UNDERSTANDING MORE QUANTUM PHYSICS:

QUANTUM STATES OF ATOMS

D0706389

MICHAEL A. MORRISON

University of Oklahoma

THOMAS L. ESTLE

Rice University

NEAL F. LANE

Rice University

PRENTICE HALL, Upper Saddle River, NJ 07458

Library of Congress Cataloging-in-Publication Data

MORRISON, MICHAEL A., (date)
Understanding more quantum physics / Michael A. Morrison, Thomas L. Estle, Neal F. Lane.
p. cm.
Reprint of part I of: Quantum states of atoms, molecules, and solids. c1976.

Includes bibliographical references and index.
1. Quantum theory. I. Estle, Thomas L. (Thomas Leo), (date).
II. Lane, Neal F. III. Title.
QC174.12.M68 1991 91-2755
530.1′2—dc20 CIP
ISBN 0-13-928300-5

Acquisitions Editor: Tim Bozik
Production Editor: Debra A. Wechsler
Cover design: Ben Santora
Prepress buyer: Paula Massenaro
Manufacturing buyer: Lori Bulwin

© 1991 by Prentice-Hall, Inc.

A Pearson Education Company
Upper Saddle River, NJ 07458

All rights reserved. No part of this book may be reproduced, in any form or by any means, without permission in writing from the publisher.

Printed in the United States of America
10 9 8 7 6 5 4 3 2

ISBN 0-13-928300-5

Prentice-Hall International (UK) Limited, London
Prentice-Hall of Australia Pty. Limited, Sydney
Prentice-Hall Canada Inc., Toronto
Prentice-Hall Hispanoamericana, S.A., Mexico
Prentice-Hall of India Private Limited, New Delhi
Prentice-Hall of Japan, Inc., Tokyo
Pearson Education Asia Pte. Ltd., Singapore
Editoria Prentice-Hall do Brasil, Ltda., Rio De Janeiro

Contents

M. A. M.— *To my parents, Janice and Alfred Morrison*
T. L. E.— *To my parents, V. Lambert and Ruby O. Estle*
N. F. L.— *To my parents, Hattie and Walter Lane, my wife, Joni Sue, and my children, Christy and John*

Preface

This book completes the story begun in the first volume of *Understanding Quantum Physics* (1990), the story of quantum mechanics. It can be used in the second semester of a two-semester introductory quantum mechanics course at the junior or senior undergraduate level following a semester devoted to *Understanding Quantum Physics* or any other undergraduate quantum text that covers the material summarized in Chapter 1: the time-dependent and time-independent Schroedinger equations, the solution of these equations for simple one-dimensional systems (square wells, harmonic oscillators, etc.), and the basic definitions of the mathematical constructs of quantum mechanics (expectation value, Hamiltonian, commutator, and the like). *Understanding More Quantum Physics* was designed for students of physics, chemistry, or engineering who have had such an introduction to the fundamentals of quantum mechanics and some college math and physics.

This book is a reprint of Part I of *Quantum States of Atoms, Molecules, and Solids* by Michael A. Morrison, Neal F. Lane, and Thomas L. Estle, published by Prentice Hall in 1976, now out of print. In that book this material served as essential groundwork for the study (in Parts II and III) of molecules and solids. This stand-alone book can still be used in that capacity. Except for a few minor differences of notation and units between this book and the first volume of *Understanding Quantum Physics,* the two meld smoothly. In the future, this book will, in turn, be supplanted by the second volume of *Understanding Quantum Physics,* which will cover the same topics and a bit more besides.

The pedagogical philosophy Neal Lane, Tom Estle, and I used in writing this text is identical to my current thinking about how to explain quantum mechanics, and most of what we wrote in the Preface to *Quantum States of Atoms, Molecules, and Solids* still holds. We have presented major concepts and problem-solving techniques in as simple a context as possible, trying wherever possible to avoid nonessential mathematical or physical complications. Thus we sometimes introduce new ideas via simple one-dimensional models, as in the introduction to the quantum mechanics of many-particle systems in Chapter 8. This strategy is predicated on our conviction that you will find it easier to tackle mathematically and physically sophisticated problems once you grasp the underlying principles of the problem.

In each chapter you will find a selection of "exercises" and, at the end of each chapter, a wide-ranging selection of problems. The former tend to be straightforward manipulations of the sort usually prefaced by the egregious phrase "it can be shown that. . . ." We have kept the number of such exercises to a minimum and included steps in derivations in the text where the truth of the matter is that "it can be shown (but only with difficulty) that. . . ." By contrast, the problems at the end of each chapter will help you understand the principles presented in the text, isolate and clarify possible points of confusion, and extend the primary text material to other topics. We mean these problems to be learning tools and so have loaded them with hints, intermediate steps, and other forms of guidance.

We developed and tested this material over a six-year period at Rice University, and at least one group of students worked every exercise and problem. Each problem comes with a "rating" of one to four stars: * = easy; ** = average; *** = a little challenging, but don't panic; **** = pretty difficult. These ratings, which are based on evaluations by students, should guide you to problems at an appropriate level of difficulty.

Like *Understanding Quantum Physics, Quantum States of Atoms, Molecules, and Solids* was born of our enthusiasm for quantum physics and our belief that learning about this subject is interesting, exciting, and should be (dare I say it?) fun. I hope this sense of the wonder and excitement of quantum physics is communicated to you and that you will leave this book eager to turn to other treatments (there are a lot of good, varied quantum books out there!) and to the fascinating applications of quantum mechanics to atoms, molecules, solids, and the rest of the universe.

Michael A. Morrison
Norman, Oklahoma
January, 1991

UNDERSTANDING MORE QUANTUM PHYSICS:

QUANTUM STATES OF ATOMS

1

Recollections of Quantum Theory: A Survey Chapter

"How often have I said to you that when you have eliminated the impossible, whatever remains, however improbable, *must be the truth."*
Arthur Conan Doyle, Sherlock Holmes to Dr. Watson in "The Sign of Four," The Complete Sherlock Holmes *(Garden City, N.Y.: Doubleday & Company).*

Throughout the nineteenth century great strides were made in the study of classical physics, and by the early 1900s a complacent mood had settled over physicists. Newtonian mechanics had been carried forward by Hamilton and Lagrange to the point where it appeared not only correct but also quite beautiful. Moreover, the electromagnetic theory of James Clark Maxwell masterfully dealt with problems involving electric and magnetic fields. Little did most of the scientists of this era realize that this mood was about to be shattered, that a series of experiments and comparisons of observations to theory would shortly reveal gaping holes in physical theory as it applied to the microscopic realm, and that a series of papers by Einstein, Schroedinger, de Broglie, Heisenberg, and others was about to burst on the scene causing shock waves whose reverberations would be felt throughout the twentieth century. For theirs was to become a new physics called *quantum mechanics* and was destined to revise radically man's thinking about the physical world and his relation to it.

However, a history of the development of quantum physics cannot be presented here; such a digression would lead us too far from our goals. Nor shall we seek to present a pedagogically sound discourse on the principles of quantum theory; a number of other books have already done so.[1] Rather we shall briefly survey the principal concepts, methods, and results of elementary nonrelativistic quantum theory. These results are not to be taken lightly, for they constitute the theoretical foundation on which the rest of this book is constructed. This brief treatment is simply a review and summary of important material, most of which should already be familiar to the reader and which can be skimmed by anyone who is not enthralled by summaries. We shall refer to and explicate results from this chapter as needed in later work.

1.1 THE WAVE-PARTICLE DUALITY

At the heart of quantum physics is a revolutionary concept of the nature of matter. Until the advent of quantum mechanics, particles had been thought of as entities possessing well-defined positions, momenta, energies, and so on, and as obeying the laws of classical mechanics. Then came the wave theory of matter and its consequences as set forth by Heisenberg, who demonstrated that the nature of the universe prohibits us from determining simultaneously the position and momentum of a particle to an aribitrary degree of precision. This disconcerting notion is expressed mathematically by the *Heisenberg uncertainty principle*:[2]

$$\Delta p_x \, \Delta x \gtrsim \hbar, \qquad \Delta p_y \, \Delta y \gtrsim \hbar, \qquad \Delta p_z \, \Delta z \gtrsim \hbar, \tag{1.1}$$

where Δp_x is the uncertainty in the x component of linear momentum, Δx is the uncertainty in the x component of position, and so on. The constant $\hbar$ is

$$\hbar \equiv \frac{\mathrm{h}}{2\pi}, \tag{1.2}$$

where h is *Planck's constant*,[3]

$$\mathrm{h} = 6.6257 \times 10^{-27} \text{ erg-sec}. \tag{1.3}$$

[1]See the Suggested Readings at the end of this chapter for references to histories and elementary texts.

[2]A similar relationship holds for energy and time, $\Delta E \, \Delta t \gtrsim \hbar$.

[3]See Appendix 1 for a table of fundamental constants.

This principle, together with experimental observations, forces us to consider a particle of matter (e.g., an electron) as having a dual nature: it has both wavelike and corpuscular properties. In fact, it is not strictly a "particle" at all; it is a *quantum mechanical system.*[4] As first articulated by de Broglie, this point of view insists that a *wavelength* λ and a *frequency* (angular) ω must be ascribed to the particle. The wavelength is related to the momentum of the particle by the *de Broglie relation,*

$$\lambda p = h, \tag{1.4}$$

or, equivalently,

$$p = \hbar k, \tag{1.5}$$

where k is the *wave number*, defined by

$$k = \frac{2\pi}{\lambda}. \tag{1.6}$$

The energy of the particle is related to its frequency by the *Einstein relation,*

$$E = \hbar\omega. \tag{1.7}$$

1.2 THE SCHROEDINGER WAVE EQUATION

This new concept of matter necessitates a revision of the ways physicists describe the location and motion of particles. No longer, for example, does it make sense to describe a particle in classical terms by precise specification of its position at each instant in time. Instead it is necessary to introduce a *wave function,* necessarily a complex function, which in some sense contains all that it is possible to know about a quantum mechanical state of the system. In general, this wave function depends on position and time[5] and is written $\Psi(\mathbf{r}, t)$.

Not all complex functions of space and time are valid wave functions. To be *physically admissible,* the function and its first derivative must be finite, continuous, and single valued throughout all space. Moreover, it must be *square-integrable*—that is, it must satisfy

$$\int |\Psi(\mathbf{r}, t)|^2 \, d\mathbf{r} = \int \Psi^*(\mathbf{r}, t)\Psi(\mathbf{r}, t) \, d\mathbf{r} < \infty, \tag{1.8}$$

[4]Although a structureless particle is used as the "system" in this chapter, the results quoted can (and will) be generalized to more complex systems, such as atoms, molecules, and solids.

[5]As we shall see in Chapter 6, the wave function can also depend on other quantities (e.g., spin).

where Ψ^* denotes the complex conjugate of Ψ and $\int \cdots d\mathbf{r}$ means integration over all space. In this text all the functions studied will satisfy these properties.

Although it is not possible to say with complete certainty where the particle is (if, indeed, "where" has any meaning in our new view of the universe), its probable location can be calculated. Thus we introduce the *probability density*, or probability per unit volume, $\rho(\mathbf{r}, t)$, defined as

$$\rho(\mathbf{r}, t) \equiv \Psi^*(\mathbf{r}, t)\Psi(\mathbf{r}, t). \tag{1.9}$$

In quantum mechanics, $\rho(\mathbf{r}, t)\, d\mathbf{r}$ is the probability that the particle will be found at time t in a volume element $d\mathbf{r}$ centered on the point $\mathbf{r}$.

Since the particle must clearly be somewhere in order to exist at all, the probability density must satisfy

$$\int \rho(\mathbf{r}, t)\, d\mathbf{r} = \int \Psi^*(\mathbf{r}, t)\Psi(\mathbf{r}, t)\, d\mathbf{r} = 1. \tag{1.10}$$

A wave function that satisfies this condition is said to be *normalized.*[6]

Thus the position of the particle is described by its wave function. The motion of the particle is described by its wave equation, the quantum mechanical equivalent of a Newtonian equation of motion. In particular, the appropriate quantum mechanical wave equation is the *time-dependent Schroedinger equation.* For a particle of mass m with potential energy $V(\mathbf{r}, t)$, this equation is

$$\mathcal{H}\Psi(\mathbf{r}, t) = i\hbar\frac{\partial}{\partial t}\Psi(\mathbf{r}, t), \tag{1.11}$$

where $\mathcal{H}$ is the Hamiltonian operator (see Sec. 1.3). The *Hamiltonian* $\mathcal{H}$ is the operator equivalent of the total energy of the particle and is equal to the sum of the kinetic and potential energies,

$$\mathcal{H} = T + V(\mathbf{r}, t), \tag{1.12}$$

where the kinetic energy is

$$T = \frac{p^2}{2m} \tag{1.13}$$

for a particle of mass m. The operator for the linear momentum $\mathbf{p}$ in this equation is

$$\mathbf{p} = -i\hbar\, \nabla. \tag{1.14}$$

[6]Clearly, if this condition is violated to the extent that the integral in Eq. (1.10) is equal to some finite number, say M, normalization can be enforced by simply redefining $\Psi(\mathbf{r}, t)$ to incorporate a multiplicative constant $1/\sqrt{M}$.

Therefore we can write

$$T = -\frac{\hbar^2}{2m}\nabla^2, \tag{1.15}$$

∇^2 being the Laplacian operator, which, in cartesian coordinates, is

$$\nabla^2 = \frac{\partial^2}{\partial x^2} + \frac{\partial^2}{\partial y^2} + \frac{\partial^2}{\partial z^2}. \tag{1.16}$$

In accordance with the probabilistic interpretation of quantum mechanics introduced in Eq. (1.9), the motion of the particle may be further described by its *probability flux density* $\mathbf{j}(\mathbf{r}, t)$. This quantity is defined as

$$\mathbf{j}(\mathbf{r}, t) \equiv -\frac{i\hbar}{2m}[\Psi^*(\mathbf{r}, t)\,\nabla\Psi(\mathbf{r}, t) - \Psi(\mathbf{r}, t)\,\nabla\Psi^*(\mathbf{r}, t)]; \tag{1.17}$$

$\mathbf{j}(\mathbf{r}, t)\cdot d\mathbf{a}$ is the probability per unit time that the particle passes through the element of surface area da, where

$$d\mathbf{a} = \hat{\mathbf{n}}\, da, \tag{1.18}$$

$\hat{\mathbf{n}}$ being a unit vector directed normal to the surface element.

An important relationship between probability density $\rho(\mathbf{r}, t)$ and probability flux density $\mathbf{j}(\mathbf{r}, t)$ is the *equation of continuity*,

$$\frac{\partial}{\partial t}\rho(\mathbf{r}, t) = -\nabla\cdot\mathbf{j}(\mathbf{r}, t). \tag{1.19}$$

This equation expresses the conservation of probability with time. It can be used, for example, to show that the normalization of a wave function is time independent.

As the preceding remarks suggest, the wave function $\Psi(\mathbf{r}, t)$ is all important to the theory of quantum mechanics. Indeed, most of our efforts in this text will be directed at obtaining wave functions by solution of the Schroedinger equation (1.11). If the potential energy does not explicitly depend on time —that is, if $V(\mathbf{r}, t) = V(\mathbf{r})$—then special solutions of this equation can be obtained by employing the mathematical procedure of separation of variables. The time dependence factors out of these wave functions:

$$\Psi(\mathbf{r}, t) = e^{-i(E/\hbar)t}\psi_E(\mathbf{r}), \tag{1.20}$$

where E is the energy of the particle. The new function $\psi_E(\mathbf{r})$ is labeled by the energy E. It depends only on $\mathbf{r}$ and satisfies the *time-independent Schroedinger equation*

$$\mathcal{H}\psi_E(\mathbf{r}) = E\psi_E(\mathbf{r}). \tag{1.21}$$

Each function $\psi_E(\mathbf{r})$ satisfying Eq. (1.21) corresponds to a state of the particle for which the energy is precisely known and does not change with time. The full wave function $\Psi(\mathbf{r}, t)$ for the state is given by Eq. (1.20). Such a state is called a *stationary state*; we say that the energy in a stationary state is *sharp* (i.e., well defined). For a given system, there are, in general, many stationary states having negative energy ($E < 0$) which are necessarily *bound states*, and an infinity of stationary states having positive energy ($E > 0$) which are not bound states and are usually referred to as *continuum states.*

In general, a system whose potential energy is independent of time will be found in a mixture of stationary states, and its energy will not be sharp. The wave function $\Psi(\mathbf{r}, t)$ for such a state can be constructed by forming a linear combination of all stationary-state wave functions,

$$\Psi(\mathbf{r}, t) = \sum a_E \psi_E(\mathbf{r}) e^{-i(E/\hbar)t}, \tag{1.22}$$

where the sum is taken over all stationary-state wave functions. [If there exist continuum solutions as well, they must be included in the sum of Eq. (1.22) by integration over the continuous region of the spectrum.] In this summation, a_E is a constant independent of $\mathbf{r}$ and t; it is the expansion coefficient for the stationary state of energy E.

1.3 EIGENVALUE EQUATIONS AND OPERATORS

The time-independent Schroedinger equation (1.21) is one of a special class of equations called *eigenvalue equations.*[7] This equation states that an operator $\mathcal{H}$ acts on a function $\psi_E(\mathbf{r})$ to give the function itself multiplied by a constant E. The function in such an equation is called an *eigenfunction* and the constant an *eigenvalue.* It is common practice to label the eigenfunction by its eigenvalue as we have done in Eq. (1.21).

The solution of the time-independent Schroedinger equation yields a set of eigenfunctions of the Hamiltonian operator together with the corresponding eigenvalues. These eigenvalues fall into two classes: discrete and continuous. The collection of all the eigenvalues of the Hamiltonian operator is called the *energy spectrum* of the system.

If two or more distinct eigenfunctions have the same eigenvalue, the eigenfunctions are said to be *degenerate.* Eigenfunctions with different eigenvalues are said to be *nondegenerate.*

Just as the time-independent Schroedinger equation is typical of a class of equations (eigenvalue equations), the Hamiltonian that appears in this equation is one of a class of mathematical entities called linear operators. In

[7]In German, "eigen" means "characteristic"; so eigenvalue equations are "characteristic equations".

general, an *operator* transforms one function into another that may be of different form; a familiar operator is the differential operator d/dx. A *linear operator* is an operator that transforms the sum of two or more functions into the sum of transformed functions. All operators that arise in quantum mechanics are linear, and we will restrict our discussion to such operators from here on. Linear operators of importance in quantum mechanics are the Hamiltonian $\mathcal{H}$, the orbital angular momentum $\mathbf{L}$, and the linear momentum $\mathbf{p}$. We shall encounter these operators in our study of the central force problem (see Chapters 2 and 3).

We write an eigenvalue equation for an operator A,

$$A\varphi_a = a\varphi_a, \tag{1.23}$$

where φ_a is an eigenfunction of the operator and a is the eigenvalue corresponding to the eigenfunction φ_a. We choose to label φ with its eigenvalue—as φ_a. Alternately, we may introduce an index, say i, to distinguish different eigenfunctions and eigenvalues—φ_i and a_i. The latter technique is especially useful if the eigenvalues of A are discrete; then the index simply takes on all integral values $i = 1, 2, \ldots, N$, where N is the number of discrete eigenvalues. Throughout this book the convention that seems most convenient at the time will be adopted.

The importance of the set of eigenvalues a_i obtained by solution of Eq. (1.23) is that in a single observation of the dynamical observable[8] A on a system, the only result that can be obtained is one of the eigenvalues of the operator corresponding to A. For example, in a single observation of the energy of a system, the only value we could get would be one of the energy eigenvalues E obtained by solution of the time-independent Schroedinger equation. This result provides an important link between theory and experiment and is the basis of the quantum mechanical theory of measurement.[9,10]

The important operators in quantum mechanics have a special property with profound implications for the corresponding eigenfunctions and eigen-

[8]The term *observable*, which may be unfamiliar, simply refers to any physically measurable quantity. Typical observables are energy, linear momentum, position, and angular momentum. We shall use the same notation to refer to a dynamical observable as to the corresponding quantum mechanical operator.

[9]The word *measurement* has a special meaning in quantum physics. Specifically, by a "measurement of A on a system," we refer to a large number of single observations carried out on the members of an ensemble of *identical* systems—that is, systems with identical Hamiltonians and identical histories. Thus the wave function $\Psi(\mathbf{r}, t)$ is the same for each system in the ensemble before the measurement is made. In general, the process of observing will disturb the states of the members of the ensemble in a way which cannot be controlled.

[10]See, for example, David Bohm, *Quantum Theory* (Englewood Cliffs, N.J.: Prentice-Hall, 1951), Part VI.

values: they are Hermitian. By definition, an operator $A(\mathbf{r}, t)$ is *Hermitian* if it satisfies the equation

$$\int f^*(\mathbf{r}) A(\mathbf{r}, t) g(\mathbf{r})\, d\mathbf{r} = \int [A(\mathbf{r}, t) f(\mathbf{r})]^* g(\mathbf{r})\, d\mathbf{r}, \tag{1.24}$$

where f and g are arbitrary functions and $A(\mathbf{r}, t)$ operates only on $f(\mathbf{r})$ in the integral on the right-hand side. We often write integrals such as the ones in Eq. (1.24) as $\langle f(\mathbf{r}) | A(\mathbf{r}, t) | g(\mathbf{r}) \rangle$, it being understood that the Hermitian operator A can operate to the right or A^* can operate to the left. This integral is called a *matrix element.*

A special significance is ascribed to Hermitian operators in quantum mechanics; one of the basic postulates of quantum theory is that *for every dynamical observable there exists a corresponding Hermitian operator.*

The solutions of the eigenvalue equation for a Hermitian operator satisfy the following properties:

1. The eigenvalues are real.
2. The eigenfunctions corresponding to distinct eigenvalues are orthogonal.
3. The eigenfunctions form a complete set.

Thus, if A is a Hermitian operator, the first property ensures that all its eigenvalues a_i are real. From the second property, we know that

$$\int \varphi_i^*(\mathbf{r}) \varphi_j(\mathbf{r})\, d\mathbf{r} = 0, \qquad i \neq j, \tag{1.25}$$

where $\varphi_i(\mathbf{r})$ and $\varphi_j(\mathbf{r})$ are eigenfunctions corresponding to distinct eigenvalues[11] a_i and a_j $(a_i \neq a_j)$. If we further specify that the eigenfunctions of A are normalized [satisfy Eq. (1.10)], we can combine this fact with their orthogonality and write

$$\int \varphi_i^*(\mathbf{r}) \varphi_j(\mathbf{r})\, d\mathbf{r} = \delta_{ij}, \tag{1.26}$$

where δ_{ij} is the Kronecker delta, defined as

$$\delta_{ij} = \begin{cases} 0 & \text{if } i \neq j \\ 1 & \text{if } i = j. \end{cases} \tag{1.27}$$

[11] Although Eq. (1.25) does not hold as written if $a_i = a_j$, there is a procedure whereby linear combinations of the eigenfunctions in the original set $\{\varphi_i\}$ can be formed in such a way that the new eigenfunctions are orthonormal. This is called the Gram-Schmidt orthogonalization process. See Elmer Anderson, *Modern Physics and Quantum Mechanics* (Philadelphia: W. B. Saunders Co., 1971), Sec. 6.4.

A set of functions that satisfies Eq. (1.26) is said to be *orthonormal.* To understand the significance of the third property, we must first define "complete set." A *complete set* of functions is defined by the property that any other function of the same coordinates may be expanded in terms of the elements of the set. Thus the set of eigenfunctions of A, $\{\varphi_i\}$, is complete. Because the Hamiltonian is linear, we can expand, say, an energy eigenfunction $\psi_E(\mathbf{r})$ in terms of the set $\{\varphi_i\}$:

$$\psi_E(\mathbf{r}) = \sum_{i=1}^{N} c_i \varphi_i(\mathbf{r}), \tag{1.28}$$

where the numbers c_i are called *expansion coefficients.* [In writing Eq. (1.28), we have assumed that there are N eigenfunctions of A in the set.] Since the set of eigenfunctions of A is orthonormal, we can determine the coefficients by simple integration:

$$c_i = \int \varphi_i^*(\mathbf{r})\psi_E(\mathbf{r})\, d\mathbf{r}. \tag{1.29}$$

These coefficients, obtained by expanding an energy eigenfunction in a complete set of eigenfunctions of some observable, contain information about the probability that a particular eigenvalue of that observable is obtained in a single observation on a system. Suppose, for example, that we carry out a series of observations of A on a large ensemble (collection) of identical systems; in each observation we get one of the eigenvalues in the set $\{a_i\}$. The probability of obtaining a particular eigenvalue, say the jth one, in a single observation of A on a member system with energy E in the ensemble is

$$c_j^* c_j = |c_j|^2 = |\langle\varphi_j|\psi_E\rangle|^2. \tag{1.30}$$

Once again our theory is linked to the real world of experimental physics by the theory of measurement.

1.4 COMMUTING OPERATORS

The state of a quantum mechanical system is usually characterized by the eigenvalues of a number of operators (such as energy and linear momentum). An important question involves the relationship of these operators to one another. Specifically, do they "commute"? Two operators A and B are said to *commute* if the order in which they operate is immaterial—that is, if

$$AB = BA. \tag{1.31}$$

This relationship is usually written in terms of the *commutator* of A and B, defined as

$$[A, B] \equiv AB - BA. \tag{1.32}$$

Thus A and B commute if their commutator is zero: $[A, B] = 0$. Another quantity that is useful in the study of operators is the *anticommutator*, defined as

$$[A, B]_+ \equiv AB + BA. \tag{1.33}$$

Now, suppose that we write an eigenvalue equation for the operator A:

$$A\varphi_a = a\varphi_a. \tag{1.34}$$

An important theorem of quantum mechanics states that if φ_a is an eigenfunction[12] of A with eigenvalue a and if A commutes with B, then φ_a is also an eigenfunction of the operator B with eigenvalue, say, b. Thus if $[A, B] = 0$, then we can write

$$B\varphi_{a,b} = b\varphi_{a,b}. \tag{1.35}$$

Stated succinctly, the theorem reads: *Operators which commute possess simultaneous eigenfunctions.* The two eigenvalues a and b correspond to allowed values for the two dynamical observables whose operators are A and B, respectively. These two numbers provide convenient labels for the eigenfunction φ and are sometimes called *quantum numbers*.

Much of this theory will seem less abstract when we consider specific physical systems involving several operators and examine eigenfunctions and quantum numbers. For example, in our study of the hydrogen atom we shall introduce an operator L^2 for the square of the orbital angular momentum of the electron about the nucleus. According to our theorem, if $[\mathcal{H}, L^2] = 0$, then the energy eigenfunctions of the Hamiltonian—the solutions $\psi_E(\mathbf{r})$ of Schroedinger's equation for the hydrogen atom—can also be eigenfunctions of L^2 and thus can be characterized by the eigenvalue of L^2 as well as by the energy E.

In quantum mechanics, an operator which commutes with the Hamiltonian of a system is said to be a *constant of the motion*.

1.5 THE EXPECTATION VALUE

Let us return now to the expansion of the energy eigenfunction $\psi_E(\mathbf{r})$ in terms of a set of nondegenerate eigenfunctions of the operator A that is complete in the coordinates $\mathbf{r}$,

$$\psi_E(\mathbf{r}) = \sum_{i=1}^{N} c_i \varphi_i(\mathbf{r}). \tag{1.36}$$

[12] We are considering only the nondegenerate situation in this section; thus, each eigenvalue a is associated with one and only one eigenfunction φ_a. See the Suggested Readings at the end of the chapter for discussions of the degenerate case.

From the last section we know that if A commutes with the Hamiltonian—that is, if

$$[A, \mathcal{H}] = 0, \tag{1.37}$$

then $\psi_E(\mathbf{r})$ will be an eigenfunction of A as well as of $\mathcal{H}$ with a distinct eigenvalue, say, a_j. That is,

$$A\psi_E(\mathbf{r}) = a_j\psi_E(\mathbf{r}). \tag{1.38}$$

Therefore all the coefficients in the expansion of Eq. (1.36) will be zero except the jth one, which, if all the functions $\{\varphi_i\}$ are normalized, will be equal to unity. Expressing this mathematically, we write

$$c_i = \delta_{ij}. \tag{1.39}$$

In this event, the observable A is sharp; in any measurement of A on the system in the state ψ_E, we obtain precisely the value a_j.

Of course, in general, not all operators commute with the Hamiltonian of a system. If A does not commute with $\mathcal{H}$, then the simple relation of Eq. (1.39) does not hold, and the energy eigenfunction $\psi_E(\mathbf{r})$ is said to be a mixture of eigenfunctions of A. If such is the case, we cannot be so precise about the results of a measurement of A; we saw in Sec. 1.3 that we could only predict statistical results of making a large number of observations on an ensemble of identical systems.

Thus we can say something about the statistical average of such a measurement. This average corresponds to the expectation value of the operator. For an operator A, the *expectation value* of A is defined as

$$\langle A \rangle \equiv \int \psi_E^*(\mathbf{r}) A \psi_E(\mathbf{r})\, d\mathbf{r}, \tag{1.40}$$

where we have assumed that A is independent of time and that the system is in a stationary state.[13] Notice that the expectation value of A is defined with respect to a particular stationary state.

If A is the Hamiltonian operator, $\mathcal{H}$, the expectation value has a particularly simple form; from the time-independent Schroedinger equation it follows that, for a stationary state of energy E, the expectation value of the Hamiltonian is equal to the energy E. If, on the other hand, A is not the

[13]In general, the expectation value is

$$\langle A \rangle = \int \Psi^*(\mathbf{r}, t) A \Psi(\mathbf{r}, t)\, d\mathbf{r},$$

where the full wave function $\Psi(\mathbf{r}, t)$, which satisfies the time-dependent Schroedinger equation, is used. See David Saxon, *Elementary Quantum Mechanics* (San Francisco: Holden-Day, 1968), pp. 91 ff.

Hamiltonian, we can still calculate $\langle A \rangle$ provided that we know the coefficients $\{c_i\}$ in the expansion of Eq. (1.36). Substituting this expansion into the definition of $\langle A \rangle$, we find

$$\langle A \rangle = \sum_{i=1}^{N} |c_i|^2 a_i; \tag{1.41}$$

that is, the expectation value of the operator A is the *weighted* sum of the eigenvalues of A. The weighting factor for the ith eigenvalue a_i is the square of the ith coefficient in the expansion of the energy eigenfunction in terms of eigenfunctions of A.

Clearly, if A commutes with $\mathcal{H}$ as in Eq. (1.37), then the expectation value of A is equal to the appropriate eigenvalue of A—for example,

$$\langle A \rangle = a_j. \tag{1.42}$$

This is merely another way of saying that the observable A is sharp.

1.6 CONCLUDING REMARKS

Some of the results of this chapter may appear quite alien and abstract, particularly the ones expressed in terms of "some operator" rather than a specific operator. This situation should not be alarming; we shall be applying and examining many of these results throughout the remaining chapters until, by the end, they will be old and familiar friends.

The elementary application of most of these ideas is to one-dimensional systems such as potential wells and barriers, simple harmonic potentials, and so on. These are important problems, however unphysical they may seem, for they illustrate the fundamental principles of quantum physics in particularly simple contexts. Indeed, we shall make extensive use of one-dimensional systems in our study of atoms, molecules, and solids as we try to "model" these systems in such a way as to extract their fundamental physics.

Several of the references presented at the conclusion of this chapter discuss applications of elementary quantum mechanics to one-dimensional systems like the ones mentioned above. If these applications are not familiar, a short review would probably be useful. In addition, Prob. 1.1 is recommended for review purposes. However, we are primarily interested in the applications of quantum theory to more realistic physical systems, a study that begins in the next chapter.

SUGGESTED READINGS

An enormous number of so-called introductory quantum mechanics texts are currently available. Therefore, we list here only a few personal favorites with which we are most familiar and which seem particularly useful to the student of our book.

The reader who desires additional references is referred to the Bibliography at the end of the book or to his neighborhood bookstore.

A number of histories of the evolution of quantum theory are available; two of the most enjoyable are

GAMOW, GEORGE, *Thirty Years that Shook Physics*. New York: Doubleday, 1966.

CROPPER, WILLIAM H., *The Quantum Physicists and an Introduction to Their Physics*. London: Oxford University Press, 1970.

A very useful introductory text is

BOCKHOFF, FRANK J., *Elements of Quantum Theory*. Reading, Mass.: Addison-Wesley, 1969.

Of particular interest are Chapters 4 to 6, which lead into quantum mechanics, and Chapters 7 and 8, which deal with simple examples of elementary quantum theory. At a slightly more advanced level are

SAXON, DAVID S., *Elementary Quantum Mechanics*. San Francisco: Holden-Day, 1968.

ANDERSON, ELMER E., *Modern Physics and Quantum Mechanics*. Philadelphia: W. B. Saunders Co., 1971.

The latter text contains a very good chapter on the formal structure of quantum mechanics (Chapter 7). This topic is also treated with clarity in Saxon's book, which contains a wealth of information on the momentum representation of quantum states.

Two volumes that are old enough to be called classics and yet are so well written that they should be recommended are

PAULING, LINUS, and E. B. WILSON, *Introduction to Quantum Mechanics*. New York: McGraw-Hill, 1935.

EYRING, HENRY, JOHN WALTER, and GEORGE E. KIMBALL, *Quantum Chemistry*. New York: Wiley, 1944.

Both books deal with quantum chemistry and contain sections on atomic and molecular physics as well as explications of elementary quantum mechanics.

Finally, we should mention

ROJANSKY, VLADIMIR, *Introductory Quantum Mechanics*. Englewood Cliffs, N.J.: Prentice-Hall, 1938.

This book, although old, was clearly written with the reader in mind and has provided help for many a perplexed student in time of stress. See especially the discussions of problems in one dimension, Chapters 2 to 8, 10, and 11.

PROBLEMS

1.1 Review of Simple One-Dimensional Systems (**)

Consider a one-dimensional symmetric potential energy [i.e., $V(x) = V(-x)$] of the form

$$V(x) = V_0\,|x|^n.$$

If $n = 2$, this is the potential energy of a simple harmonic oscillator. If $n = -1$, it is the one-dimensional analog of the coulomb potential energy. Finally, if n is a large number, then $V(x)$ approximates an infinite square well or "box."

(a) Let us assume that the solutions for very large n are essentially those of a particle in a "box" defined by

$$V(x) = \begin{cases} 0, & |x| < 1 \\ \infty, & |x| > 1. \end{cases}$$

What are the energies of the stationary states for this system? As energy increases, do the states get closer together or farther apart? Sketch the wave functions for the lowest four states. Discuss the contributions of the kinetic energy and the potential energy to the total energy.

(b) Now consider the case $n = 2$. What are the stationary-state energies? As energy increases, do the states get closer together or farther apart? Sketch the wave-functions for the lowest four states. Explain, in terms of the relative importance of the kinetic and potential energies in determining the total energy, why the energy spectrum differs from that in part (a).

The first excited state is an odd state [i.e., $\psi(x) = -\psi(-x)$] with a single node at $x = 0$. If the wavefunction for this state were more extended (so that the peaks in the probability were farther apart), would the total energy be lower? Would the kinetic energy be lower? The potential energy?

(c) For $n = -1$, we have the one-dimensional analog of the coulomb potential energy. However, in one dimension this potential energy has too strong a singularity for our purposes, and it is actually necessary to truncate the potential in order to obtain discrete levels [see L. Haines and D. Roberts, *American Journal of Physics* **37**, 1145(1969)]. For our purposes, we will simply draw analogies from three dimensions. Thus we have a series of levels specified by the principal quantum number n. All but the lowest ($n = 1$) level are degenerate. In three dimensions this degeneracy results from the various nonzero values of the electron's orbital angular momentum. The equivalent in one dimension is the double degeneracy of an odd state with $n - 1$ nodes and of an even state with n nodes (these are analogous to the np and ns states in three dimensions). Assuming, also by analogy with three dimensions, that the levels get closer together as the energy increases, explain why this is so in terms of the kinetic and potential energies.

(d) Ignoring, for the moment, the obvious problem of n going from a positive to a negative value, explain what qualitative changes occur in the energy spectrum as n is varied. In addition, explain, in terms of the changes in shape of $V(x)$, why this occurs. Would you expect similar results in two or three dimensions? What arguments can you advance to justify your conclusions?

2

Solution of the Central Force Problem

Polonius: "Though this be madness, yet there is method in't."

William Shakespeare, Hamlet, *ii. 2*

In this chapter we shall consider the solutions of the time-independent nonrelativistic Schroedinger equation for a single particle

$$\mathcal{H}\psi_E(\mathbf{r}) = E\psi_E(\mathbf{r}), \tag{2.1}$$

for a special class of problems. Many three-dimensional systems of interest possess *spherical symmetry*; that is, all directions in space are equivalent. The potential energy of such a system will depend only on the radial distance from a suitably chosen origin:

$$V(\mathbf{r}) = V(r). \tag{2.2}$$

Then the Hamiltonian in Eq. (2.1) is

$$\mathcal{H} = T + V(r). \tag{2.3}$$

Problems dealing with such systems are called *central force problems.* One example of a system that satisfies Eq. (2.2) and is of great interest is the one-electron atom; the potential energy for an electron in the field of a nucleus of charge Ze is

$$V(r) = -\frac{Ze^2}{r}. \tag{2.4}$$

We shall study this system in Chapter 3.

2.1 CENTRAL FORCES

The central force problem in classical mechanics is probably familiar.[1] The classical force is related to the potential energy by

$$\mathbf{F} = -\nabla V. \tag{2.5}$$

Therefore any force derived from a potential energy that satisfies Eq. (2.2) is radially directed and independent of the angular variables θ and φ of spherical coordinates. Hence it is called a *central force.*

In classical mechanics, certain properties related to the conservation of angular momentum are characteristic of the solutions of any central force problem. We shall see that in quantum mechanics similar properties are common to the solution of any central force problem. Therefore if we can at least partially solve the general quantum-mechanical central force problem, our labor will be considerably reduced when we take up a specific problem with a particular $V(r)$.

Let us briefly review some of the properties of central force motion in classical mechanics. A particle of mass m_1 under the influence of a central force always moves in a plane in such a way that the total energy E and the orbital angular momentum L have vanishing time derivatives—they are constants of the motion. They are given by

$$L = m_1 r^2 \dot{\theta} = \text{constant} \tag{2.6}$$

and

$$E = \frac{1}{2} m_1 (\dot{r}^2 + r^2 \dot{\theta}^2) + V(r) = \text{constant}, \tag{2.7}$$

where the dot (e.g., $\dot{\theta}$) indicates the first time derivative (e.g., $d\theta/dt$). The expression for the total energy can be written in the suggestive form

$$E = \frac{1}{2} m_1 \dot{r}^2 + \frac{1}{2m_1 r^2} L^2 + V(r) = \text{constant}, \tag{2.8}$$

[1]See, for example, Keith Symon, *Mechanics*, 3rd ed. (Reading, Mass.: Addison-Wesley, 1971), Sec. 3.13.

where the first term is the radial kinetic energy, the second term is the angular kinetic energy, and $V(r)$ is the potential energy. It is useful to define an *effective potential energy*,

$$V_{\text{eff}}(r) \equiv \frac{1}{2m_1r^2}L^2 + V(r), \tag{2.9}$$

in terms of which the total energy can be written as

$$E = \frac{1}{2}m_1\dot{r}^2 + V_{\text{eff}}(r). \tag{2.10}$$

The effective potential helps us to visualize and understand the properties of the orbit of the particle. An analogous effective potential will be introduced in our study of the quantum-mechanical central force problem.

Before explicitly considering this problem, let us write down potential energies for three commonly encountered central force problems; we shall return to these problems in this and subsequent chapters. Of paramount importance in atomic physics is the *coulomb potential energy* of Eq. (2.4). The *harmonic spherical well* is useful in constructing and studying models in atomic and molecular physics. The corresponding potential energy is

$$V(r) = \gamma r^2, \tag{2.11}$$

where γ is a constant. Finally, we should mention the *spherical well*, which is the first central force problem we shall solve in detail (see Example 2.1); the potential energy is given by

$$V(r) = \begin{cases} -V_0, & 0 < r < a \\ 0, & r \geq a. \end{cases} \tag{2.12}$$

Keeping these examples in mind, we now turn to the general central force problem in quantum mechanics.

2.2 THE SCHROEDINGER EQUATION FOR THE CENTRAL FORCE PROBLEM

Consider a particle of mass m_1 in a spherically symmetric force field. We know that the potential energy satisfies Eq. (2.2), so the Hamiltonian eigenvalue equation (2.1) can be written

$$\left[-\frac{\hbar^2}{2m_1}\nabla^2 + V(r) - E\right]\psi_E(\mathbf{r}) = 0. \tag{2.13}$$

The fact that there is no preferred direction in space for the system may have led you to expect the energy eigenfunctions to be independent of θ and φ. Not true! However, the special nature of the potential energy will enable us to find angular-dependent eigenfunctions of a particularly simple form.

The spherical symmetry of $V(r)$ suggests that we work in spherical coordinates; this coordinate system is shown in Fig. 2.1. We must first express

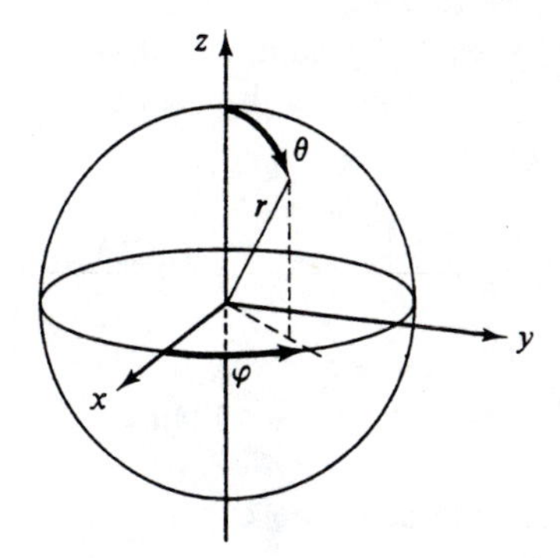

Figure 2.1 Spherical coordinate system.

the Laplacian ∇^2 in spherical coordinates. In cartesian coordinates, this operator is

$$\nabla^2 = \frac{\partial^2}{\partial x^2} + \frac{\partial^2}{\partial y^2} + \frac{\partial^2}{\partial z^2}. \tag{2.14}$$

The transformation between the two systems is given by

$$x = r \sin\theta \cos\varphi, \tag{2.15a}$$

$$y = r \sin\theta \sin\varphi, \tag{2.15b}$$

$$z = r \cos\theta, \tag{2.15c}$$

and it is possible to use Eqs. (2.15) to show that

$$\nabla^2 = \frac{1}{r^2}\frac{\partial}{\partial r}\left(r^2 \frac{\partial}{\partial r}\right) + \frac{1}{r^2 \sin\theta}\frac{\partial}{\partial\theta}\left(\sin\theta \frac{\partial}{\partial\theta}\right) + \frac{1}{r^2 \sin^2\theta}\frac{\partial^2}{\partial\varphi^2}. \tag{2.16}$$

This result enables us to write the Hamiltonian eigenvalue equation (2.13) in spherical coordinates:

$$\left\{-\frac{\hbar^2}{2m_1}\left[\frac{1}{r^2}\frac{\partial}{\partial r}\left(r^2 \frac{\partial}{\partial r}\right) + \frac{1}{r^2 \sin\theta}\frac{\partial}{\partial\theta}\left(\sin\theta \frac{\partial}{\partial\theta}\right) + \frac{1}{r^2 \sin^2\theta}\frac{\partial^2}{\partial\varphi^2}\right] + V(r) - E\right\}\psi_E(\mathbf{r}) = 0. \tag{2.17}$$

If we manipulate this equation a bit, we can rewrite it as

$$-\frac{\hbar^2}{2m_1}\frac{1}{r^2}\frac{\partial}{\partial r}\left(r^2\frac{\partial}{\partial r}\right)\psi_E(\mathbf{r}) + \frac{1}{2m_1 r^2}\left\{-\hbar^2\left[\frac{1}{\sin\theta}\frac{\partial}{\partial\theta}\left(\sin\theta\frac{\partial}{\partial\theta}\right) + \frac{1}{\sin^2\theta}\frac{\partial^2}{\partial\varphi^2}\right]\right\}\psi_E(\mathbf{r}) + [V(r) - E]\psi_E(\mathbf{r}) = 0. \tag{2.18}$$

The form of this equation bears a strong resemblance to that of the classical expression for the total energy, Eq. (2.8). Let us exploit this similarity and tentatively identify the operator corresponding to the square of the orbital angular momentum with the factors in braces in Eq. (2.18):

$$L^2 = -\hbar^2\left[\frac{1}{\sin\theta}\frac{\partial}{\partial\theta}\left(\sin\theta\frac{\partial}{\partial\theta}\right) + \frac{1}{\sin^2\theta}\frac{\partial^2}{\partial\varphi^2}\right]. \tag{2.19}$$

Of course, in no way have we proven that the right-hand side of Eq. (2.19) is the operator for the square of the orbital angular momentum. We shall do so in Sec. 2.3, where we examine angular momentum in more depth.

Accepting this reasonable identification for the moment, we can write the Laplacian in terms of L^2,

$$\nabla^2 = \frac{1}{r^2}\frac{\partial}{\partial r}\left(r^2\frac{\partial}{\partial r}\right) - \frac{1}{\hbar^2 r^2}L^2; \tag{2.20}$$

then the time-independent Schroedinger equation becomes

$$\boxed{\left[-\frac{\hbar^2}{2m_1}\frac{1}{r^2}\frac{\partial}{\partial r}\left(r^2\frac{\partial}{\partial r}\right) + \frac{1}{2m_1 r^2}L^2 + V(r) - E\right]\psi_E(\mathbf{r}) = 0.}$$

Hamiltonian eigenvalue equation for a central force problem (2.21)

Separation of Radial and Angular Coordinates

Equation (2.21) is a second-order partial differential equation in three variables, r, θ, and φ, and as such is rather formidable. To make the task of solving this equation easier, let us seek solutions for which the angular motion, the dependence of ψ on θ and φ, is separated from the radial motion, the dependence of ψ on r. Such a function will have the form

$$\psi_E(\mathbf{r}) = R(r)Y(\theta, \varphi), \tag{2.22}$$

where $R(r)$ is the *radial function* and $Y(\theta, \varphi)$ is the *angular function*.[2] This is

[2]We do not suggest that all solutions of Eq. (2.21) are of this form. However, any valid solution of Eq. (2.21) can be written as a linear combination of functions of the form of Eq. (2.22).

the familiar method of separation of variables.[3] Substituting Eq. (2.22) into (2.21), we obtain

$$Y(\theta, \varphi)\left[-\frac{\hbar^2}{2m_1}\frac{1}{r^2}\frac{d}{dr}\left(r^2\frac{d}{dr}\right) + V(r) - E\right]R(r) + \frac{1}{2m_1 r^2}R(r)L^2Y(\theta, \varphi) = 0, \tag{2.23}$$

where we have used the fact that L^2 has no effect on a function of r only. To see the separation of variables more clearly, let us multiply Eq. (2.23) by $-[2m_1r^2/\hbar^2][R(r)\,Y(\theta, \varphi)]^{-1}$:

$$\underbrace{\frac{r^2}{R(r)}\left\{\frac{1}{r^2}\frac{d}{dr}\left(r^2\frac{d}{dr}\right) + \frac{2m_1}{\hbar^2}[E - V(r)]\right\}R(r)}_{\text{function of } r \text{ only}} - \underbrace{\frac{1}{\hbar^2 Y(\theta, \varphi)}L^2Y(\theta,\varphi)}_{\text{function of } \theta \text{ and } \varphi \text{ only}} = 0. \tag{2.24}$$

Since the radial coordinate r varies independently of the angular coordinates θ and φ, Eq. (2.24) has nontrivial solutions if and only if

$$\frac{r^2}{R(r)}\left\{\frac{1}{r^2}\frac{d}{dr}\left(r^2\frac{d}{dr}\right) + \frac{2m_1}{\hbar^2}[E - V(r)]\right\}R(r) = \alpha \tag{2.25}$$

and

$$-\frac{1}{\hbar^2}\frac{1}{Y(\theta, \varphi)}L^2Y(\theta, \varphi) = -\alpha, \tag{2.26}$$

for some *separation constant* α. For obvious reasons, we call (2.25) the *radial equation* and (2.26) the *angular equation*. Notice that this method works because the potential energy $V(r)$ is independent of θ and φ.

One term in the radial equation contains the potential energy, so the radial motion certainly depends on the particular central force problem being considered and cannot be determined until $V(r)$ is specified. However, *the angular equation is independent of the potential energy*. Therefore once we have solved Eq. (2.26) for $Y(\theta, \varphi)$, we can use this function to describe the angular motion for any central force problem. Then for any spherically symmetric potential energy—for example, Eqs. (2.4), (2.11), or (2.12)—we merely solve the radial equation for $R(r)$, and Eq. (2.22) gives us $\psi_E(\mathbf{r})$.

Let us look more closely at the angular equation. Multiplying through by $-\hbar^2 Y(\theta, \varphi)$, we obtain

$$L^2Y(\theta, \varphi) = \hbar^2\alpha Y(\theta, \varphi), \tag{2.27}$$

which is in the form of an eigenvalue equation [see Eq. (1.23)]. Thus we conclude that $Y(\theta, \varphi)$ is an eigenfunction of the operator L^2 with eigenvalues

[3] If this method is not familiar, refer to a good differential equations text. See, for example, Earl Rainville and Phillip Bedient, *Elementary Differential Equations* (New York: Macmillan, 1969), pp. 361–364.

$\hbar^2\alpha$. Admittedly, as yet we know very little about L^2. In particular, we must verify that it is the quantum mechanical operator corresponding to the square of the orbital angular momentum. We shall now try to discover the physical significance and properties of L^2. Having done so, we shall be ready to tackle (and solve) Eq. (2.27).

2.3 ORBITAL ANGULAR MOMENTUM AND THE L^2 EIGENVALUE PROBLEM

Classical Angular Momentum

Consider a classical particle of mass m_1. Suppose that the linear momentum of the particle is $\mathbf{p} = m_1\mathbf{v}$ and its position is $\mathbf{r}$, defined relative to a specified origin. The angle between $\mathbf{p}$ and $\mathbf{r}$ is β, as shown in Fig. 2.2. Then the orbital

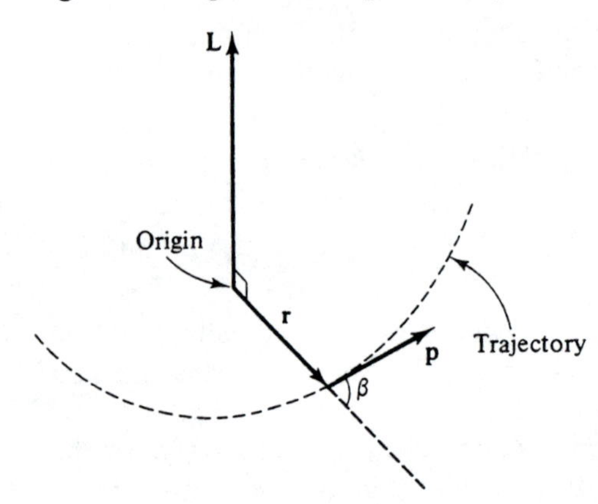

Figure 2.2 Quantities used in the definition of the classical orbital angular momentum **L**. The particle moves along the trajectory shown with linear momentum $\mathbf{p} = m_1\mathbf{v}$ a distance **r** from the origin.

angular momentum about the origin specified is

$$\mathbf{L} = \mathbf{r} \times \mathbf{p}, \tag{2.28}$$

where **L** is a vector perpendicular to **r** and **p** of magnitude

$$|\mathbf{L}| = rp \sin \beta; \tag{2.29}$$

or

$$|\mathbf{L}| = rm_1 v \sin \beta. \tag{2.30}$$

The Transition to Quantum Mechanics

We now wish to obtain the quantum mechanical operator L^2 that corresponds to the square of the observable orbital angular momentum. We begin by deriving the operator equivalent of **L** from Eq. (2.28). The linear momentum operator is [see Eq. (1.14)]

$$\mathbf{p} = -i\hbar\,\nabla, \tag{2.31}$$

or

$$\mathbf{p} = -i\hbar\left(\hat{\mathbf{x}}\frac{\partial}{\partial x} + \hat{\mathbf{y}}\,\frac{\partial}{\partial y} + \hat{\mathbf{z}}\frac{\partial}{\partial z}\right), \tag{2.32}$$

where $\hat{\mathbf{x}}$, $\hat{\mathbf{y}}$, and $\hat{\mathbf{z}}$ are unit vectors in the x, y, and z directions. In cartesian coordinates, the components of the classical angular momentum $\mathbf{L}$ are

$$L_x = yp_z - zp_y, \tag{2.33a}$$

$$L_y = zp_x - xp_z, \tag{2.33b}$$

$$L_z = xp_y - yp_x. \tag{2.33c}$$

Using Eq. (2.32) for the components of $\mathbf{p}$, we obtain Hermitian operators for the components of $\mathbf{L}$:

$$L_x = -i\hbar\left(y\frac{\partial}{\partial z} - z\frac{\partial}{\partial y}\right), \tag{2.34a}$$

$$L_y = -i\hbar\left(z\frac{\partial}{\partial x} - x\frac{\partial}{\partial z}\right), \tag{2.34b}$$

$$L_z = -i\hbar\left(x\frac{\partial}{\partial y} - y\frac{\partial}{\partial x}\right). \tag{2.34c}$$

These are the three cartesian components of the orbital angular momentum operator.

> **Exercise 2.1** Show that the operator L_z is Hermitian and thus has real eigenvalues.

We chose to make the transition from classical observables to quantum mechanical operators in cartesian coordinates because of the simplicity of the required equations. However, for purposes of the central force problem, we prefer to use spherical coordinates. From the transformation equation, Eq. (2.15), it follows that the components of $\mathbf{L}$ can be written as

$$L_x = i\hbar\left(\sin\varphi\frac{\partial}{\partial\theta} + \operatorname{ctn}\theta\cos\varphi\frac{\partial}{\partial\varphi}\right), \tag{2.35a}$$

$$L_y = -i\hbar\left(\cos\varphi\frac{\partial}{\partial\theta} - \operatorname{ctn}\theta\sin\varphi\frac{\partial}{\partial\varphi}\right), \tag{2.35b}$$

$$L_z = -i\hbar\frac{\partial}{\partial\varphi}. \tag{2.35c}$$

We shall have more to say about these important operators shortly. For the present, let us note that only derivatives with respect to θ and φ appear and that the component of orbital angular momentum along the polar axis ($\hat{z}$ axis) of the spherical coordinate system has a particularly simple form. This latter fact is simply a consequence of Eq. (2.15c) and has no special physical significance, since the choice of the polar axis was arbitrary.

We can now calculate the operator for L^2. By squaring each of Eqs. (2.35), we find

$$L^2 = L_x^2 + L_y^2 + L_z^2 = -\hbar^2\left[\frac{1}{\sin\theta}\frac{\partial}{\partial\theta}\left(\sin\theta\frac{\partial}{\partial\theta}\right) + \frac{1}{\sin^2\theta}\frac{\partial^2}{\partial\varphi^2}\right]. \tag{2.36}$$

This result is identical with the identification postulated in Eq. (2.19). We have therefore proven that the operator artificially introduced into the Schroedinger equation in Sec. 2.2 is, in fact, the operator for the square of the orbital angular momentum.

From this fact it follows that the eigenvalue equation (2.27), which resulted from the separation of radial and angular coordinates in the central force problem, is precisely the eigenvalue problem for the square of the orbital angular momentum. Hence the eigenvalues $\hbar^2\alpha$ are the only values that can be obtained in a measurement of L^2. Since $Y(\theta, \varphi)$ is the angular part of the energy eigenfunction $\psi_E(\mathbf{r})$, we conclude that the eigenfunction $\psi_E(\mathbf{r})$ is simultaneously an eigenfunction of the operators $\mathcal{H}$ and L^2. This is consistent with the fact that these two operators commute,

$$[\mathcal{H}, L^2] = 0, \tag{2.37}$$

which tells us that L^2 is a constant of the motion for *any* central force problem. In general, if an observation of the square of the orbital angular momentum is made on a spherically symmetric system in some stationary state, the result will be one of the eigenvalues $\hbar^2\alpha$.[4] We say that L^2 is sharp. Since each $\psi_E(\mathbf{r})$ will be identified in this fashion with a particular eigenvalue of L^2, we could use the L^2 eigenvalue, in addition to the energy eigenvalue E, to label $\psi(\mathbf{r})$ and thus to distinguish one state function from another.

2.4 SOLUTION OF THE L^2 EIGENVALUE PROBLEM

Separation of Angular Coordinates

Using the expression for L^2 in spherical coordinates, Eq. (2.36), we can rewrite the L^2 eigenvalue equation as

$$\left[\frac{1}{\sin\theta}\frac{\partial}{\partial\theta}\left(\sin\theta\frac{\partial}{\partial\theta}\right) + \frac{1}{\sin^2\theta}\frac{\partial^2}{\partial\varphi^2} + \alpha\right]Y(\theta, \varphi) = 0. \tag{2.38}$$

[4]This statement does not hold if accidental degeneracies are present. Such is the case, for example, in the one-electron atom (see Chapter 3).

To solve this equation, we again employ the method of separation of variables. Thus we seek solutions of the form

$$Y(\theta, \varphi) = \Theta(\theta)\Phi(\varphi). \tag{2.39}$$

Inserting Eq. (2.39) into (2.38) and multiplying through by $\sin^2 \theta / Y(\theta, \varphi)$, we find

$$\underbrace{\frac{1}{\Theta(\theta)}\left[\sin\theta \frac{d}{d\theta}\left(\sin\theta \frac{d}{d\theta}\right) + \alpha \sin^2\theta\right]\Theta(\theta)}_{\text{function of } \theta \text{ only}} + \underbrace{\frac{1}{\Phi(\varphi)}\frac{d^2}{d\varphi^2}\Phi(\varphi)}_{\text{function of } \varphi \text{ only}} = 0. \tag{2.40}$$

Introducing a separation constant m^2 (we choose m^2 rather than, say, m, for later convenience), we obtain two new eigenvalue equations, a θ-equation

$$\left[-\frac{1}{\sin\theta}\frac{d}{d\theta}\left(\sin\theta\frac{d}{d\theta}\right) + \frac{m^2}{\sin^2\theta}\right]\Theta(\theta) = \alpha\Theta(\theta) \tag{2.41}$$

and a φ-equation

$$\frac{d^2}{d\varphi^2}\Phi(\varphi) = -m^2\Phi(\varphi). \tag{2.42}$$

All we need do is solve these equations separately and multiply the resulting functions to form $Y(\theta, \varphi)$.

Solution of the φ-Equation

Equation (2.42) is familiar from the elementary theory of ordinary differential equations. Two linearly independent solutions[5] are $e^{im\varphi}$ and $e^{-im\varphi}$. We can therefore write the φ-dependence of the Hamiltonian eigenfunction as

$$\boxed{\Phi_m(\varphi) = \frac{1}{\sqrt{2\pi}}e^{im\varphi}, \qquad m = 0, \pm 1, \pm 2, \ldots} \tag{2.43}$$

where we have appended the subscript m to $\Phi(\varphi)$ and where the single-valuedness of Φ requires that m be an integer.

> **Exercise 2.2** Show that the physical requirement that $\Phi(\varphi)$ must be single valued with respect to the variable φ leads to the restriction that m take on only integral values.

[5] Equally valid linearly independent solutions are $\sin m\varphi$ and $\cos m\varphi$. We choose to use the exponential forms here.

The factor of $1/\sqrt{2\pi}$ in Eq. (2.43) was introduced so that $\Phi(\varphi)$ would be normalized:

$$\int_0^{2\pi} |\Phi_m(\varphi)|^2 \, d\varphi = 1. \tag{2.44}$$

[We could have waited until we had obtained $R(r)$ and $\Theta(\theta)$ and then normalized $\psi_E(\mathbf{r})$. However, independent normalization of the three functions turns out to be more convenient.] It is easy to check that the set of solutions $\{\Phi_m\}$ is orthonormal,

$$\int_0^{2\pi} \Phi_m^*(\varphi)\Phi_{m'}(\varphi) \, d\varphi = \delta_{mm'}. \tag{2.45}$$

Solution of the θ-Equation

Unfortunately, the θ-equation (2.41) is more difficult to solve. However, it can be put in the form of a special type of ordinary differential equation, the solutions of which are well known.[6] This is the *associated Legendre equation*:

$$\frac{d}{d\xi}\left[(1-\xi^2)\frac{d}{d\xi}\Theta(\xi)\right] + \left(\alpha - \frac{m^2}{1-\xi^2}\right)\Theta(\xi) = 0. \tag{2.46}$$

Exercise 2.3 By defining the variable

$$\xi = \cos\theta, \tag{2.47}$$

derive Eq. (2.46) from (2.41).

If we now rewrite the separation constant α as

$$\alpha = \ell(\ell+1), \tag{2.48}$$

Eq. (2.46) becomes

$$\frac{d}{d\xi}\left[(1-\xi^2)\frac{d}{d\xi}\Theta(\xi)\right] + \left[\ell(\ell+1) - \frac{m^2}{1-\xi^2}\right]\Theta(\xi) = 0, \tag{2.49}$$

Some further manipulation leads to the equivalent equation

$$\left[(1-\xi^2)\frac{d^2}{d\xi^2} - 2\xi\frac{d}{d\xi} + \ell(\ell+1) - \frac{m^2}{1-\xi^2}\right]\Theta(\xi) = 0. \tag{2.50}$$

The solutions of Eq. (2.50) are the *associated Legendre polynomials* of the

[6] Here is where we encounter our first "special function." From time to time you may wish to seek out more information on a particular special function than we have space to present; several useful references are mentioned at the end of this chapter.

first and second kind,[7] $P_\ell^m(\xi)$ and $Q_\ell^m(\xi)$, where ℓ is an integer that can take on the values

$$\ell = |m|, |m| + 1, |m| + 2, \ldots . \tag{2.51}$$

Of the two linearly independent solutions, $P_\ell^m(\xi)$ is well behaved for all finite values of ξ, but $Q_\ell^m(\xi)$ has a singularity at $\xi = \pm 1$ (i.e., the function becomes infinite at this point). The values of θ corresponding to $\xi = +1$ and $\xi = -1$ are $\theta = 0$ and $\theta = \pi$, respectively. Since the wave function must be finite at all physical points, the functions $Q_\ell^m(\xi)$ are physically unacceptable solutions. We are left with the associated Legendre polynomials of the first kind, $P_\ell^m(\xi)$, as the solutions to the θ-equation.

For reasonably small values of ℓ and m, the associated Legendre polynomials are rather simple functions; several of them appear in Table 2.1.

Table 2.1
Legendre polynomials for $\ell = 0$, 1, 2, and 3 and corresponding associated Legendre polynomials.

ℓ	$P_\ell(\cos\theta)$	m	$P_\ell^m(\cos\theta)$
0	$P_0 = 1$	0	$P_0^0 = 1$
1	$P_1 = \cos\theta$	0	$P_1^0 = \cos\theta$
		1	$P_1^1 = \sin\theta$
2	$P_2 = \frac{1}{2}(3\cos^2\theta - 1)$	0	$P_2^0 = \frac{1}{2}(3\cos^2\theta - 1)$
		1	$P_2^1 = 3\sin\theta\cos\theta$
		2	$P_2^2 = 3\sin^2\theta$
3	$P_3 = \frac{1}{2}(5\cos^3\theta - 3\cos\theta)$	0	$P_3^0 = \frac{1}{2}(5\cos^3\theta - 3\cos\theta)$
		1	$P_3^1 = \frac{3}{2}\sin\theta\,(5\cos^2\theta - 1)$
		2	$P_3^2 = 15\sin^2\theta\cos\theta$
		3	$P_3^3 = 15\sin^3\theta$

Also included in Table 2.1 are the *Legendre polynomials* $P_\ell(\xi)$. They may be used to generate the associated Legendre polynomials by the differential formula

$$P_\ell^m(\xi) = (1 - \xi^2)^{m/2} \frac{d^m}{d\xi^m} P_\ell(\xi), \qquad m \geq 0, \quad \ell \geq m, \tag{2.52}$$

which permits us to derive any $P_\ell^m(\xi)$ given $P_\ell(\xi)$. The latter can be generated by use of the *Rodrigues formula*,

$$P_\ell(\xi) = \frac{1}{2^\ell \ell!} \frac{d^\ell}{d\xi^\ell} (\xi^2 - 1)^\ell . \tag{2.53}$$

[7]These polynomials may be obtained by using the method of series expansion to solve Eq. (2.50). See, for example, W. W. Bell, *Special Functions for Scientists and Engineers* (London: D. Van Nostrand Co. Ltd., 1968), Chaps. 1 and 3.

Exercise 2.4 (a) Use Eq. (2.53) to show that

$$P_4(\cos\theta) = \tfrac{1}{8}(35\cos^4\theta - 30\cos^2\theta + 3).$$

(b) Use Eq. (2.52) and the fact that

$$P_5(\cos\theta) = \tfrac{1}{8}(63\cos^5\theta - 70\cos^3\theta + 15\cos\theta)$$

to derive expressions for $P_5^1(\cos\theta)$ and $P_5^3(\cos\theta)$.

Inspection of Table 2.1 reveals that for $\ell = 0, 1, 2,$ and 3, $P_\ell(\xi)$ is $\begin{Bmatrix}\text{even}\\ \text{odd}\end{Bmatrix}$ for $\begin{Bmatrix}\text{even}\\ \text{odd}\end{Bmatrix}$ values of ℓ, a useful property that is true in general and that can be derived from Eq. (2.53).

Notice that the value of m in Eq. (2.52) is restricted to nonnegative integers. It is convenient to define $P_\ell^m(\cos\theta)$ for a negative value of m by the relation

$$P_\ell^{-m}(\xi) = P_\ell^m(\xi), \qquad m \geq 0. \tag{2.54}$$

Since we would like the solutions of the θ-equation (2.50) to be normalized, we shall define real functions by

$$\boxed{\Theta_{\ell m}(\theta) = (-1)^{(m+|m|)/2}\left[\frac{(2\ell+1)}{2}\frac{(\ell-|m|)!}{(\ell+|m|)!}\right]^{1/2} P_\ell^m(\cos\theta),} \tag{2.55}$$

which is valid for all m. These are normalized solutions of Eq. (2.50):

$$\int_0^\pi \Theta_{\ell m}(\theta)\Theta_{\ell' m}(\theta)\sin\theta\, d\theta = \delta_{\ell\ell'}. \tag{2.56}$$

In Eq. (2.55), the phase convention has been chosen so that

$$\Theta_{\ell,-m}(\theta) = (-1)^m \Theta_{\ell m}(\theta), \qquad m \geq 0. \tag{2.57}$$

Spherical Harmonics

This completes the task of solving the eigenvalue equation (2.27). The full eigenfunctions of L^2 are simply products of the functions $\Theta(\theta)$ and $\Phi(\varphi)$; that is,

$$Y_{\ell m}(\theta, \varphi) = \Theta_{\ell m}(\theta)\Phi_m(\varphi), \tag{2.58}$$

where we have labeled the function Y with the integers ℓ and m. These functions, which are also encountered in the study of boundary value problems in electrostatics, are called *spherical harmonics*; several spherical harmonics are presented in Table 2.2.[8] They have the property

Table 2.2
Some normalized spherical harmonics.

ℓ	m	$Y_{\ell m}(\theta, \varphi)$
0	0	$Y_{00} = \frac{1}{\sqrt{4\pi}}$
1	0	$Y_{10} = \left(\frac{3}{4\pi}\right)^{1/2} \cos\theta$
	± 1	$Y_{1\pm 1} = \mp\left(\frac{3}{8\pi}\right)^{1/2} \sin\theta\, e^{\pm i\varphi}$
2	0	$Y_{20} = \left(\frac{5}{16\pi}\right)^{1/2} (3\cos^2\theta - 1)$
	± 1	$Y_{2\pm 1} = \mp\left(\frac{15}{8\pi}\right)^{1/2} \sin\theta\cos\theta\, e^{\pm i\varphi}$
	± 2	$Y_{2\pm 2} = \left(\frac{15}{32\pi}\right)^{1/2} \sin^2\theta\, e^{\pm 2i\varphi}$
3	0	$Y_{30} = \left(\frac{7}{16\pi}\right)^{1/2} (5\cos^3\theta - 3\cos\theta)$
	± 1	$Y_{3\pm 1} = \mp\left(\frac{21}{64\pi}\right)^{1/2} \sin\theta\,(5\cos^2\theta - 1) e^{\pm i\varphi}$
	± 2	$Y_{3\pm 2} = \left(\frac{105}{32\pi}\right)^{1/2} \sin^2\theta\cos\theta\, e^{\pm 2i\varphi}$
	± 3	$Y_{3\pm 3} = \mp\left(\frac{35}{64\pi}\right)^{1/2} \sin^3\theta\, e^{\pm 3i\varphi}$

$$Y_{\ell,-m}(\theta, \varphi) = (-1)^m Y^*_{\ell m}(\theta, \varphi). \tag{2.59}$$

Moreover, it follows from Eqs. (2.45) and (2.56) that the spherical harmonics form an orthonormal set,

$$\int Y^*_{\ell m}(\theta, \varphi)\, Y_{\ell' m'}(\theta, \varphi)\, d\hat{\mathbf{r}} = \delta_{\ell\ell'}\delta_{mm'}, \tag{2.60}$$

where $d\hat{\mathbf{r}} = \sin\theta\, d\theta\, d\varphi$. We shall look at graphs of some spherical harmonics in the next chapter.

Notice that for each value of ℓ there are $2\ell + 1$ spherical harmonics $Y_{\ell m}(\theta, \varphi)$, each with a different value of m in the range $-\ell \leq m \leq \ell$. These are degenerate eigenfunctions of L^2; they are labeled by the same value of ℓ.

[8] For a more extensive list of spherical harmonics, see Linus Pauling and E. B. Wilson: *Introduction to Quantum Mechanics* (New York: McGraw-Hill, 1935), pp. 134–135.

The spherical harmonics may look rather trivial, but they contain a wealth of physics. For one thing, they completely describe the angular motion for *any* central force problem. We need only solve the radial equation (2.25) for $R(r)$ and multiply it by one of the spherical harmonics to obtain an energy eigenfunction $\psi_E(\mathbf{r})$ for the problem [recall Eq. (2.22)].

Moreover, the spherical harmonics are the eigenfunctions of the square of the orbital angular momentum. Summarizing the results obtained above, we write

$$\boxed{L^2 Y_{\ell m}(\theta, \varphi) = \hbar^2 \ell(\ell + 1) Y_{\ell m}(\theta, \varphi) \qquad \text{for } \ell = 0, 1, 2, \ldots; -\ell \leq m \leq \ell,} \tag{2.61}$$

where the conditions on ℓ and m are equivalent to those expressed in Eqs. (2.51) and (2.43).

We have pointed out that since $\psi_E(\mathbf{r})$ is simultaneously an eigenfunction of L^2 and $\mathcal{H}$, we can use the eigenvalues of L^2—namely, $\hbar^2\ell(\ell + 1)$—to label $\psi_E(\mathbf{r})$. Clearly, all that must be specified is ℓ. Thus ℓ labels the eigenvalues of the Hermitian operator L^2 and is a quantum number, the *orbital angular momentum quantum number*. (Although ℓ is sometimes loosely called the "orbital angular momentum," the actual magnitude of the orbital angular momentum is $\hbar\sqrt{\ell(\ell + 1)}$.)

The L_z Eigenvalue Problem

Still more useful information can be extracted from the spherical harmonics. To obtain it, let us operate on $Y_{\ell m}(\theta, \varphi)$ with the operator for the z component of orbital angular momentum L_z. This operator, given in Eq. (2.35c), operates only on functions of φ; we find that

$$\boxed{L_z Y_{\ell m}(\theta, \varphi) = m\hbar Y_{\ell m}(\theta, \varphi) \qquad \text{for } \ell = 0, 1, 2, \ldots; -\ell \leq m \leq \ell.} \tag{2.62}$$

Therefore the spherical harmonics satisfy *two* eigenvalue equations and are eigenfunctions of both the square of the orbital angular momentum *and* the z component of the orbital angular momentum. Since m labels the eigenvalues of L_z, it is also a quantum number, the so-called *magnetic quantum number*.

Commutation Relations

Since the spherical harmonics, which form a complete set in the variables θ and φ, satisfy eigenvalue equations for L^2 and L_z, we expect that these operators commute. Indeed, it is easy to demonstrate that[9]

$$[L^2, L_z] = 0. \tag{2.63}$$

The components of **L** also satisfy a number of other commutation relations:

$$[L^2, L_x] = [L^2, L_y] = 0, \tag{2.64}$$

$$[L_x, L_y] = i\hbar L_z, [L_y, L_z] = i\hbar L_x, \quad [L_z, L_x] = i\hbar L_y. \tag{2.65}$$

Equation (2.63) reaffirms what we already know: that we can find simultaneous eigenfunctions of L^2 and L_z. Equations (2.65) tell us that these functions cannot *also* be eigenfunctions of L_x or L_y, since L_x and L_y do not commute with L_z. However, we know from Eq. (2.64) that functions do exist that are simultaneous eigenfunctions of L_x and L^2 *or* of L_y and L^2; they are not these spherical harmonics, of course.[10]

Exercise 2.5 Show that $[L_x, L^2] = 0$ and $[L_y, L_z] = i\hbar L_x$.

Consequently, in an experiment we can simultaneously determine the energy, the magnitude of orbital angular momentum, and the component of **L** along one and only one spatial axis. If we define the z direction along that axis, then we cannot also determine L_x or L_y; if we insist on specifying L_x, we must give up knowledge about L_z. These are, of course, nonclassical features of the theory and are really consequences of the Heisenberg uncertainty principle.

Raising and Lowering Operators

It is useful to define two new operators: L_+, the *raising operator*, and L_-, the *lowering operator*. They are defined as

$$L_\pm \equiv L_x \pm iL_y, \tag{2.66}$$

where the operators L_x and L_y correspond to the x and y components of orbital angular momentum. Since L_x and L_y are Hermitian, the raising and lowering operators satisfy

$$L_+^\dagger = L_- \quad \text{and} \quad L_-^\dagger = L_+, \tag{2.67}$$

[9]To do so, you will need the easily verified commutator relationships

$$[x, p_x] = [y, p_y] = [z, p_z] = i\hbar$$
$$[x, p_y] = [y, p_x] = \cdots = 0.$$

[10]In fact, it is possible to form linear combinations of the $Y_{\ell m}(\theta, \varphi)$ involving different values of m that are eigenfunctions of L_x and L^2 but not of L_z (see Prob. 2.7).

where $L_+^\dagger$ is the Hermitian conjugate (adjoint) of L_+.[11] Other properties of L_+ and L_- are considered in Prob. 2.5.

Exercise 2.6 Using Eqs. (2.34) and (2.35) for L_x and L_y, write expressions for L_+ and L_- in cartesian and spherical coordinates.

These new operators can be used to generate spherical harmonics having different values of m. The result of operation on $Y_{\ell m}(\theta, \varphi)$ with L_+ and L_- is [see Prob. 2.5b]

$$L_\pm Y_{\ell m}(\theta, \varphi) = \hbar\sqrt{(\ell \mp m)(\ell \pm m + 1)}\, Y_{\ell, m\pm 1}(\theta, \varphi). \tag{2.68}$$

Two applications of the raising and lowering operators will be found in Probs. 2.6 and 2.7.

2.5 FULL EIGENFUNCTIONS OF THE CENTRAL FORCE PROBLEM

Solutions of the Hamiltonian Eigenvalue Problem

Although much has been achieved by the preceding analysis, additional work remains to be done. Still before us is the spectre of the radial equation (2.25), which must be solved. Using Eq. (2.48) for α, we can rewrite this equation as

$$\left\{\frac{1}{r^2}\frac{d}{dr}\left(r^2\frac{d}{dr}\right) - \frac{\ell(\ell+1)}{r^2} + \frac{2m_1}{\hbar^2}[E_{n\ell} - V(r)]\right\} R_{n\ell}(r) = 0. \tag{2.69}$$

We have labeled the energy and the radial function with two quantum numbers: n and ℓ. E and $R(r)$ depend on ℓ, the orbital angular momentum quantum number, because $\mathcal{H}$ contains L^2 [see Eq. (2.21)]. The other quantum number n is a new one; it is an integer that we shall use to distinguish different linearly independent solutions of Eq. (2.69). Thus for fixed ℓ we have a set of energies and radial functions: $E_{n\ell}$ and $R_{n\ell}(r)$ for $n = 0, 1, 2, \ldots$.

If we can somehow solve Eq. (2.69) for $R_{n\ell}(r)$ and $E_{n\ell}$, we can obtain a set of valid stationary-state eigenfunctions by multiplying by the spherical harmonics:

$$\boxed{\psi_{n\ell m}(\mathbf{r}) = R_{n\ell}(r) Y_{\ell m}(\theta, \varphi).} \tag{2.70}$$

[11]The *Hermitian conjugate* of an operator A is written $A^\dagger$ and is defined by $\int \psi_1^* A^\dagger \psi_2 d\mathbf{r} = \int (A\psi_1)^* \psi_2 d\mathbf{r}$ for arbitrary admissable ψ_1 and ψ_2. [See David Saxon, *Elementary Quantum Mechanics* (San Francisco: Holden-Day, 1968), pp. 88 ff.]

Each function in the set is an eigenfunction of the mutually commuting operators $\mathcal{H}$, L^2, and L_z and is labeled by the appropriate quantum numbers. For each n and ℓ there are $2\ell + 1$ functions $\psi_{n\ell m}(\mathbf{r})$, each of which has the same energy $E_{n\ell}$. These are degenerate eigenfunctions of the Hamiltonian; the state with this value of $E_{n\ell}$ is said to be $(2\ell + 1)$-fold degenerate.

In some cases, additional so-called accidental degeneracies can occur. For example, if $V(r)$ is the coulomb potential energy, some eigenfunctions with different values of ℓ turn out to be degenerate. We shall study this situation in Chapter 3; it is not a general property of solutions to the central force problem.

The Centrifugal Potential Energy

Although the radial equation (2.69) cannot be solved without specializing to a particular $V(r)$, something can be learned about the nature of its solutions by rewriting it in a slightly different form. So let us define a new radial function $\chi(r)$ by

$$\chi_{n\ell}(r) \equiv rR_{n\ell}(r). \tag{2.71}$$

Substituting Eq. (2.71) into the radial equation, we find that $\chi_{n\ell}(r)$ satisfies the equation

$$\left\{\frac{d^2}{dr^2} - \frac{\ell(\ell+1)}{r^2} + \frac{2m_1}{\hbar^2}[E_{n\ell} - V(r)]\right\}\chi_{n\ell}(r) = 0. \tag{2.72}$$

We now define an *effective potential energy* $V^{\ell}_{\text{eff}}(r)$, analogous to the classical effective potential energy of Eq. (2.9):

$$V^{\ell}_{\text{eff}}(r) \equiv V(r) + \frac{\hbar^2}{2m_1}\frac{\ell(\ell+1)}{r^2} \tag{2.73}$$

Equation (2.72) can then be written as:

$$\left\{-\frac{\hbar^2}{2m_1}\frac{d^2}{dr^2} + V^{\ell}_{\text{eff}}(r)\right\}\chi_{n\ell}(r) = E_{n\ell}\chi_{n\ell}(r). \tag{2.74}$$

But this is simply the *one-dimensional* Schroedinger eigenvalue equation for a particle of mass m_1 with position coordinate r and potential energy $V^{\ell}_{\text{eff}}(r)$. The eigenfunctions are $\chi_{n\ell}(r)$, and the eigenvalues are the energies $E_{n\ell}$. Consequently, we can use our intuition developed in the study of various simple one-dimensional problems (see Prob. 1.1) to predict features of the solution of the radial equation. To illustrate, let us consider a particular central force problem.

Example 2.1
The Spherical Well

The second term in $V^{\ell}_{\text{eff}}(r)$ is often called the *centrifugal potential energy.* Let us explore the effect of different choices of ℓ in the centrifugal potential energy by looking at the bound states ($E < 0$) of a spherical well. The potential energy for this system is

$$V(r) = \begin{cases} -V_0, & 0 < r < a \\ 0, & r \geq a. \end{cases} \tag{2.75}$$

The form of this function is shown in Fig. 2.3(a). In Fig. 2.3(b) we have drawn $V^{\ell}_{\text{eff}}(r)$ for $\ell = 0$, 1, 2, and the radial functions $\chi_{n\ell}(r)$ for the lowest energy for $\ell = 0$ and 1. Several interesting physical features can be seen immediately. For example, as ℓ increases, the energy increases and the peak of probability distribution shifts to larger r.

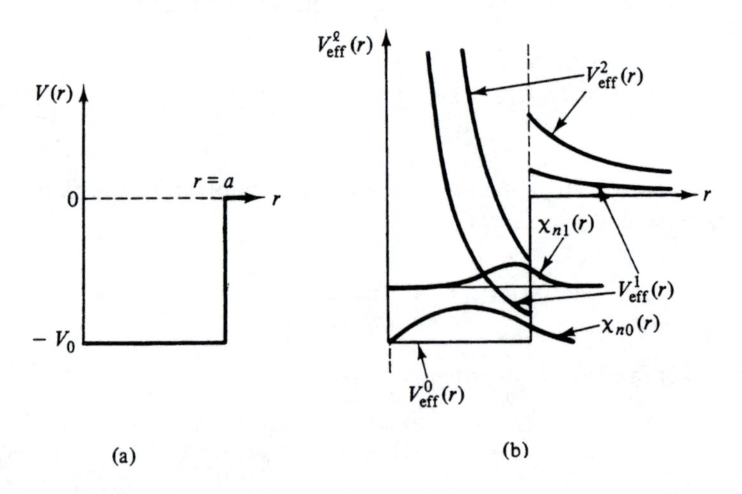

Figure 2.3 (a) The spherical potential well; (b) the effective potential energy for $\ell = 0, 1, 2$ and the radial function $\chi_{n\ell}(r)$ for the lowest energy level for $\ell = 0$ and $\ell = 1$. Also shown are the peaks in probability density for each state.

We already know the angular dependence of the stationary-state eigenfunctions $\psi_{n\ell m}(\mathbf{r})$ for this system; it is described by the spherical harmonics $Y_{\ell m}(\theta, \varphi)$. To obtain the radial function, we simply solve Eq. (2.72) with

$V(r)$ given in Eq. (2.75) and divide the resulting function $\chi_{n\ell}(r)$ by r. Let us consider the solution of this equation for the case $\ell = 0$, $m = 0$.[12]

We begin by defining two regions separated by the discontinuity in the potential energy at $r = a$. The first region contains all r in the interval $0 < r < a$; the solutions of Eq. (2.72) in this region will be denoted $\chi_n^<(r)$. The second region is $r > a$; the solutions here will be denoted $\chi_n^>(r)$. The equations for these functions can be written

$$\frac{d^2\chi_n^<(r)}{dr^2} + k_1^2\chi_n^<(r) = 0, \qquad 0 < r < a \tag{2.76a}$$

and

$$\frac{d^2\chi_n^>(r)}{dr^2} - k_2^2\chi_n^>(r) = 0, \qquad r > a, \tag{2.76b}$$

where we have introduced real positive wave numbers k_1 and k_2, defined by

$$k_1 = \sqrt{\frac{2m_1}{\hbar^2}(E_n + V_0)} \tag{2.77}$$

and

$$k_2 = \sqrt{-\frac{2m_1}{\hbar^2}E_n}. \tag{2.78}$$

(Recall that the energies of interest lie in the range $-V_0 < E_n < 0$.) The general solutions of these familiar second-order differential equations are

$$\chi_n^<(r) = A\sin k_1r + B\cos k_1r, \qquad 0 < r < a, \tag{2.79a}$$

$$\chi_n^>(r) = Ce^{k_2r} + De^{-k_2r}, \qquad r > a, \tag{2.79b}$$

where the coefficients are as-yet-arbitrary constants.

These purely mathematical solutions do not satisfy the boundary conditions on $\chi_n(r)$ demanded by the physical interpretation of the wave function, which was discussed in Sec. 1.2. In particular, we must require that

1. $\chi_n(r)$ be finite for all r, including the limit $r \longrightarrow \infty$.
2. $\chi_n(r) \longrightarrow 0$ in the limit $r \longrightarrow 0$.
3. χ_n and its first derivative be continuous at all values of r, in particular at $r = a$.

Condition (1) can be satisfied by the general solutions of Eqs. (2.79) only if $C = 0$. Condition (2) follows from applying the requirement that ψ be finite at $r = 0$. [We can also obtain this condition by viewing the problem as equivalent to a one-dimensional problem with $V = \infty$ at $r = 0$; see Eq.

[12] We shall drop the subscripts 0 from χ_{n0}, E_{n0}, ψ_{n00}, etc. in order to avoid clutter. Always keep in mind that the results obtained are only valid for $l = 0$, $m = 0$.

(2.75).] It can be satisfied only if $B = 0$. Condition (3) can be written

$$\chi_n^<(a) = \chi_n^>(a) \tag{2.80a}$$

and

$$\frac{d}{dr}\chi_n^<(r)\bigg|_{r=a} = \frac{d}{dr}\chi_n^>(r)\bigg|_{r=a}. \tag{2.80b}$$

We shall use these two equations to derive a transcendental equation for the energies E_n and to determine one of the two remaining unknown coefficients A and D. The second of these coefficients will be determined by requiring that $R_n(r)$ be normalized,—that is,

$$\int_0^\infty |R_n(r)|^2 r^2\, dr = 1, \tag{2.81}$$

so that the full eigenfunction $\psi_n(\mathbf{r})$ will be normalized. The normalization condition for $\chi_n(r)$ follows from substituting $\chi_n(r) = rR_n(r)$ into Eq. (2.81):

$$\int_0^\infty |\chi_n(r)|^2\, dr = 1. \tag{2.82}$$

This is merely the normalization requirement for the corresponding one-dimensional problem.

Applying the conditions of Eqs. (2.80) to the general solutions, Eqs. (2.79), we find

$$A \sin k_1 a = De^{-k_2 a} \tag{2.83a}$$

$$k_1 A \cos k_1 a = -k_2 De^{-k_2 a}. \tag{2.83b}$$

We now divide Eq. (2.83b) by (2.83a) and thus obtain a transcendental equation for the energy eigenvalues,

$$k_1 \cot k_1 a = -k_2. \tag{2.84}$$

Once Eq. (2.84) is solved for the allowed values of k_1 (and hence for k_2 and E_n), Eq. (2.83a) or (2.83b) can be used to relate D to A. Then all that remains is to calculate one of the coefficients A or D by requiring that the wave function ψ_n be normalized.

Exercise 2.7 Find A by normalizing $R_n(r)$. Write complete expressions for $\psi_n(\mathbf{r})$ for $0 < r < a$ and $r > a$.

To solve Eq. (2.84), we first rewrite it as

$$\tan k_1 a = -\frac{k_1}{k_2}. \tag{2.85}$$

This relation can be satisfied only by values of k_1a for which $\tan k_1a < 0$; that is, by values of k_1a in quadrants II or IV of a unit circle: $\pi/2 \leq k_1a \leq \pi$ or $3\pi/2 \leq k_1a \leq 2\pi$. (Of course, there are infinitely many other allowed values of k_1a in quadrants II or IV, separated from these values by increments of 2π—for example, $5\pi/2 \leq k_1a \leq 3\pi$.)

The easiest way to locate the values of k_1a that solve Eq. (2.85) is to graph this equation. However, it is preferable to make a simple transformation first. As can easily be verified, in quadrants II or IV the following trigonometric identity holds:

$$\tan\theta = -\frac{\alpha}{\beta} \Longleftrightarrow \sin\theta = \frac{\pm\alpha}{\sqrt{\alpha^2+\beta^2}} \tag{2.86}$$

with + in II and − in IV. (In this relation, α and β are positive real numbers.) Hence Eq. (2.85) can be transformed into

$$\sin k_1a = \frac{\pm k_1}{\sqrt{k_1^2+k_2^2}} = \frac{\pm k_1a}{\gamma}, \tag{2.87}$$

where we have introduced the parameter

$$\gamma = \sqrt{\frac{2m_1V_0a^2}{\hbar^2}}. \tag{2.88}$$

If we restrict the argument of the sine function to quadrant II, the range $\pi/2$ to π, we can write Eq. (2.87) as

$$\gamma\sin(k_1a - n\pi) = k_1a. \tag{2.89}$$

The odd values of the integer n will produce the negative sign necessary in quadrant IV, and the 2π periodicity of $\sin k_1a$ is embodied in the even values of n.

Let us now plot $y = \gamma\sin(k_1a - n\pi)$ versus $k_1a - n\pi$ for values of $k_1a - n\pi$ in the second quadrant. The solutions of Eq. (2.89)—and hence the eigenvalues—are simply obtained by finding the intersections of this curve with the lines $y = (k_1a - n\pi) + n\pi$; there will be one such line for each integer $n > 0$. (A safe upper limit on n is the largest integer that is less than or equal to γ/π.) For the case $\gamma = 3\pi$, we obtain the graph shown in Fig. 2.4. There are three bound states corresponding to $k_1a = 2.83$, 5.64, and 8.34. If we express the energies in units of $\hbar^2/2m_1a^2$, these roots correspond to energies of −80.8, −57.1, and −19.6, respectively.[13]

Why do we bother with the transformation to Eq. (2.89)? The answer is that the curves are easier to draw and allow accurate graphical solution.

[13]In these units, the depth of the well is −88.8.

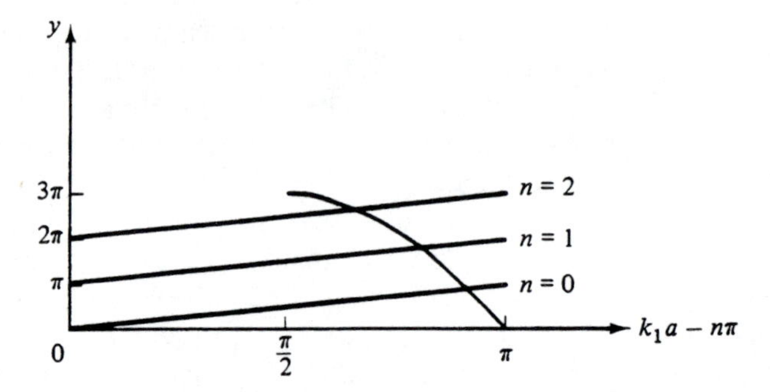

Figure 2.4 Graphical solution of the spherical square well problem (for $\gamma = 3\pi$) showing three bound states.

The graphs so obtained also facilitate understanding the effects of adjusting the well parameters. For example, we see from Fig. 2.4 and Eq. (2.89) that the number of bound states increases as γ increases—that is, as either the width or the depth of the well increases. In solving more complicated problems (e.g., in molecular physics), it is important to be able to produce such graphs easily and to glean from them insight into the physical properties of the system.

Of course, we could have chosen innumerable other central force problems instead of the spherical well. In all such problems, the angular dependence of the energy eigenfunctions is described by the spherical harmonics, and the radial dependence is obtained by solving the appropriate radial equation. Not all radial equations are as easy to solve as Eqs. (2.76). In particular, the one of greatest interest to us, the radial equation for a one-electron atom, is a bit more formidable. But it, too, can be solved; we shall do so in the next chapter.

SUGGESTED READINGS

Almost any book on "mathematical methods" will contain a chapter or two dealing with special functions. Two such books are

MARGENAU, HENRY, and GEORGE M. MURPHY, *The Mathematics of Physics and Chemistry*, 2nd ed. London: D. Van Nostrand Co. Ltd., 1956.

MATHEWS, JON, and R. L. WALKER, *Mathematical Methods of Physics*, 2nd ed. New York: W. A. Benjamin, 1970.

An especially useful text on this subject is

BELL, W. W., *Special Functions for Scientists and Engineers*. London: D. Van Nostrand Co. Ltd., 1968.

PROBLEMS

2.1 Spherical Well Eigenfunctions for $\ell = 0$ (**)

Consider again the spherical potential well of Example 2.1. Suppose that the well has depth and width such that $V_0 a^2 = 9/2$ in units of $\hbar^2/\text{m}$. Consider the case of $\ell = 0$.

(a) Write down radial wave functions for $0 < r < a$ and $r > a$ in terms of k_1 and k_2 (the wave numbers inside and outside the well, respectively). Write your wave function for $r > a$ in terms of sinh and cosh functions. Determine all but one of the unknown coefficients in these functions by requiring that $\chi^<(r = 0) = 0$ and that $\chi(r)$ and $d\chi/dr$ be continuous at $r = a$. Now apply the additional restriction that the radial wave function have unit slope at the origin: $(d\chi^</dr)|_{r=0} = 1$. (This is just an alternate way to normalize the function.)

(b) Suppose that you are given three values of k_1 for this system: $k_1 = 0.79$, 0.76, and 0.73. Sketch semiquantitatively[14] the radial wave function for each of these values of k_1. To what energies do these values correspond? Give a thorough explanation of the behavior of these three wave functions for $r > a$, using physical rather than mathematical arguments.

(c) Based on your sketches in part (b), which of these values of the energy is closest to an actual bound-state eigenvalue for this system?

(d) Use the transcendental equation (2.84), which can also be derived from your expression for $\chi^>(r)$ in part (a), to see how far from an actual eigenvalue your choice of k_1 in part (c) is. (Do not go through an iterative procedure. Simply see how well the equation is satisfied by your choice of k_1.)

2.2 Infinite Potential Well ($\ell = 0$ States) (**)

A special case of the spherical potential well of Example 2.1 is the infinite potential well defined by the potential energy

$$V(r) = \begin{cases} 0, & 0 < r < a \\ \infty, & r > a. \end{cases}$$

Suppose that a particle of mass m_1 has this potential energy. Consider only the $\ell = 0$ states.

(a) Write an expression for $\psi_{n\ell m}(\mathbf{r})$ for $0 < r < a$.

(b) Defining a wave number $k_{n\ell}^2 = 2m_1 E_{n\ell}/\hbar^2$, write a differential equation for $\chi_{n\ell}^<(r)$, where $\chi_{n\ell}(r) = rR_{n\ell}(r)$ [see Eq. (2.72)]. Find a real, general solution to your equation and determine the two "unknown" coefficients by applying boundary conditions and normalization. Qualitatively compare the form of your result with the eigenfunctions of the one-dimensional infinite potential well.

[14]A semiquantitative sketch is made by calculating just enough points to enable you to draw a reasonably accurate curve.

(c) For a well of width $a = 3a_0$, write explicit expressions for $\psi_{n00}(\mathbf{r})$ for the lowest two states.[15] Sketch the corresponding radial functions, $R_{10}(r)$ and $R_{20}(r)$, semiquantitatively as a function of $k_{n0}r$.

(d) Obtain an explicit expression for the energy eigenvalues E_{n0}. Calculate E_{n0} for a well of width $a = 3a_0$ for the lowest four states.

2.3 Infinite Potential Well ($\ell \neq 0$) (***)

[This problem does not require that you have done Prob. 2.2. You will, however, need a book on special functions. See Suggested Readings for this chapter.]

Consider a particle of mass m_1 with the infinite potential energy of Prob. 2.2. We shall study states with $\ell \neq 0$.

(a) Defining $k_{n\ell}^2 = 2m_1E_{n\ell}/\hbar^2$, write a differential equation for $\chi_{n\ell}^{<}(r)$, where $\chi_{n\ell}(r) = rR_{n\ell}(r)$ [see Eq. (2.72)]. This equation is not yet in a "standard form." To derive an equation in such a form, define a new variable $z \equiv k_{n\ell}r$ and a function $\eta_{n\ell}(z)$ by

$$\eta_{n\ell}(z) \equiv (k_{n\ell}r)^{-1/2}\chi_{n\ell}^{<}(r).$$

Derive a differential equation for $\eta_{\ell n}(z)$.

(b) Using a table of special functions, find a general solution to your equation for $\eta_{n\ell}(z)$. Obtain an expression for $\chi_{n\ell}^{<}(r)$, expressing your result in terms of the spherical Bessel functions $j_\ell(z)$ and $n_\ell(z)$. Use the properties of these functions and the boundary conditions of this problem to determine $R_{n\ell}^{<}(r)$ to within a multiplicative normalization constant.

(c) Consider a well of width $a = 3a_0$.[15] Choose the normalization constant to be 1. For $\ell = 1$, write explicit expressions for $R_{n\ell}^{<}(r)$ and $\psi_{n\ell m}(\mathbf{r})$ for all allowed values of m. Plot semiquantitatively $R_{11}^{<}(r)$ as a function of $k_{11}r$.

(d) Derive a transcendental equation for $E_{n\ell}$. How many bound states exist for each value of ℓ? Consider again the well of width $a = 3a_0$ and obtain the three lowest energies for $\ell = 1$.

2.4 Working with Spherical Harmonics (*)

Consider the spherical harmonic $Y_{21}(\theta, \varphi)$ (see Table 2.2).

(a) Show by explicit differentiation that this function is an eigenfunction of L^2 with eigenvalue $6\hbar^2$.

(b) Show by explicit differentiation that this function is an eigenfunction of L_z with eigenvalue $\hbar$.

(c) Verify that this function is normalized.

(d) Show that this function is orthogonal to $Y_{10}(\theta, \varphi)$.

2.5 Raising and Lowering Operators for Orbital Angular Momentum (*)

In this problem we shall prove some useful properties of the raising and lowering operators L_+ and L_- introduced in Eqs. (2.66) to (2.68).

[15] The constant a_0 is the first Bohr radius of the hydrogen atom, $a_0 \simeq 0.529$ Å [see Eq. (3.40) and Appendix 1].

(a) Prove the following commutation relations

(i) $[L^2, L_\pm] = 0$
(ii) $[L_z, L_\pm] = \pm\hbar L_\pm$
(iii) $[L_+, L_-] = 2\hbar L_z$

and the identities

(iv) $L_+L_- = L^2 - L_z(L_z - \hbar)$
(v) $L_-L_+ = L^2 - L_z(L_z + \hbar)$
(vi) $L^2 = \frac{1}{2}(L_+L_- + L_-L_+) + L_z^2$
$= \frac{1}{2}[L_+, L_-]_+ + L_z^2,$

where the anticommutator $[A, B]_+$ is defined in Eq. (1.33).

(b) Use the properties of L_+ and L_- to show that the result of operation on an orbital angular momentum eigenfunction (i.e., a spherical harmonic) by L_+ or L_- is given by

(vii) $L_\pm Y_{\ell m}(\theta, \varphi) = \hbar\sqrt{(\ell \mp m)(\ell \pm m + 1)}\, Y_{\ell m\pm 1}(\theta, \varphi)$
$= \hbar\sqrt{\ell(\ell + 1) - m(m \pm 1)}\, Y_{\ell m\pm 1}(\theta, \varphi).$

Show that this can be written as a matrix element,

(viii) $\langle Y_{\ell m\pm 1}(\theta, \varphi) | L_\pm | Y_{\ell m}(\theta, \varphi)\rangle = \hbar\sqrt{(\ell \mp m)(\ell \pm m + 1)}$
$= \hbar\sqrt{\ell(\ell + 1) - m(m \pm 1)},$

where the matrix element notation is defined by

$$\langle f(\mathbf{r}) | A | g(\mathbf{r})\rangle \equiv \int f^*(\mathbf{r}) A g(\mathbf{r})\, d\mathbf{r}$$

for arbitrary functions $f(\mathbf{r})$ and $g(\mathbf{r})$ and an operator A. (This is called the Dirac bracket notation.)

[HINT: The following suggested procedure may prove helpful.

Assume: $L_\pm Y_{\ell m}(\theta, \varphi) = C^\pm_{\ell m} Y_{\ell m\pm 1}$

Consider: $\langle L_\pm Y_{\ell m} | L_\pm Y_{\ell m}\rangle = |C^\pm_{\ell m}|^2$

Show: $\langle L_\pm Y_{\ell m} | L_\pm Y_{\ell m}\rangle = \langle Y_{\ell m} | L_\mp L_\pm Y_{\ell m}\rangle$

and evaluate the latter and so obtain $|C^\pm_{\ell m}|^2$ and $C^\pm_{\ell m}$ (aside from arbitrary phase).]

2.6 Applications of the Angular Momentum Raising and Lowering Operators: 1 (**)

[This problem makes use of some of the results of Prob. 2.5.]

It is often convenient to be able to generate the eigenfunctions of L^2 and L_z (i.e., the spherical harmonics) when extensive tables are not at hand. One procedure is to remember the fairly simple result for $m = 0$

$$Y_{\ell 0}(\theta, \varphi) = \left(\frac{2\ell + 1}{4\pi}\right)^{1/2} P_\ell(\cos\theta)$$

and then apply the operators L_+ or L_- to obtain the functions for larger or smaller values of m. Carry out this procedure to obtain Y_{21} and Y_{22}, starting with

$$P_2(\cos\theta) = \tfrac{1}{2}(3\cos^2\theta - 1).$$

Choose your phase to agree with our convention and check your results against those in Table 2.2.

2.7 Application of the Angular Momentum Raising and Lowering Operators: 2 (***)

Recall that since L_x and L_y commute with L^2, it is possible to find simultaneous eigenfunctions of L_x and L^2 or of L_y and L^2, but not of L_x and L_y or of L_x and L_z. For the special case $\ell = 1$, obtain the normalized simultaneous eigenfunctions of L_x and L^2 and of L_y and L^2. Find the corresponding eigenvalues of L_x and L_y. [HINT: Consider an expansion in spherical harmonics and make use of the properties of the raising and lowering operators.]

2.8 Constants of the Motion and Symmetry Operations (****)

In quantum mechanics, a constant of the motion is a dynamical variable (or observable) whose operator commutes with the system Hamiltonian. A symmetry element is defined as a transformation of variables (e.g., rotation, reflection) represented by an operator that commutes with the Hamiltonian. In this problem we will consider the eight dynamical variables: T, p_x, p_y, p_z, L_x, L_y, L_z, and L^2 for a particle in three dimensions. We will also consider the five symmetry operations: $\boldsymbol{i}$ (inversion $\mathbf{r} \longrightarrow -\mathbf{r}$), σ_x (reflection in yz plane), σ_y (reflection in xz plane), σ_z (reflection in xy plane), and $R_z(\varphi_0)$ (rotation about $\hat{z}$ axis by angle φ_0—i.e., $\varphi \longrightarrow \varphi + \varphi_0$).

(a) Commuting operators, whether dynamical variables or symmetry operations, are often useful in characterizing states of a quantum mechanical system. Consider the eight observables and five symmetry operations given above and find which of the 40 commutators do not vanish. [HINT: A 5×8 table with zeros and x's (in place of nonzero commutators) will provide a convenient representation.]

(b) Consider a system consisting of a particle in a three-dimensional potential field $V(x, y, z)$, which may, in general, depend on all spatial variables. For a particular case—for example, "central field"—we write $V(r)$, the implication being that V is independent of θ and φ. Give all constants of motion and symmetry elements for each of the following cases [in the case of $R_z(\varphi)$, be sure to specify which angles φ are allowed].

(i) "Free particle": $V =$ constant.
(ii) "Central potential": $V = V(r)$.
(iii) "Cylindrically symmetric potential": $V = V(r, \theta)$.
(iv) An electron in the field of two protons located on the $\hat{z}$ axis at $z = \pm a$.
(v) "Square potential tube":

$$V = V(x, y) = \begin{cases} \infty, & |x| > a \text{ or } |y| > a \\ 0, & |x| < a \text{ and } |y| < a. \end{cases}$$

(vi) An electron in the field of four protons located in the xy plane at $(x = a,\ y = a)$, $(x = a,\ y = -a)$, $(x = -a,\ y = a)$, and $(x = -a,\ y = -a)$.

(vii) "Rectangular potential tube":

$$V = V(x, y) = \begin{cases} \infty, & |x| > a \text{ or } |y| > b, \\ 0, & |x| < a \text{ and } |y| < b, \end{cases} \quad a \neq b.$$

(viii) "Cylindrical potential tube":

$$V = V(x, y) = \begin{cases} \infty, & \sqrt{x^2 + y^2} > a \\ 0, & \sqrt{x^2 + y^2} < a. \end{cases}$$

(ix) "Potential box":

$$V = V(x, y, z) = \begin{cases} \infty, & |x| > a, |y| > b, \text{ or } |z| > c \\ 0, & |x| < a, |y| < b, \text{ and } |z| < c. \end{cases}$$

2.9 Rigid Rotator (***)

When we eventually study the structure and spectra of molecules, it will be a welcome surprise to find that the rotation of most diatomic molecules may be described quantum mechanically by the rigid rotator, a particularly simple system. We may define the rigid rotator to be a rigid massless rod of length R_0 which has point masses at its ends. Only rotation about a perpendicular bisector of the rod is of interest in this problem. (See Fig. 2.5.)

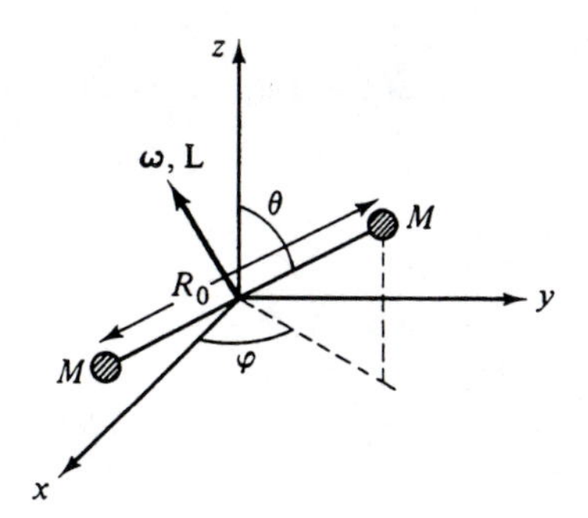

Figure 2.5

(a) *Classical treatment:* Show that by transforming to center-of-mass and relative coordinates, the total classical energy of a rigid rotator may be written

$$E = \frac{1}{2}(2M)v_{\text{cm}}^2 + \frac{1}{2I}L^2,$$

where **L** is the relative angular momentum of the two masses and I is the moment of inertia $I = \mu R_0^2$, μ being the reduced mass (see Sec. 3.2). We shall disregard center-of-mass motion.

(b) *Rotation in the plane—quantum mechanical treatment:* Consider the rather artificial problem of a rigid rotator confined to rotate so that both masses remain in the xy plane (planar rotation). Construct the Hamiltonian $\mathcal{H}$ and solve the time-independent Schroedinger equation for the energy eigenvalues and the corresponding orthonormal set of wave functions. Discuss the degeneracy of the energy levels. Show that L_z is a constant of the motion (L_x and L_y have no meaning here, since the system is artificially constrained) and find simultaneous eigenfunctions of $\mathcal{H}$ and L_z. Show how to construct eigenfunctions of $\mathcal{H}$ that are not eigenfunctions of L_z. Calculate the value of $\langle L_z \rangle$ for these eigenfunctions. Indicate appropriate choices of quantum numbers in each case.

(c) *General rotation in space—quantum mechanical treatment:* Now we deal with the more realistic problem of unconstrained rotation. Again, construct the Hamiltonian (ignoring center-of-mass motion) and solve the time-independent Schroedinger equation for the eigenvalues and the corresponding orthonormal set of wave functions. Show that it is appropriate to introduce two quantum numbers, say j and k, where j alone labels the energy levels. Draw an energy level diagram[16] and indicate the degeneracy of the levels. Show that L^2 and $\mathbf{L}$ are constants of the motion and indicate which choices of wave functions are simultaneous eigenfunctions of L^2 and L_z and which choices are eigenfunctions of L^2 alone. Is this actually a central field problem?

(d) *Diatomic* CO *molecule:* Let us consider the rigid rotator as a model for the rotation of the CO molecule, where $R_0 = 1.13$ Å. Calculate the numerical values for the first three energy levels ($j = 0, 1, 2$) in ergs, electron volts, and Rydbergs (see Appendix 2 for definition of units). Calculate the wavelength and frequency of light that would result from a $j = 2 \longrightarrow j = 1$ transition and a $j = 1 \longrightarrow j = 0$ transition. Is this visible, infrared, microwave, etc.?

[16]That is, draw a sketch showing the relative location of the energy levels. See Fig. 3.3 for the hydrogen-atom energy level diagram.

3
The One-Electron Atom

Apparently there is colour, apparently sweetness, apparently bitterness; actually there are only atoms and the void.

Democritus, 420 B.C.

Our study of atoms commences in this chapter and will ultimately lead to the analysis of atomic systems containing numerous electrons. Such multielectron systems are sufficiently complicated that we cannot solve them exactly; instead we must develop approximation methods. Rather than go immediately into the derivation of these methods, we shall begin with a system that can be solved exactly: the simple one-electron atom. In fact, the quantum mechanical problem of the one-electron (or "hydrogenic") atom can be solved by using the techniques of the last chapter.

Hydrogen (nuclear charge $Z = 1$) and its isotopes deuterium and tritium are the only neutral one-electron atomic systems. However, the results obtained in this chapter will also apply to all hydrogenic atomic ions—He^{+} ($Z = 2$), Li^{++} ($Z = 3$), and so on. Each of these systems consists of a nucleus (of charge Ze) and an electron (of charge $-e$); we are dealing with two-body problems. As we shall see, the Schroedinger equation for the system can be reduced to that of a central force problem and then solved.

3.1 THE SCHROEDINGER EQUATION FOR THE ONE-ELECTRON ATOM

The coordinates for a one-electron atom are sketched in Fig. 3.1. The nonrelativistic Hamiltonian describing this system is

$$\mathcal{H} = T + V, \tag{3.1}$$

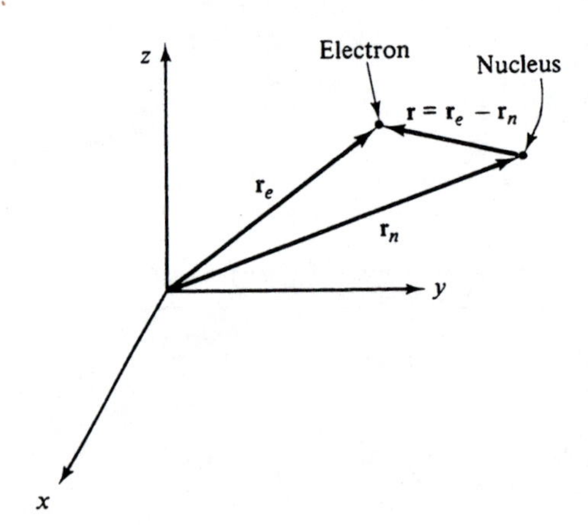

Figure 3.1 Electronic and nuclear coordinates for the one-electron atom; **r** is the coordinate of the electron relative to the nucleus.

where T is the total kinetic energy of the system and V is the potential energy of interaction between the two particles. The kinetic energy is the sum of two terms, one for each particle; that is,

$$T = T_e + T_n = -\frac{\hbar^2}{2m_e}\nabla_e^2 - \frac{\hbar^2}{2m_n}\nabla_n^2, \tag{3.2}$$

where m_e is the mass of the electron and m_n is the mass of the nucleus. In Eq. (3.2), ∇_e^2 acts only on the coordinates of the electron, $\mathbf{r}_e$, and ∇_n^2 acts only on the coordinates of the nucleus, $\mathbf{r}_n$. The potential energy is due to the coulomb force;[1] for a hydrogenic atom of nuclear charge Ze, it is simply

$$V(\mathbf{r}_e, \mathbf{r}_n) = -\frac{Ze^2}{|\mathbf{r}_e - \mathbf{r}_n|}. \tag{3.3}$$

The time-dependent Schroedinger equation is

$$\mathcal{H}\Psi(\mathbf{r}_e, \mathbf{r}_n, t) = i\hbar\frac{\partial}{\partial t}\Psi(\mathbf{r}_e, \mathbf{r}_n, t). \tag{3.4}$$

[1] In this chapter we consider the electron and the nucleus to be nonrelativistic, structureless point charges and ignore a whole host of more subtle effects that could be included in our theory (magnetic interactions, electron spin, nuclear spin). We shall return to some of them later (see Chapters 5 to 9).

The solution of this equation, subject to prescribed initial conditions, gives us the time-dependent wave function $\Psi(\mathbf{r}_e, \mathbf{r}_n, t)$, which contains locked within it all that man can know about the system. Because $V(\mathbf{r}_e, \mathbf{r}_n)$ is independent of time, the Schroedinger equation is separable in space and time (see Sec. 1.2); the stationary-state wave functions are

$$\Psi(\mathbf{r}_e, \mathbf{r}_n, t) = \psi_E(\mathbf{r}_e, \mathbf{r}_n)e^{-i(E/\hbar)t}, \tag{3.5}$$

where E is the eigenvalue of the Hamiltonian and $\psi_E(\mathbf{r}_e, \mathbf{r}_n)$ satisfies

$$\boxed{\mathcal{H}\psi_E(\mathbf{r}_e, \mathbf{r}_n) = E\psi_E(\mathbf{r}_e, \mathbf{r}_n).} \tag{3.6}$$

Time-independent Schroedinger equation for a one-electron atom

We must solve this equation for the stationary-state eigenfunctions $\psi_E(\mathbf{r}_e, \mathbf{r}_n)$ such that $|\psi_E(\mathbf{r}_e, \mathbf{r}_n)| < \infty$ for all $\mathbf{r}_e$ and $\mathbf{r}_n$. Moreover, these functions must be single valued and continuous everywhere and must possess continuous first derivatives. Let us see if we can reduce Eq. (3.6) to a more tractable form.

3.2 SEPARATION OF CENTER-OF-MASS AND RELATIVE MOTION

Written out in full, the time-independent Schroedinger equation for a one-electron atom is

$$\left(-\frac{\hbar^2}{2m_e}\nabla_e^2 - \frac{\hbar^2}{2m_n}\nabla_n^2 - \frac{Ze^2}{|\mathbf{r}_e - \mathbf{r}_n|}\right)\psi_E(\mathbf{r}_e, \mathbf{r}_n) = E\psi_E(\mathbf{r}_e, \mathbf{r}_n). \tag{3.7}$$

We can simplify this equation by noting that the nucleus is more massive than the electron.[2] Calling on our classical intuition, we expect the nucleus to "move much more slowly" than the electron. Therefore an (reasonable) approximate way to handle Eq. (3.7) is to treat the nucleus as an "infinitely massive point" and take the origin of coordinates at this point.

Approximation of Infinite Nuclear Mass

Let us define the coordinate of the electron's motion relative to the nucleus:

$$\mathbf{r} \equiv \mathbf{r}_e - \mathbf{r}_n. \tag{3.8}$$

Setting $m_n = \infty$, we can rewrite Eq. (3.7) as

$$\left(-\frac{\hbar^2}{2m_e}\nabla^2 - \frac{Ze^2}{r}\right)\psi_E(\mathbf{r}) = E\psi_E(\mathbf{r}). \tag{3.9}$$

[2]The ratio of nuclear mass to electron mass for hydrogen is $m_n/m_e \simeq 1836$.

This equation, which describes the motion of a particle of mass m_e in a coulomb field, is identical to the central force equation (2.13) with $V(r)$ equal to the coulomb potential energy (2.4). Thus the angular dependence of the wave function is known, and there is reason to believe that it is possible to solve the corresponding radial equation.

Of course, the nucleus is not really infinitely massive, and we must accept the fact that the preceding approximation will not yield exact results. However, this simple exercise does provide an idea of how to handle the two-body equation (3.7): reduce it to a central force problem. (We shall return to the approximation of infinite nuclear mass shortly and see how good an approximation it is.)

The approach suggested is familiar from classical mechanics. The classical total energy for two interacting particles can be "separated" by a change of variables to center-of-mass and relative coordinates.[3] But the Hamiltonian is simply the operator equivalent of the classical energy, so it is reasonable to hope that such a separation will work in the quantum-mechanical two-body problem.

Separation of Variables

The mathematical method of separation of variables was discussed in Chapter 2, where we separated the radial and angular dependence of the Hamiltonian eigenfunction for a central force problem (see Sec. 2.2). We shall apply the same method here. Let us introduce *center-of-mass coordinates* $\mathbf{r}_{\mathrm{cm}}$, defined by

$$x_{\mathrm{cm}} \equiv \frac{m_e x_e + m_n x_n}{m_n + m_e}, \qquad y_{\mathrm{cm}} \equiv \frac{m_e y_e + m_n y_n}{m_n + m_e}, \qquad z_{\mathrm{cm}} \equiv \frac{m_e z_e + m_n z_n}{m_n + m_e}, \tag{3.10}$$

and *relative coordinates* $\mathbf{r}$, defined by

$$x \equiv x_e - x_n, \qquad y \equiv y_e - y_n, \qquad z \equiv z_e - z_n. \tag{3.11}$$

We now write the Hamiltonian, Eq. (3.1), in rectangular coordinates x_e, y_e, z_e and x_n, y_n, z_n and use the chain rule of differentiation to convert it to center-of-mass and relative coordinates. We seek solutions to the Schroedinger equation of the form

$$\psi_E(\mathbf{r}_{\mathrm{cm}}, \mathbf{r}) = \psi_{\mathrm{cm}}(\mathbf{r}_{\mathrm{cm}})\psi(\mathbf{r}). \tag{3.12}$$

[3] See Keith R. Symon, *Mechanics*, 3rd ed. (Reading, Mass.: Addison-Wesley, 1971), Chap. 4.

Therefore we substitute this form for ψ_E into the time-independent Schroedinger equation, using the Hamiltonian just derived, and carry out the separation of variables. We obtain a center-of-mass equation

$$\left(-\frac{\hbar^2}{2M}\nabla^2_{\rm cm} - E_{\rm cm}\right)\psi_{\rm cm}(\mathbf{r}_{\rm cm}) = 0. \tag{3.13}$$

Center-of-mass equation for a one-electron atom

and a relative equation

$$\left(-\frac{\hbar^2}{2\mu}\nabla^2 - \frac{Ze^2}{r} - E_{\rm rel}\right)\psi(\mathbf{r}) = 0. \tag{3.14}$$

Equation of relative motion for a one-electron atom

In Eq. (3.13), M is the total mass of the system

$$M \equiv m_e + m_n, \tag{3.15}$$

and $\nabla^2_{\rm cm}$ operates only on the center-of-mass coordinates. Similarly, in Eq. (3.14), μ is the *reduced mass* of the system

$$\mu \equiv \frac{m_e m_n}{m_e + m_n}, \tag{3.16}$$

and ∇^2 operates only on the relative coordinates. The energies in these equations satisfy

$$E_{\rm cm} + E_{\rm rel} = E. \tag{3.17}$$

Exercise 3.1 Carry out the steps outlined above and so derive Eqs. (3.13) and (3.14).

Equation (3.13) is simply the Schroedinger equation for a free particle of mass M with coordinates $\mathbf{r}_{\rm cm}$ and energy $E_{\rm cm}$. Equation (3.14) is a Schroedinger equation for a "particle" of mass μ with coordinates $\mathbf{r}$, moving in a coulomb field with energy $E_{\rm rel}$. Notice that this equation is the same as (3.9) except that m_e has been replaced by μ. We can write the reduced mass as[4]

$$\mu \simeq m_e\left(1 - \frac{m_e}{m_n}\right).$$

[4]To obtain this result, use the binomial expansion

$$(1 \pm x)^{-1} = 1 \mp x + \cdots,$$

which is valid for $|x| < 1$.

Since m_e/m_n is of the order of 10^{-3} or less, it is clear that, to a good approximation, $\mu \simeq m_e$.

Equation (3.14), which describes the relative motion of the electron with respect to the nucleus, is of the same form as the general central force equation (2.13). Although E in Eq. (3.17) is the total energy of the hydrogenic atom, the solution of (3.14) entirely determines the energy spectrum of the atom.

Let us look at each of the new equations (3.13) and (3.14) and see what, if anything, is known about their solutions.

3.3 THE CENTER-OF-MASS MOTION

The functions that satisfy the center-of-mass equation (3.13) are the stationary-state wave functions of a free particle of mass M. They may be chosen as *plane-wave eigenfunctions*[5]

$$\psi_{\text{cm}}(\mathbf{r}_{\text{cm}}) = \frac{1}{\sqrt{(2\pi)^3}} e^{i\mathbf{k}\cdot\mathbf{r}_{\text{cm}}}$$

corresponding to a continuous range of energies

$$E_{\text{cm}} = \frac{\hbar^2}{2M}k^2 = \frac{\hbar^2}{2M}(k_x^2 + k_y^2 + k_z^2),$$

where $\mathbf{k}$ is the wave vector introduced in Chapter 1. Since this function is an eigenfunction of the linear momentum operator $\mathbf{p}_{\text{cm}}$ with eigenvalue $\hbar\mathbf{k}$ and of the free particle Hamiltonian, both the energy and the linear momentum of the state it describes are sharp. Therefore since we know exactly the linear momentum of the center of mass (i.e., $\Delta\mathbf{p} = 0$), the Heisenberg uncertainty principle [Eq. (1.1)] forbids us from knowing anything about its location (except, of course, that it is somewhere). In the laboratory we usually know something about where the atom is, even if only that our hydrogen atoms are in a gas that is confined to a bottle or are restricted to a beam of some sort. Hence it is more appropriate to describe the center-of-mass of the atom by a *wave packet*

$$\psi_{\text{cm}}(\mathbf{r}_{\text{cm}}) = \frac{1}{\sqrt{(2\pi)^3}} \int \varphi(\mathbf{k}) e^{i\mathbf{k}\cdot\mathbf{r}_{\text{cm}}}\, d\mathbf{k}, \tag{3.18}$$

[5]See Elmer E. Anderson, *Modern Physics and Quantum Mechanics* (Philadelphia: W. B. Saunders Co., 1971), Chap. 4, for a discussion of free-particle eigenfunctions and wave packets.

where $\varphi(\mathbf{k})$ is the momentum amplitude.[6] Then the product function $\psi_{cm}(\mathbf{r}_{cm})\psi(\mathbf{r})$ can be made to characterize the stationary states of a hydrogen atom (the center of mass of which is partially localized in space) by allowing the values of $\mathbf{k}$ to be spread in some manner specified by the precise form of $\varphi(\mathbf{k})$.

And so we see that the center-of-mass motion of the atom is quite simple and can be described by free-particle wave packets. Let us now turn to the more interesting motion of the electron relative to the nucleus. This motion is given by the solutions of the relative equation (3.14). Moreover, the eigenvalues obtained by solution of this equation adequately describe the energy spectrum of the hydrogenic atom as observed spectroscopically.

3.4 SOLUTION OF THE RELATIVE EQUATION

Since the relative equation is a central force equation, we can write it in spherical coordinates [see Eq. (2.21)] as

$$\left[-\frac{\hbar^2}{2\mu}\frac{1}{r^2}\frac{\partial}{\partial r}\left(r^2\frac{\partial}{\partial r}\right)+\frac{1}{2\mu r^2}L^2-\frac{Ze^2}{r}-E_{n\ell}\right]\psi_{n\ell m}(\mathbf{r})=0, \tag{3.19}$$

where we have labeled the wave functions and energies with the appropriate subscripts from Chapter 2. We can immediately write down the separation

$$\psi_{n\ell m}(\mathbf{r})=R_{n\ell}(r)Y_{\ell m}(\theta,\varphi). \tag{3.20}$$

The spherical harmonics $Y_{\ell m}(\theta, \varphi)$ specify the angular dependence of the energy eigenfunctions—how $\psi_{n\ell m}(\mathbf{r})$ depends on θ and φ. The radial functions $R_{n\ell}(r)$ must satisfy a radial equation; with m_1 replaced by μ and $V(r)$ given by $-Ze^2/r$, this equation, (2.69), can be written

$$\left[\frac{1}{r^2}\frac{d}{dr}\left(r^2\frac{d}{dr}\right)-\frac{\ell(\ell+1)}{r^2}+\frac{2\mu}{\hbar^2}\left(E_{n\ell}+\frac{Ze^2}{r}\right)\right]R_{n\ell}(r)=0. \tag{3.21}$$

The functions $\psi_{n\ell m}(\mathbf{r})$ are simultaneous eigenfunctions of the Hamiltonian $\mathcal{H}$, the square of the orbital angular momentum L^2, and the z component of the orbital angular momentum L_z. Although not the only stationary-state eigenfunctions of $\mathcal{H}$, they are particularly useful functions.

Solution of the Radial Equation

The radial equation for the hydrogenic atom, Eq. (3.21), can be most easily solved by transforming it into a standard differential equation.

[6]The plane-wave eigenfunctions can be obtained from the wave packet (3.18) by setting $\varphi(\mathbf{k})$ equal to a delta function $\delta_3(\mathbf{k}-\mathbf{k}')$ and integrating over $\mathbf{k}'$.

To illustrate, let us define a new variable

$$\rho \equiv \alpha r, \tag{3.22}$$

where

$$\alpha \equiv \frac{2}{\hbar}\sqrt{-2\mu E_{n\ell}}. \tag{3.23}$$

(Notice that α is real, since $E_{n\ell} < 0$ for bound states of the system.) We introduce α into Eq. (3.21) by dividing the equation by α^2 and appropriately grouping terms:

$$\left[\frac{d^2}{d(\alpha r)^2} + \frac{2}{(\alpha r)}\frac{d}{d(\alpha r)} + \frac{2\mu}{\hbar^2\alpha^2}E_{n\ell} + \frac{2\mu Ze^2}{\hbar^2\alpha(\alpha r)} - \frac{\ell(\ell+1)}{(\alpha r)^2}\right]R_{n\ell}(r) = 0. \tag{3.24}$$

If we now define

$$\gamma \equiv \frac{2\mu Ze^2}{\hbar^2\alpha}, \tag{3.25}$$

Eq. (3.24) can be written

$$\left[\frac{d^2}{d\rho^2} + \frac{2}{\rho}\frac{d}{d\rho} - \frac{1}{4} + \frac{\gamma}{\rho} - \frac{\ell(\ell+1)}{\rho^2}\right]R_{n\ell}(\rho) = 0. \tag{3.26}$$

Notice that ρ is the independent variable of the radial function in this equation.

Asymptotic Limits

We can learn something about the form of the solutions of Eq. (3.26) by considering two important limits. First, consider the limit of very large r ($\rho \longrightarrow \infty$). In this limit, Eq. (3.26) becomes

$$\left(\frac{d^2}{d\rho^2} - \frac{1}{4}\right)R_{n\ell}(\rho) = 0, \qquad \rho \longrightarrow \infty. \tag{3.27}$$

The solution to this equation that is physically admissible (finite everywhere) is[7]

$$R_{n\ell}(\rho) \underset{\rho\to\infty}{\sim} e^{-\rho/2}. \tag{3.28}$$

Second, consider $\rho \longrightarrow 0$. In this limit, Eq. (3.26) becomes

$$\frac{d^2}{d\rho^2}[\rho R_{n\ell}(\rho)] - \frac{\ell(\ell+1)}{\rho^2}[\rho R_{n\ell}(\rho)] = 0, \qquad \rho \longrightarrow 0, \tag{3.29}$$

[7] Actually, the radial function is of the form $\rho^\ell e^{-\rho/2}$ for large values of ρ, but it is the exponential factor that dominates and that causes the radial function to die off as $\rho \longrightarrow \infty$.

where we have kept only terms of second order in $1/\rho$. The solution of Eq. (3.29) that is regular at $\rho = 0$ has the limiting behavior

$$R_{n\ell}(\rho) \underset{\rho \to 0}{\sim} \rho^{\ell}. \tag{3.30}$$

These results tell us how *all* radial functions $R_{n\ell}(\rho)$ must behave for very large and very small values of ρ; they must be of the form

$$R_{n\ell}(\rho) = e^{-\rho/2}\rho^{\ell} f_{n\ell}(\rho), \tag{3.31}$$

where $f_{n\ell}(\rho)$ is some as-yet-unknown function of ρ that in the limits $\rho \longrightarrow 0$ and $\rho \longrightarrow \infty$ behaves in such a way that $R_{n\ell}(\rho)$ remains finite.

An Equation for $f_{n\ell}(\rho)$

By substituting Eq. (3.31) into (3.26) and performing some simple algebra, we can derive the following equation for $f_{n\ell}(\rho)$:

$$\left\{\rho \frac{d^2}{d\rho^2} + [2(\ell + 1) - \rho]\frac{d}{d\rho} + [\gamma - (\ell + 1)]\right\} f_{n\ell}(\rho) = 0. \tag{3.32}$$

Exercise 3.2 Derive Eq. (3.32).

Fortunately, this result is in a standard form; specifically, it is of the form

$$\left[\rho \frac{d^2}{d\rho^2} + (j + 1 - \rho)\frac{d}{d\rho} + (q - j)\right] f_{n\ell}(\rho) = 0 \tag{3.33}$$

with

$$j = 2\ell + 1 \quad \text{and} \quad q = \gamma + \ell. \tag{3.34}$$

The solutions of Eq. (3.33) are the *associated Laguerre polynomials*[8] $L_q^j(\rho)$; they are defined only for integral values of q and j satisfying the inequality

$$j \leq q. \tag{3.35}$$

Thus we have

$$f_{n\ell}(\rho) = L_{\gamma+\ell}^{2\ell+1}(\rho). \tag{3.36}$$

Since q is an integer, so is γ. It is conventional to choose γ as the quantum number n that labels the different radial functions resulting from the solution of the radial equation for a particular value of ℓ. Of course, n also labels the energy eigenvalues $E_{n\ell}$. This label is called the *principal quantum number*.

Summarizing our results thus far, we have separated the center-of-mass and relative motion and solved each of the resulting equations. We can now

[8]See the references on special functions in the Selected Readings list for Chapter 2. These functions can also be obtained from Eq. (3.32) by the method of series expansion (see Prob. 3.1).

describe the quantum mechanical behavior of our first atom! (Its properties will be the main topic of the rest of this chapter.) The center-of-mass motion is simply that of a free particle of mass M. The relative motion is that of a particle of mass μ in a particular central potential. The angular motion of the electron is described by the spherical harmonics $Y_{\ell m}(\theta, \varphi)$; the radial motion is described by the radial functions that we just obtained. These functions can be written

$$R_{n\ell}(\rho) = N_{n\ell} e^{-\rho/2} \rho^{\ell} L_{n+\ell}^{2\ell+1}(\rho), \tag{3.37}$$

where $N_{n\ell}$ is an as-yet-undetermined normalization constant. The variable ρ is simply

$$\rho = \alpha r, \tag{3.38}$$

or

$$\rho = \frac{2Z}{na_0} r, \tag{3.39}$$

where we have used Eq. (3.25) for α (with $\gamma = n$) and have introduced the useful constant

$$a_0 \equiv \frac{\hbar^2}{\mu e^2} \simeq 0.529 \text{ Å}. \tag{3.40}$$

This constant is equal to the radius of the first Bohr orbit of hydrogen and is called the *Bohr radius*. There are some restrictions on the quantum numbers n and ℓ. In particular, since the associated Laguerre polynomials are defined only for $j \leq q$, n and ℓ must satisfy

$$2\ell + 1 \leq n + \ell \tag{3.41}$$

or

$$n \geq \ell + 1 \qquad \text{for } \ell = 0, 1, 2, 3, \ldots. \tag{3.42}$$

This is equivalent to the restriction

$$\ell \leq n - 1 \qquad \text{for } n = 1, 2, 3, \ldots \tag{3.43}$$

which, as we shall see, is more convenient. We shall normalize[9] $R_{n\ell}(r)$ and study the associated Laguerre polynomials in the next section.

Before doing so, notice that we can now write a convenient expression for the energy eigenvalues appearing in the relative equation (3.19). Using Eqs. (3.23) and (3.25) and $\gamma = n$, we obtain

$$E_n = -\frac{\mu Z^2 e^4}{2\hbar^2 n^2}. \tag{3.44}$$

[9] Since the spherical harmonics $Y_{\ell m}(\theta, \varphi)$ that multiply these radial functions are themselves normalized [see Eq. (2.60)], we want to independently normalize $R_{n\ell}(r)$. Then the full Hamiltonian eigenfunction $\psi_{n\ell m}(\mathbf{r})$ will also be normalized.

Notice that the orbital-angular-momentum quantum number ℓ does not appear in this result; in this nonrelativistic theory, the energy eigenvalues for the hydrogen atom do not explicitly depend on ℓ. We shall have more to say about this important fact shortly.

3.5 PROPERTIES OF THE RADIAL FUNCTIONS

The associated Laguerre functions $L_q^j(\rho)$ that form a part of the radial functions $R_{n\ell}(\rho)$ are rather simple polynomials for small n and ℓ. Several are presented in Table 3.1.

Table 3.1
Associated Laguerre polynomials $L_q^j(\rho)$ for $q = 0, 1, 2, 3$.

q	j	$L_q^j(\rho)$
0	0	$L_0^0(\rho) = 1$
1	0	$L_1^0(\rho) = 1 - \rho$
	1	$L_1^1(\rho) = -1$
2	0	$L_2^0(\rho) = 2 - 4\rho + \rho^2$
	1	$L_2^1(\rho) = 2\rho - 4$
	2	$L_2^2(\rho) = 2$
3	0	$L_3^0(\rho) = 6 - 18\rho + 9\rho^2 - \rho^3$
	1	$L_3^1(\rho) = -18 + 18\rho - 3\rho^2$
	2	$L_3^2(\rho) = 18 - 6\rho$
	3	$L_3^3(\rho) = -6$

In general, the associated Laguerre functions are defined as the functions $L_q^j(\rho)$ that satisfy the *Laguerre differential equation*

$$\left[\rho\frac{d^2}{d\rho^2} + (j + 1 - \rho)\frac{d}{d\rho} + (q - j)\right]L_q^j(\rho) = 0, \qquad (3.45)$$

where j and q are any real numbers such that $j \leq q$. If j and q are integers, we call the functions $L_q^j(\rho)$ the associated Laguerre polynomials; these polynomials appear in Table 3.1.

The associated Laguerre polynomials $L_q^j(\rho)$ are related to the simpler *Laguerre polynomials* $L_q(\rho)$ by the equation

$$L_q^j(\rho) = \frac{d^j}{d\rho^j}L_q(\rho). \qquad (3.46)$$

The Laguerre polynomials can be generated via

$$L_q(\rho) = e^{\rho} \frac{d^q}{d\rho^q}(\rho^q e^{-\rho}). \tag{3.47}$$

Notice that both the Laguerre polynomials $L_q(\rho)$ and the associated Laguerre polynomials $L_q^j(\rho)$ are real functions of ρ; hence the radial function $R_{n\ell}(\rho)$ is real. It follows from Eq. (3.46) that

$$L_q(\rho) = L_q^0(\rho). \tag{3.48}$$

The Laguerre polynomials satisfy the useful recursion relation

$$L_{q+1}(\rho) = (2q + 1 - \rho)L_q(\rho) - q^2 L_{q-1}(\rho). \tag{3.49}$$

Exercise 3.3 Use Eq. (3.47) to generate $L_1(\rho)$ and $L_2(\rho)$. Use your expression for $L_2(\rho)$ and Eq. (3.46) to generate $L_2^1(\rho)$. Make a rough sketch of these three functions.

We can normalize the radial functions $R_{n\ell}(\rho)$ by making use of the convenient identity[10]

$$\int_0^\infty e^{-\rho}\rho^{2\ell}[L_{n+\ell}^{2\ell+1}(\rho)]^2 \rho^2 \, d\rho = \frac{2n[(n+\ell)!]^3}{(n-\ell-1)!}. \tag{3.50}$$

Then $R_{n\ell}(r)$ is given by

$$R_{n\ell}(r) = N_{n\ell} e^{-\rho/2} \rho^\ell L_{n+\ell}^{2\ell+1}(\rho); \tag{3.51}$$

by requiring that $R_{n\ell}(r)$ be normalized,

$$\int_0^\infty [R_{n\ell}(r)]^2 r^2 \, dr = 1, \tag{3.52}$$

we obtain

$$N_{n\ell} = -\left\{\frac{\alpha^3(n-\ell-1)!}{2n[(n+\ell)!]^3}\right\}^{1/2}, \tag{3.53}$$

where

$$\alpha = \frac{2Z}{na_0} \tag{3.54}$$

and we have taken the negative square root as a matter of convention.[11]

Exercise 3.3 Derive Eq. (3.53), using Eq. (3.50).

[10]For a proof of this relation, see Linus Pauling and E. B. Wilson, *Introduction to Quantum Mechanics* (New York: McGraw-Hill, 1935), pp. 132 ff.

[11]It makes no difference if $R_{n\ell}(\rho)$ is multiplied by a minus sign because $(-1) = e^{i\pi}$, a phase factor, and all wave functions are indeterminate to within a phase factor.

The Radial Functions

Thus we can finally write an expression for the *normalized radial functions for a one-electron atom*:

$$R_{n\ell}(r) = -\left\{\left(\frac{2Z}{na_0}\right)^3 \frac{(n-\ell-1)!}{2n[(n+\ell)!]^3}\right\}^{1/2} e^{-\rho/2}\rho^{\ell}L_{n+\ell}^{2\ell+1}(\rho), \quad (3.55)$$

Radial wave function of the one-electron atom

where

$$\rho = \frac{2Z}{na_0}r. \quad (3.56)$$

Explicit expressions for the radial eigenfunctions for $n = 1$, 2, and 3 appear in Table 3.2.[12]

Table 3.2
Normalized radial functions for the one-electron atom for $n = 1$, 2, and 3.

n	ℓ	$R_{n\ell}(r)$
1	0	$2\left(\frac{Z}{a_0}\right)^{3/2} e^{-Zr/a_0}$
2	0	$\left(\frac{Z}{2a_0}\right)^{3/2}\left(2 - \frac{Zr}{a_0}\right)e^{-Zr/2a_0}$
2	1	$\frac{1}{\sqrt{3}}\left(\frac{Z}{2a_0}\right)^{3/2}\left(\frac{Zr}{a_0}\right)e^{-Zr/2a_0}$
3	0	$\frac{2}{3}\left(\frac{Z}{3a_0}\right)^{3/2}\left(3 - \frac{2Zr}{a_0} + \frac{2Z^2r^2}{9a_0^2}\right)e^{-Zr/3a_0}$
3	1	$\frac{2\sqrt{2}}{9}\left(\frac{Z}{3a_0}\right)^{3/2}\left(\frac{2Zr}{a_0} - \frac{Z^2r^2}{3a_0^2}\right)e^{-Zr/3a_0}$
3	2	$\frac{4}{27\sqrt{10}}\left(\frac{Z}{3a_0}\right)^{3/2}\left(\frac{Z^2r^2}{a_0^2}\right)e^{-Zr/3a_0}$

Like the θ- and φ-functions obtained in Chapter 2, the radial functions satisfy an orthonormality relation

$$\int_0^\infty R_{n\ell}(r)R_{n'\ell}(r)r^2\,dr = \delta_{nn'}, \quad (3.57)$$

This point can be verified by using the integral properties of the associated Laguerre polynomials.

Graphs of the radial functions for $n = 1$, 2, and 3 are presented in Fig. 3.2. These functions are quite important and should be examined carefully. Notice, for example, that the number of nodes in the radial function $R_{n\ell}(\rho)$

[12]A more complete table for $n = 1$ through $n = 6$ may be found in Linus Pauling and E. B. Wilson, *Introduction to Quantum Mechanics* (New York: McGraw-Hill, 1935), pp. 135–136.

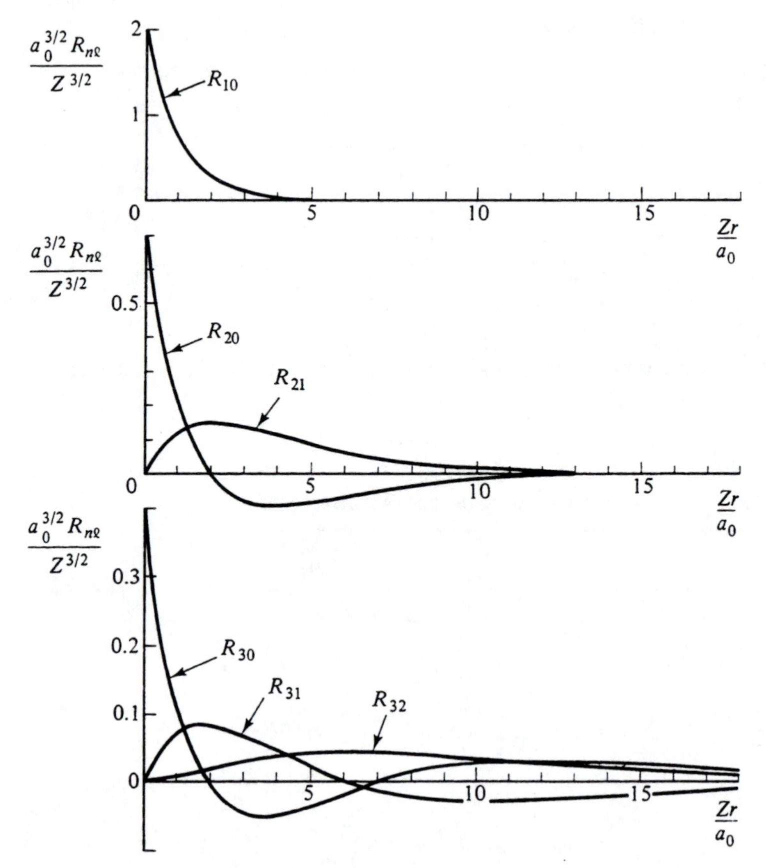

Figure 3.2 Hydrogenic radial functions $R_{n\ell}(r)$ for $n = 1, 2$, and 3. (See Table 3.2 for functional forms.) (From R. B. Leighton, *Principles of Modern Physics.* New York: McGraw-Hill, 1959.)

is $n - \ell - 1$. Also notice that each radial function for $\ell = 0$ approaches a nonzero constant as $r \longrightarrow 0$, whereas each function for $\ell > 0$ approaches zero as $r \longrightarrow 0$.

3.6 FULL EIGENFUNCTIONS OF THE ONE-ELECTRON ATOM HAMILTONIAN

We set out at the beginning of this chapter to solve the time-independent Schroedinger equation for the one-electron atom, and we have achieved that goal. The normalized stationary-state wave functions are

$$\psi_{n\ell m}(\mathbf{r}) = R_{n\ell}(r)Y_{\ell m}(\theta, \varphi), \tag{3.58}$$

where $R_{n\ell}(r)$ is given by Eq. (3.55) and the $Y_{\ell m}(\theta, \varphi)$ are the spherical harmonics of Eq. (2.58). These functions for $n = 1, 2$, and 3 are listed in Table

Table 3.3
Energy eigenfunctions for the hydrogenic atoms for $n = 1, 2,$ and 3.

$$\psi_{100} = \frac{1}{\sqrt{\pi}}\left(\frac{Z}{a_0}\right)^{3/2} e^{-Zr/a_0}$$

$$\psi_{200} = \frac{1}{4\sqrt{2\pi}}\left(\frac{Z}{a_0}\right)^{3/2}\left(2 - \frac{Zr}{a_0}\right)e^{-Zr/2a_0}$$

$$\psi_{210} = \frac{1}{4\sqrt{2\pi}}\left(\frac{Z}{a_0}\right)^{3/2}\frac{Zr}{a_0}e^{-Zr/2a_0}\cos\theta$$

$$\psi_{21\pm1} = \frac{\mp1}{8\sqrt{\pi}}\left(\frac{Z}{a_0}\right)^{3/2}\frac{Zr}{a_0}e^{-Zr/2a_0}\sin\theta\, e^{\pm i\varphi}$$

$$\psi_{300} = \frac{1}{81\sqrt{3\pi}}\left(\frac{Z}{a_0}\right)^{3/2}\left(27 - 18\frac{Zr}{a_0} + 2\frac{Z^2r^2}{a_0^2}\right)e^{-Zr/3a_0}$$

$$\psi_{310} = \frac{\sqrt{2}}{81\sqrt{\pi}}\left(\frac{Z}{a_0}\right)^{3/2}\left(6 - \frac{Zr}{a_0}\right)\frac{Zr}{a_0}e^{-Zr/3a_0}\cos\theta$$

$$\psi_{31\pm1} = \frac{\mp1}{81\sqrt{\pi}}\left(\frac{Z}{a_0}\right)^{3/2}\left(6 - \frac{Zr}{a_0}\right)\frac{Zr}{a_0}e^{-Zr/3a_0}\sin\theta\, e^{\pm i\varphi}$$

$$\psi_{320} = \frac{1}{81\sqrt{6\pi}}\left(\frac{Z}{a_0}\right)^{3/2}\frac{Z^2r^2}{a_0^2}e^{-Zr/3a_0}(3\cos^2\theta - 1)$$

$$\psi_{32\pm1} = \frac{\mp1}{81\sqrt{\pi}}\left(\frac{Z}{a_0}\right)^{3/2}\frac{Z^2r^2}{a_0^2}e^{-Zr/3a_0}\sin\theta\cos\theta\, e^{\pm i\varphi}$$

$$\psi_{32\pm2} = \frac{1}{162\sqrt{\pi}}\left(\frac{Z}{a_0}\right)^{3/2}\frac{Z^2r^2}{a_0^2}e^{-Zr/3a_0}\sin^2\theta\, e^{\pm 2i\varphi}$$

3.3. From the orthonormality relations of the individual product functions [Eqs. (3.57) and (2.60)], we conclude that the one-electron ("hydrogenic") eigenfunctions satisfy the orthonormality relation

$$\int \psi^*_{n\ell m}(\mathbf{r})\psi_{n'\ell'm'}(\mathbf{r})\, d\mathbf{r} = \delta_{nn'}\,\delta_{\ell\ell'}\,\delta_{mm'}, \tag{3.59}$$

where, as usual, $d\mathbf{r} = r^2\, dr \sin\theta\, d\theta\, d\varphi$.

Quantum Numbers

Let us summarize what we have learned about the quantum numbers for the one-electron atom. It is important to remember that quantum numbers serve two purposes. First, they provide a convenient way to distinguish different quantum states. Second, they contain information about physical properties of the states, since they label eigenvalues of various constants of the motion for the system. So far we have found three quantum numbers for the one-electron atom: the principal quantum number n, which labels the energies and radial wave functions, the orbital-angular-momentum quantum number ℓ, which specifies the magnitude of $\mathbf{L}$ via the L^2 eigenvalue equation, and the magnetic quantum number m, which specifies the z component of $\mathbf{L}$ via the

L_z eigenvalue equation. The allowed values of these quantum numbers are restricted—that is,

$$\boxed{\begin{aligned} n &= 1, 2, 3, \ldots, \\ \ell &= 0, 1, \ldots, n-1, \\ m &= -\ell, -\ell+1, \ldots, \ell-1, \ell. \end{aligned}} \tag{3.60}$$

For example, we can speak of the "$n = 2$, $\ell = 1$, $m = 0$" state of a hydrogenic atom. There is a more conventional way to label the atomic states of a one-electron atom: *spectroscopic notation*. Spectroscopists refer to the aforementioned state as a "$2p_0$" state, using a letter to represent the value of ℓ and a subscript to indicate the value of m. Similarly, the "$n = 3$, $\ell = 2$, $m = +1$" state is called the "$3d_1$" state. The letter designations for $\ell = 0$, 1, 2, 3, . . . are given in Table 3.4; we shall use this terminology throughout the remainder of the book.

Table 3.4
Spectroscopic notation.

ℓ	0	1	2	3	4	5	6	. . .
Spectroscopic designation	*s*	*p*	*d*	*f*	*g*	*h*	*i*	. . .

One-Electron Energies

Recall that as a consequence of our solution of the radial equation, we obtained a simple expression for the energies E_n,

$$\boxed{E_n = -\frac{\mu Z^2 e^4}{2\hbar^2 n^2}, \qquad n = 1, 2, \ldots .} \tag{3.61}$$

Energy eigenvalues for a one-electron atom

This gives the energy (in gaussian CGS units) of the $(n\ell m)$ state of a hydrogenic atom with nuclear charge Z. Introducing the first Bohr radius a_0 [see Eq. (3.40)] into (3.61), we can rewrite this result in the form:

$$E_n = -\frac{Z^2}{n^2}\left(\frac{e^2}{2a_0}\right), \qquad n = 1, 2, \ldots . \tag{3.62}$$

Evaluating the constants in this equation according to Appendix 1, we obtain

$$E_n = -(13.596)\frac{Z^2}{n^2}\ \text{eV}. \tag{3.63}$$

Although these units (electron volts) are standard CGS units, we shall also have occasion to use *atomic units* (see Appendix 2); in atomic units the hydrogenic energy becomes

$$E_n = -\frac{Z^2}{2n^2} \text{ Hartrees,} \tag{3.64}$$

$$E_n = -\frac{Z^2}{n^2} \text{ Rydbergs,} \tag{3.65}$$

where 1 Hartree $= e^2/a_0 \simeq 27.2$ eV and 1 Rydberg $= e^2/2a_0 \simeq 13.6$ eV.

Notice that for a given value of n there exist functions $\psi_{n\ell m}(\mathbf{r})$ for all $\ell = 0, 1, 2, \ldots, n-1$ and corresponding $m = -\ell, -\ell+1, -\ell+2, \ldots, \ell-2, \ell-1, \ell$. From Eq. (3.61) it follows that all these functions correspond to the same energy; they are degenerate. For a fixed value of n there are n allowed values of ℓ, and for each ℓ there are $2\ell+1$ values of m. Hence the state $(n\ell m)$ is n^2-fold degenerate, and there are n^2 linearly independent wave functions with energy E_n.

This "high degeneracy" is a special property of the solutions of Schroedinger's equation for a coulomb potential energy, $-Ze^2/r$. The fact that for fixed n and ℓ there are $2\ell+1$ degenerate functions, each with a different magnetic quantum number, should come as no surprise; this is true of all central force problems. However, it is not always true that states with the same n but different ℓ are degenerate.[13]

The energies of a system can be presented diagrammatically via an *energy level diagram*. We have drawn such a diagram for the hydrogen atom ($Z = 1$) in Fig. 3.3, where special care has been taken to indicate the high degree of degeneracy.

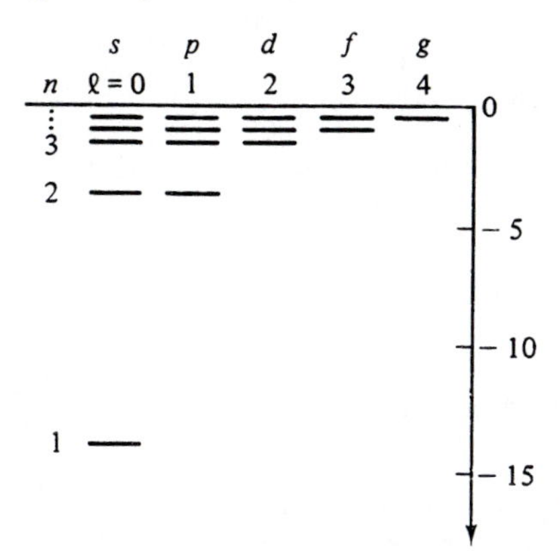

Figure 3.3 Energy level diagram for the hydrogen atom ($Z = 1$) in the approximation that the potential energy is "pure coulomb".

[13] In fact, this degeneracy disappears when we take into account the neglected interactions alluded to in footnote 1. Only in the approximation that the hydrogen atom potential energy in Eq. (3.1) is "pure coulomb" does it appear.

In the remainder of this chapter we shall focus on the properties of the hydrogenic eigenfunctions $\psi_{n\ell m}(\mathbf{r})$ and relate them to physical observables.

3.7 PROPERTIES OF THE HYDROGENIC WAVE FUNCTIONS

Recall from Chapter 1 that we relate wave functions to physical observables by examining expectation values and probabilities. For example, if a hydrogenic atom is in the stationary state $(n\ell m)$, then the average radial distance of the electron from the nucleus is given by the expectation value [see Eq. (1.40)]

$$\langle r \rangle_{n\ell m} = \int \psi^*_{n\ell m}(\mathbf{r}) r \psi_{n\ell m}(\mathbf{r})\, d\mathbf{r}, \tag{3.66}$$

where the subscript on $\langle r \rangle$ reminds us that it is defined with respect to state $(n\ell m)$. This expression can be evaluated analytically; using Eqs. (3.55) and (3.58), plus the properties of the associated Laguerre polynomials and the spherical harmonics, we obtain

$$\langle r \rangle_{n\ell m} = \frac{a_0}{2Z}[3n^2 - \ell(\ell + 1)]. \tag{3.67}$$

Exercise 3.4 Derive Eq. (3.67) for some specific case.

Expectation values of some other operators are given in Appendix 3.

This expectation value provides information about the statistical average of results of several position observations on the atom. The probability that the electron will be found at position $\mathbf{r}$ in volume element $d\mathbf{r}$ is given by the probability density [Eq. (1.9)],

$$\rho_{n\ell m}(\mathbf{r})\, d\mathbf{r} = \psi^*_{n\ell m}(\mathbf{r})\psi_{n\ell m}(\mathbf{r})\, d\mathbf{r}; \tag{3.68}$$

$\rho_{n\ell m}(\mathbf{r})$ is independent of time, since the atom is assumed to be in a stationary state. Using Eq. (3.58), we can write this expression as

$$\rho_{n\ell m}(\mathbf{r})\, d\mathbf{r} = [R_{n\ell}(r)]^2 Y^*_{\ell m}(\theta, \varphi) Y_{\ell m}(\theta, \varphi) r^2\, dr \sin\theta\, d\theta\, d\varphi. \tag{3.69}$$

Another useful quantity, which is easier to visualize than $\rho_{n\ell m}(\mathbf{r})\, d\mathbf{r}$, is obtained by integrating Eq. (3.69) over the angles θ and φ:

$$\int_0^{2\pi}\int_0^{\pi} \rho_{n\ell m}(\mathbf{r}) r^2\, dr \sin\theta\, d\theta\, d\varphi = r^2[R_{n\ell}(r)]^2\, dr \equiv P_{n\ell}(r)\, dr, \tag{3.70}$$

where we have defined the *radial probability density* $P_{n\ell}(r)$. Then the probability of finding the electron anywhere within a shell of thickness dr located a distance r from the nucleus is $P_{n\ell}(r)\, dr$.

The radial probability densities $P_{n\ell}(r)$ for $n = 1, 2,$ and 3 are plotted in Fig. 3.4. These functions provide a "picture" of the radial distribution of the electron and, as we shall see in Chapter 9, aid us in deducing physical properties of atoms with many electrons. The little arrows in the figure point to

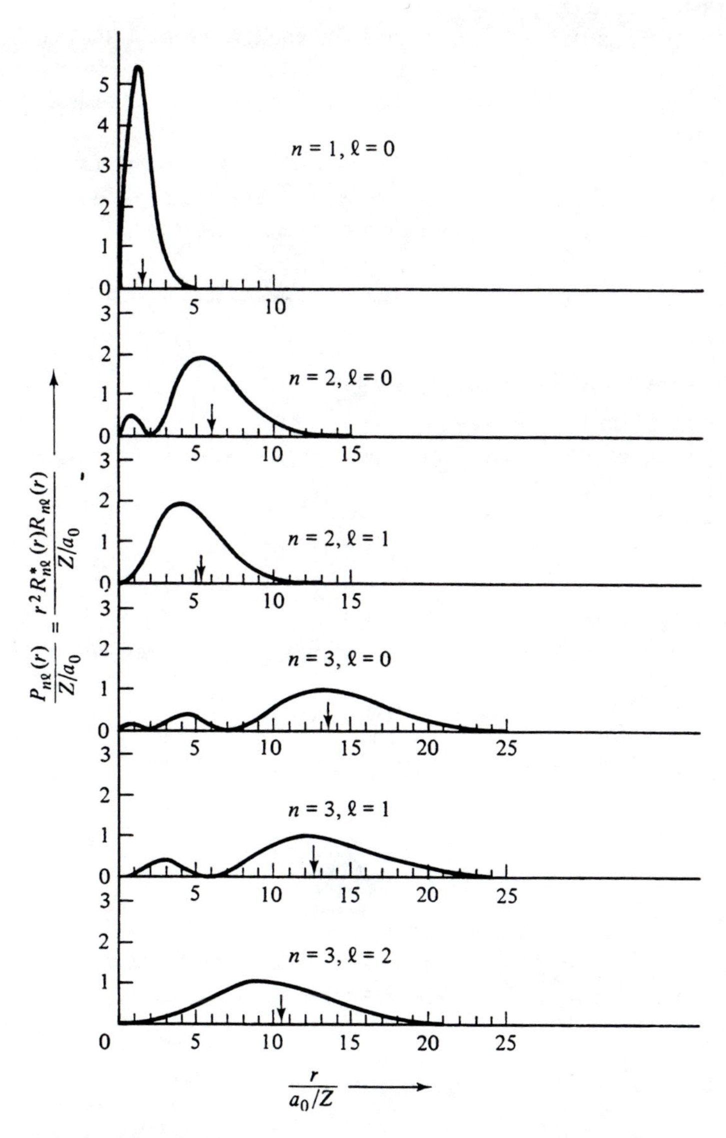

Figure 3.4 Radial probability density functions $P_{n\ell}(r)$ for the hydrogenic atom for $n = 1, 2,$ and 3. The arrows indicate $\langle r \rangle_{n\ell m}$. (From R. M. Eisberg, *Fundamentals of Modern Physics.* New York: Wiley, 1961.)

the expectation value of r for each state. In a crude sense, we could think of $\langle r \rangle$ as defining a "radius" at which the electron is most likely to be found.

In fact, we often speak of electron shells and subshells. There is one *shell* for each principal quantum number n. The shell radius for a given value of n [e.g., for the "third shell" ($n = 3$)] is taken to be the expectation value of r defined with respect to $\psi_{n00}(\mathbf{r})$, $\langle r \rangle_{n00}$.

Similarly, there is one *subshell* defined for each value of ℓ allowed by $l = 0, 1, 2, \ldots, n - 1$. The subshell radius is taken to be the expectation value of r defined with respect to $\psi_{n\ell 0}(\mathbf{r})$, $\langle r \rangle_{n\ell 0}$. Thus each shell consists of a number of subshells.

Several trends in the radial probability densities are apparent from Fig. 3.4. For example, for fixed n the expectation value $\langle r \rangle$ decreases as ℓ increases, but for fixed ℓ the expectation value increases as n increases.

Example 3.1

The Ground (1*s*) State of the Hydrogen Atom

The lowest state of the hydrogen atom ($Z = 1$) has principal quantum number $n = 1$. This is called the *ground state*. The energy of this state is [see Eqs. (3.63) and (3.65)]

$$E_1 = -\frac{e^2}{2a_0} = -13.6 \text{ eV} = -1 \text{ Rydberg}. \tag{3.71}$$

Since $n = 1$, the only allowed values of ℓ and m for the ground state are $\ell = 0$ and $m = 0$. The wave function corresponding to the energy E_1 is

$$\psi_{100}(\mathbf{r}) = R_{10}(r) Y_{00}(\theta, \varphi), \tag{3.72}$$

where

$$R_{10}(r) = 2\left(\frac{1}{a_0}\right)^{3/2} e^{-r/a_0} \tag{3.73}$$

and

$$Y_{00}(\theta, \varphi) = \frac{1}{\sqrt{4\pi}}. \tag{3.74}$$

Since $n^2 = 1$, the ground state is nondegenerate.

The radial function $R_{10}(r)$ can be found in Fig. 3.2, and the corresponding radial probability density $P_{10}(r)$ in Fig. 3.4. This density function has a peak very near the value of $\langle r \rangle$

$$\langle r \rangle_{1s} = \frac{3}{2} a_0, \tag{3.75}$$

although, in general, the peak in $P_{n\ell}(r)$ does not occur at $\langle r \rangle_{n\ell m}$.

Normally the radial probability density does not tell the whole story of the electron's spatial distribution; $\psi_{n\ell m}(\mathbf{r})$ also depends on θ and φ. However, the angular dependence of the ground state is particularly simple because of the restriction that $\ell = 0$ and $m = 0$. The ground state is spherically symmetric.

Finally, we recall that ℓ and m label the eigenvalues of the angular momentum operators L^2 and L_z, $\ell(\ell + 1)\hbar^2$ and $m\hbar$, respectively. For the ground state, we simply say that the electron possesses "zero orbital angular momentum."

Polar Plots

We have seen that plots of the radial probability density, like those in Fig. 3.4, enable us to visualize the radial dependence of the electron's probability distribution. For *s* states ($\ell = 0$, $m = 0$), the angular dependence is spherical and easy to see. In order to obtain pictures of the angular distribution of the electron for non-*s* states ($\ell \neq 0$), we draw *polar plots.*

The probability density $\rho_{n\ell m}(\mathbf{r})$ is independent of the angle φ [see Eq. (3.69)]. Therefore the probability density, which describes the spatial distribution of the electron, is equal to the radial probability density $P_{n\ell}(r)r^{-2}$ modulated by the factor $[\Theta_{\ell m}(\theta)]^2$, which depends only on θ. We can display $[\Theta_{\ell m}(\theta)]^2$ by plotting a curve as in Fig. 3.5. In a polar plot the distance from

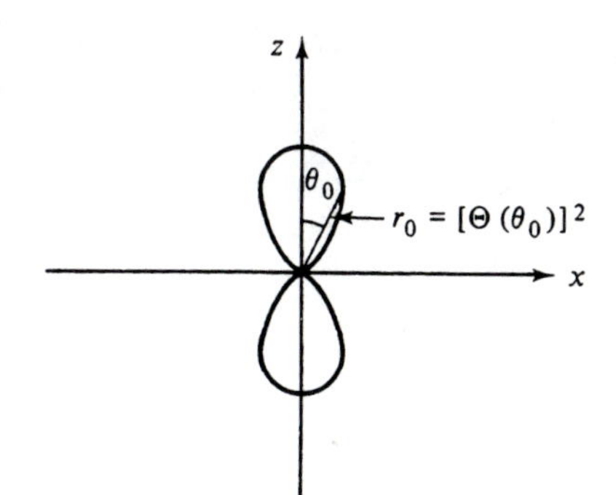

Figure 3.5 Polar plot for a *p* state ($\ell = 1$). The distance from the origin to the curve at angle $\theta = \theta_0$ is $r_0 = [\Theta(\theta_0)]^2$.

the origin of the graph to the curve at a particular value of $\theta = \theta_0$ is defined to be $[\Theta_{\ell m}(\theta_0)]^2$. The figure shows the polar plot for $\ell = 1$, $m = 0$—that is, of a p_0 state. Polar plots of other states are shown in Fig. 3.6.

A three-dimensional image of the angular dependence of $|\psi_{n\ell m}(\mathbf{r})|^2$ is obtained by rotating the polar plot for state $(n\ell m)$ through 180° about the polar axis (the $\hat{z}$ axis). The region of space swept out by the rotating curve defines a surface representing $|Y_{\ell m}(\theta, \varphi)|^2 = |\Theta_{\ell m}(\theta)\Phi_m(\varphi)|^2$. This surface possesses axial symmetry.

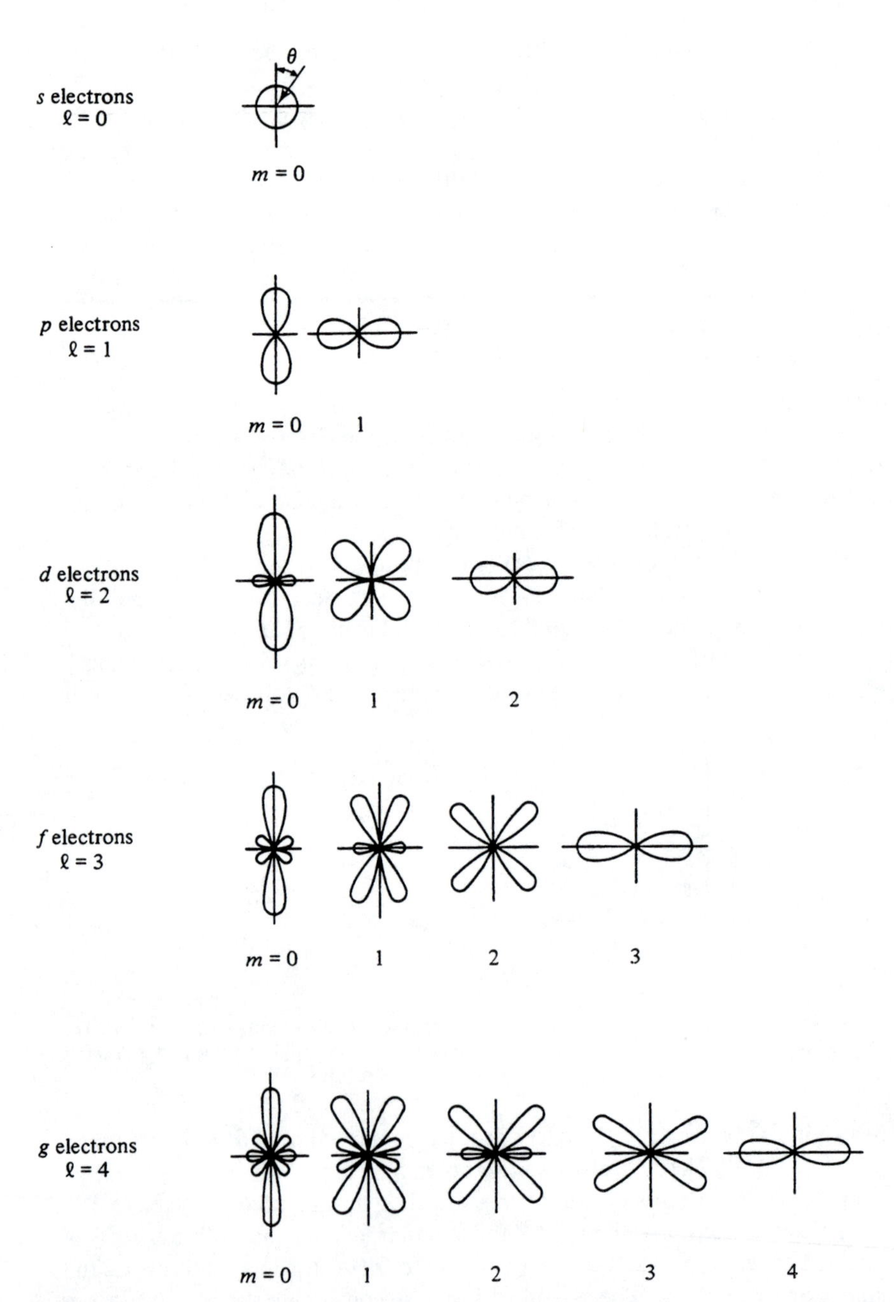

Figure 3.6 The angular probability density function $[\Theta_{\ell m}(\theta)]^2$ plotted as a function of θ for $\ell = 0, 1, 2, 3,$ and 4. The figures are not drawn to the same scale. (Adapted from F. K. Richtmyer, E. H. Kennard, and John N. Cooper, *Introduction to Modern Physics,* 6th ed. New York: McGraw-Hill, 1969.)

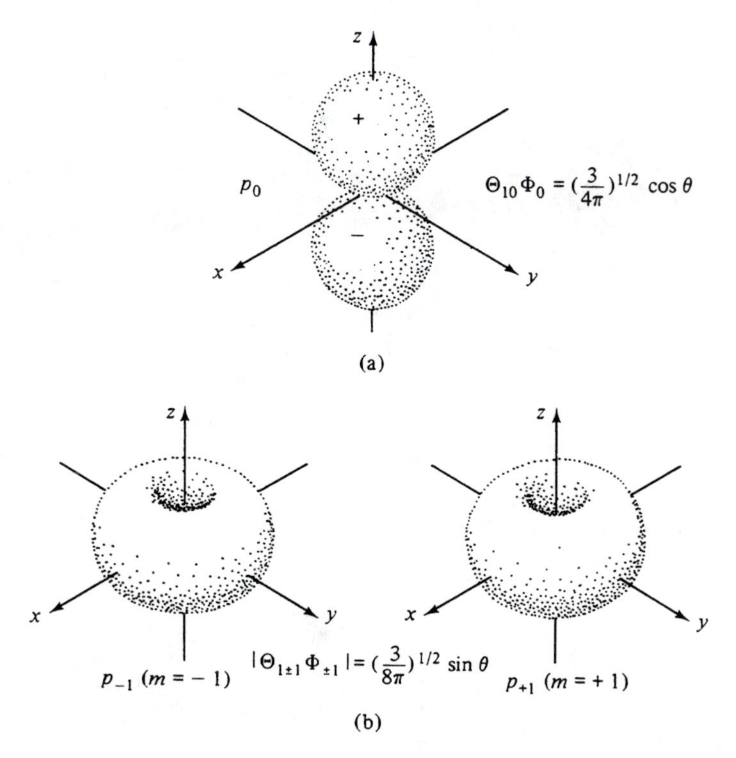

Figure 3.7 Polar plots of the spherical harmonics for $\ell = 1$. (a) Graph of Y_{10}; (b) graphs of $|Y_{1,1}|$ and $|Y_{1,-1}|$. (Adapted from Frank J. Bockoff, *Elements of Quantum Theory*. Reading, Mass: Addison-Wesley, 1969.)

Of course, the resulting surface for the state $(n\ell m)$ is merely a graphical representation of the square of the spherical harmonic $Y_{\ell m}(\theta, \varphi)$. The angular dependence of $Y_{\ell m}(\theta, \varphi)$ itself can be pictured by plotting $|Y_{\ell m}(\theta, \varphi)|$. Spherical harmonics for $\ell = 1$ (p states) are graphed in this manner in Fig. 3.7. The length of the radial coordinate at a particular $\theta = \theta_0$ and $\varphi = \varphi_0$ is $|Y_{\ell m}(\theta_0, \varphi_0)|$.

Armed with graphs of the radial probability density (Fig. 3.4) and the angular function $[\Theta_{\ell m}(\theta)]^2$ (Fig. 3.6), we can construct pictures of the full spatial probability distribution of the electron. In Fig. 3.8 we present boundary surface plots of the probability density $\rho_{n\ell m}(\mathbf{r})$ for the $1s$, $2s$, $2p_0$, and $2p_{\pm1}$ states of the hydrogen atom. (The surfaces shown in this figure enclose the regions of space in which the probability density is greater than one-tenth of its maximum value.) Such "pictures" of the probability distribution of the electron for particular states are very useful in the study of atoms and molecules.

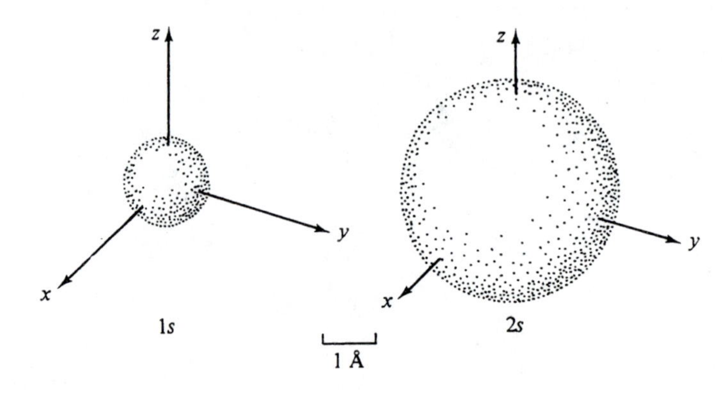

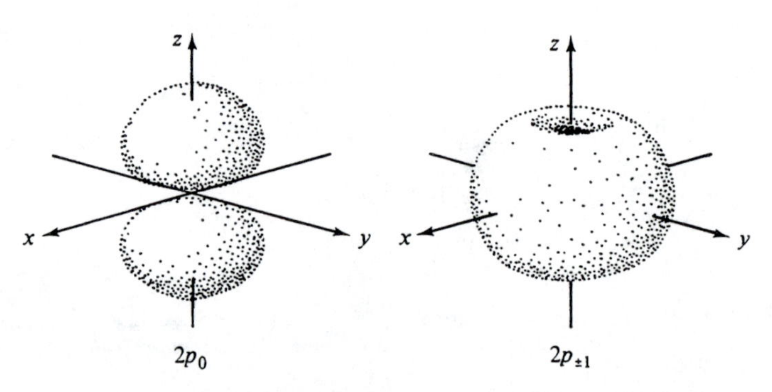

Figure 3.8 Boundary suface plots of the probability density for the 1*s*, 2*s*, $2p_0$, and $2p_{\pm 1}$ states of the hydrogen atom. (The boundary surface excludes all space in which $|\psi|^2$ is less than one-tenth of its maximum value.) (Adapted from Frank J. Bockoff, *Elements of Quantum Theory*. Reading, Mass: Addison-Wesley, 1969.)

Example 3.2

First Excited State of the Hydrogen ($Z = 1$) Atom

The energy for the $n = 2$ states of the hydrogen atom is

$$E_2 = -\frac{1}{4}\left(\frac{e^2}{2a_0}\right) = -3.40 \text{ eV} = -\frac{1}{4} \text{ Rydberg}. \tag{3.76}$$

These states are called the first excited states of the atom.

Since $n^2 = 4$, there are four degenerate eigenfunctions with this energy; these functions are listed in Table 3.5. (The explicit forms of these functions appear in Table 3.3.) Notice that the radial wave function for the $2p_0$, $2p_{+1}$

Table 3.5
Degenerate $n = 2$ eigenfunctions of the hydrogen atom.

ℓ	m	Spectroscopic notation	
0	0	$2s$	$\psi_{200}(\mathbf{r}) = R_{20}(r)Y_{00}(\theta, \varphi)$
1	0	$2p_0$	$\psi_{210}(\mathbf{r}) = R_{21}(r)Y_{10}(\theta, \varphi)$
1	+1	$2p_{+1}$	$\psi_{211}(\mathbf{r}) = R_{21}(r)Y_{11}(\theta, \varphi)$
1	−1	$2p_{-1}$	$\psi_{21-1}(\mathbf{r}) = R_{21}(r)Y_{1-1}(\theta, \varphi)$

and $2p_{-1}$ states are the same. The radial functions for $n = 2$ are plotted in Fig. 3.2; the corresponding radial probability densities appear in Fig. 3.4. From these figures we see that the radial dependence of the electron's probability distribution for the $n = 2$, $\ell = 0$ state is quite different from that of the $n = 2$, $\ell = 1$ states; the $2s$ state has two peaks, but the $2p$ states have only one peak. The node between the two peaks of the $2s$ radial function appears in the full probability density $|\psi_{200}(\mathbf{r})|^2$ for all values of θ and φ. Of course, this function is spherically symmetric, since $\ell = m = 0$.

The radius of the $n = 2$ shell is the expectation value of r; from Eq. (3.67) we find that this radius is

$$\langle r \rangle_{2s} = 6a_0; \tag{3.77}$$

the radius of the $\ell = 1$ subshell is

$$\langle r \rangle_{2p} = 5a_0. \tag{3.78}$$

Therefore, on the average, the electron is found farther from the nucleus in a $2s$ state than in a $2p$ state. (Notice that the difference between $\langle r \rangle_{2s}$ and $\langle r \rangle_{2p}$ is not as great as the difference between $\langle r \rangle_{1s}$ and $\langle r \rangle_{2s}$.)

Another important distinction can be made between the $2s$ and $2p$ radial probability densities $P_{20}(r)$ and $P_{21}(r)$. Figure 3.4 shows that at small values of r, say $r \lesssim a_0$, a small peak in the probability density occurs in the $2s$ state but not in the $2p$ state. This effect is called *penetration*; we shall encounter it again in our study of multielectron atoms (Chapter 9).

Finally, boundary surface plots of the total probability densities $|\psi_{n\ell m}(\mathbf{r})|^2$ for these $n = 2$ states are shown in Fig. 3.8.

So much time has been spent here discussing the hydrogenic functions $\psi_{n\ell m}(\mathbf{r})$ as energy eigenfunctions that we may have forgotten that they are also eigenfunctions of the orbital angular momentum operators L^2 and L_z (see

Sec. 2.4). We shall now return to this important fact and examine some of its physical consequences.

3.8 ORBITAL ANGULAR MOMENTUM OF THE HYDROGENIC ATOM

Because the coulomb potential energy is independent of θ and φ, the functions $\psi_{n\ell m}(\mathbf{r})$ satisfy the eigenvalue equations

$$L^2\psi_{n\ell m}(\mathbf{r}) = \ell(\ell + 1)\hbar^2\psi_{n\ell m}(\mathbf{r}), \tag{3.79}$$

$$L_z\psi_{n\ell m}(\mathbf{r}) = m\hbar\psi_{n\ell m}(\mathbf{r}). \tag{3.80}$$

Therefore the observables L^2 and L_z are *sharp* for a hydrogenic atom in state $(n\ell m)$, and the expectation values of the corresponding operators are simply

$$\langle L^2\rangle_{n\ell m} = \ell(\ell + 1)\hbar^2, \tag{3.81}$$

$$\langle L_z\rangle_{n\ell m} = m\hbar. \tag{3.82}$$

These results tell us, for example, that in a measurement of L_z on an ensemble of identical systems, the average value obtained is $m\hbar$. The expectation values of the other two components of $\mathbf{L}$ for the atom in state $(n\ell m)$ are

$$\langle L_x\rangle_{n\ell m} = 0, \tag{3.83}$$

$$\langle L_y\rangle_{n\ell m} = 0. \tag{3.84}$$

Vector Model

Equations (3.82), (3.83), and (3.84) suggest that we might construct a useful model of the electron's orbital angular momentum. Imagine a vector $\mathbf{L}$ that precesses about the $\hat{z}$ axis (polar axis) in such a manner that, on the average, L_x and L_y vanish but L_z is well defined. This picture is used in the so-called *vector model* of the atom. The precession of $\mathbf{L}$ in this model can be represented pictorially; the vector diagram for d states ($\ell = 2$) is drawn in Fig. 3.9. It is useful to relate this model to the polar plots of the angular dependence of the probability density that we drew in the last section. Consequently, let us look once more at Fig. 3.6 and see what trends in the polar plots can be identified.

For a given ℓ value (e.g., $\ell = 2$), the maximum in probability density shifts toward $\theta = \pi/2$ as $|m|$ increases from 0 to ℓ. By rotating the polar plots for $\ell = 2$ through all φ, we see that $|\Theta_{\ell m}(\theta)\Phi_m(\varphi)|^2$ is largest in the vicinity of the xy plane for $|m| = 2$, whereas it is largest near the $\hat{z}$ axis for

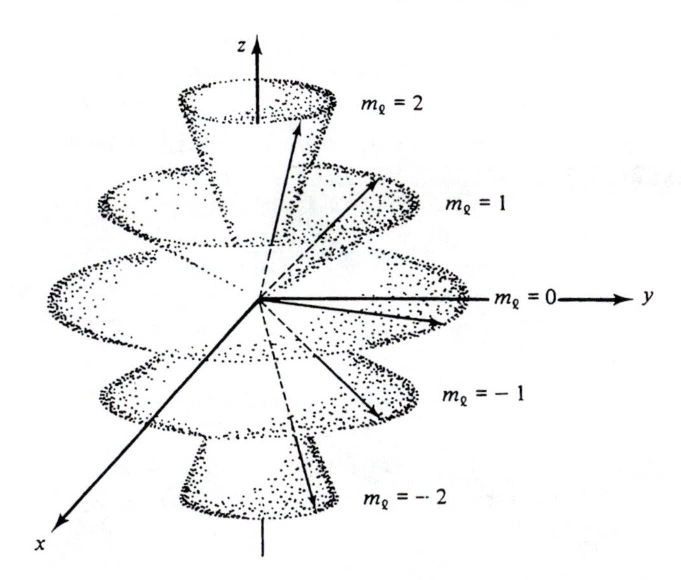

Figure 3.9 Vector model of the precession of the electron's oribital angular momentum for $\ell = 2$. We *imagine* that **L** is precessing as shown on the various cones. The length of each vector is $\sqrt{3}\hbar$.

$m = 0$. This observation is entirely consistent with the vector model, since $|m| = 2$ corresponds to the largest value of L_z (see Fig. 3.9). Classically, the maximum value of L_z arises from the largest possible electron orbits in the xy plane; clearly, it is only the $|m| = 2$ probability densities that can accommodate such orbits.

There is danger in taking the vector model too seriously. It is a mistake to conclude that we can literally follow the electron around in space and thereby pin down a well-defined orbit; we cannot. Nor can we follow **L** around in space. Actually, we do not know what the electron is doing in the classical sense; all quantum mechanics gives us is a description of the probability of an electron being found at a position **r** at time t. Nevertheless, we can obtain some feeling for the motion of the electron by studying the probability flux density.

3.9 STANDING VERSUS ROTATING WAVES

The probability flux density for a particle of mass μ is [see Eq. (1.17)]

$$\mathbf{j}(\mathbf{r}, t) = -\frac{i\hbar}{2\mu}[\Psi^*(\mathbf{r}, t)\,\nabla\Psi(\mathbf{r}, t) - \Psi(\mathbf{r}, t)\,\nabla\Psi^*(\mathbf{r}, t)]. \tag{3.85}$$

We can write $\mathbf{j}(\mathbf{r}, t)$ as

$$\mathbf{j}(\mathbf{r}, t) = \frac{\hbar}{\mu} \operatorname{Im} [\Psi^*(\mathbf{r}, t) \nabla \Psi(\mathbf{r}, t)], \tag{3.86}$$

where only the imaginary part of the quantity in brackets appears and where the gradient operator in spherical coordinates is

$$\nabla = \hat{\mathbf{r}} \frac{\partial}{\partial r} + \hat{\theta} \frac{1}{r} \frac{\partial}{\partial \theta} + \hat{\varphi} \frac{1}{r \sin \theta} \frac{\partial}{\partial \varphi}. \tag{3.87}$$

The time-dependent wave function for an electron in a hydrogenic atom in state $(n\ell m)$ is

$$\Psi_{n\ell m}(\mathbf{r}, t) = R_{n\ell}(r) \Theta_{\ell m}(\theta) \Phi_m(\varphi) e^{-(i/\hbar) E_{n\ell} t}, \tag{3.88}$$

and since $R_{n\ell}(r)$ and $\Theta_{\ell m}(\theta)$ are real functions, only the φ-component of $\Psi^* \nabla \Psi$ in Eq. (3.86) contributes to the flux density. Using Eqs. (3.87) and (3.88), we obtain

$$\mathbf{j}_{n\ell m}(\mathbf{r}, t) = \frac{\hbar}{\mu} \operatorname{Im} \left\{ \Psi^*_{n\ell m}(\mathbf{r}, t) \hat{\varphi} \frac{1}{r \sin \theta} \frac{\partial}{\partial \varphi} \Psi_{n\ell m}(\mathbf{r}, t) \right\}, \tag{3.89}$$

or

$$\boxed{\mathbf{j}_{n\ell m}(\mathbf{r}, t) = \hat{\varphi} \frac{m\hbar}{\mu r \sin \theta} \rho_{n\ell m}(\mathbf{r}).} \tag{3.90}$$

Now, $\mathbf{j}_{n\ell m}(\mathbf{r}, t)$ is the probability per unit time "flowing" through an element of surface area of unit magnitude.[14] Therefore this result explicitly shows that the electron probability "flows" (or "rotates") around the $\hat{z}$ axis. The "flow" is in the sense of increasing φ for $m > 0$ and decreasing φ for $m < 0$. The magnitude of $\mathbf{j}(\mathbf{r}, t)$ is proportional to m and is largest where the probability density is largest. The wave function $\psi_{n\ell m}(\mathbf{r})$ is called a *rotating wave*.

We have suggested that these rotating wave functions are especially useful, for they are eigenfunctions of L^2 and L_z. However, other perfectly valid stationary-state eigenfunctions of the one-electron atom Hamiltonian have angular momentum properties that differ from $\psi_{n\ell m}(\mathbf{r})$. In particular, we can construct energy eigenfunctions that do not "rotate" about the $\hat{z}$ axis by forming linear combinations of degenerate rotating waves. The new eigenfunctions are called *standing waves* and are particularly useful in molecular physics (see Chapter 14). Of course, the standing wave functions are not eigenfunctions of L_z, but sometimes this does not matter. Let us consider a specific example.

[14]We must be careful about using the word "flow" in reference to probability flux, since it is a classical term. However, it does provide a conceptual feeling for the physical significance of Eq. (3.90).

Example 3.3
Standing Waves for 2p Hydrogenic States

Of the three eigenfunctions $\psi_{n\ell m}(\mathbf{r})$ for $n = 2$, $\ell = 1$ shown in Fig. 3.8, only the $2p_0$ function is a standing wave [since $m = 0$ in Eq. (3.90) implies that $\mathbf{j}(\mathbf{r}, t) = 0$]. However, we can form standing wave functions from the $2p_1$ and $2p_{-1}$ functions by defining appropriate linear combinations. Contemplation of the spherical harmonics Y_{11} and $Y_{1,-1}$ leads us to the functions

$$\begin{aligned}\psi_{2p_y}(\mathbf{r}) &\equiv \frac{i}{\sqrt{2}}[\psi_{2p_{+1}}(\mathbf{r}) + \psi_{2p_{-1}}(\mathbf{r})] \\ &= \frac{1}{\sqrt{2}} R_{21}(r)\left(2\sqrt{\frac{3}{8\pi}}\right) \sin\theta \sin\varphi,\end{aligned} \tag{3.91}$$

$$\begin{aligned}\psi_{2p_x}(\mathbf{r}) &\equiv \frac{1}{\sqrt{2}}[\psi_{2p_{-1}}(\mathbf{r}) - \psi_{2p_{+1}}(\mathbf{r})] \\ &= \frac{1}{\sqrt{2}} R_{21}(r)\left(2\sqrt{\frac{3}{8\pi}}\right) \sin\theta \cos\varphi.\end{aligned} \tag{3.92}$$

These functions are standing waves [i.e., $\mathbf{j}(\mathbf{r}, t) = 0$]; they are eigenfunctions of $\mathcal{H}$ and L^2 but not of L_z.

Exercise 3.5 (a) Show that $\mathbf{j}(\mathbf{r}, t) = 0$ for the functions $\psi_{2p_x}(\mathbf{r})$ and $\psi_{2p_y}(\mathbf{r})$.

(b) Show that $\psi_{2p_x}(\mathbf{r})$ and $\psi_{2p_y}(\mathbf{r})$ are eigenfunctions of $\mathcal{H}$ and L^2 but not of L_x, L_y, or L_z.

(c) Show that ψ_{2p_x} and ψ_{2p_y} are normalized and orthogonal to each other.

The reason for the subscripts labeling these new functions becomes apparent when we look at the sketches of their angular dependence; see Fig. 3.10. (Notice that any linear combinations of the three functions ψ_{2p_x}, ψ_{2p_y}, and $\psi_{2p_z} = \psi_{2p_0}$ are also standing waves.)

3.10 THE ORBITAL MAGNETIC MOMENT

Let us return to the functions $\Psi_{n\ell m}(\mathbf{r}, t)$. These functions describe the motion of the electron. Since they are rotating waves, and the electron possesses a charge $-e$, our classical intuition leads us to expect that this rotational flux will have associated with it an electric current and a local magnetic field. We shall see in Chapter 7 that such is indeed the case and that the effects of this field can be observed experimentally.

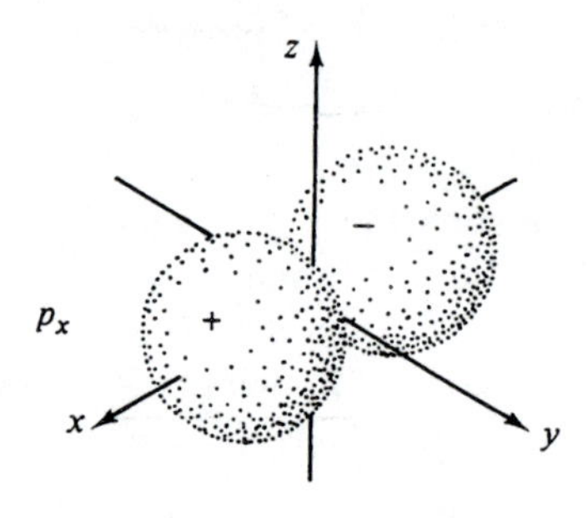

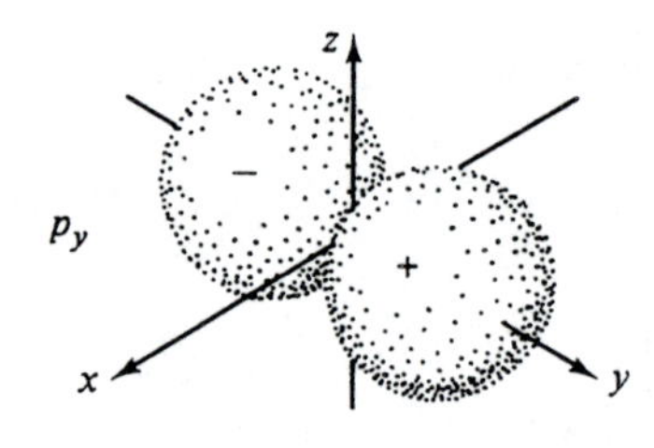

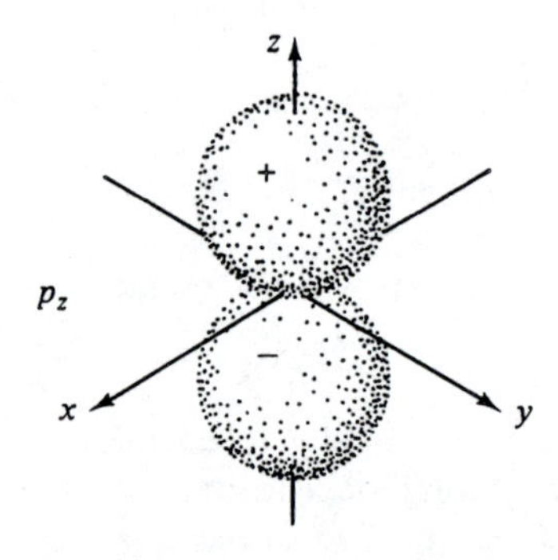

Figure 3.10 Graphs of the *angular dependence* of ψ for the p_x, p_y, and p_z standing-wave hydrogen eigenfunctions. (Adapted from Frank J. Bockoff, *Elements of Quantum Theory*, Reading, Mass: Addison-Wesley, 1969.)

In this section we shall use our expression for $\Psi_{n\ell m}(\mathbf{r}, t)$ and the probability flux density to evaluate the orbital magnetic moment due to the motion of the electron.

The contribution to the *orbital magnetic moment* due to the rotation of an element of volume dv is[15]

$$d\mathbf{M} = \frac{1}{2c}\mathbf{r} \times [-e\mathbf{j}(\mathbf{r}, t)]\, dv, \tag{3.93}$$

where $-e$ is the electronic change—that is, $e > 0$. The probability flux

[15]See Edward M. Purcell, *Electricity and Magnetism* (New York: McGraw-Hill, 1965), Chap. 10.

density $\mathbf{j}(\mathbf{r}, t)$ can be written in the useful form

$$\mathbf{j}(\mathbf{r}, t) = \frac{1}{2\mu}[\Psi^*(\mathbf{r}, t)\mathbf{p}\Psi(\mathbf{r}, t) - \Psi(\mathbf{r}, t)\mathbf{p}\Psi^*(\mathbf{r}, t)] \tag{3.94}$$

or

$$\mathbf{j}(\mathbf{r}, t) = \frac{1}{\mu}\,\mathrm{Re}\,\{\Psi^*(\mathbf{r}, t)\mathbf{p}\Psi(\mathbf{r}, t)\}, \tag{3.95}$$

where $\mathbf{p} = -i\hbar\,\nabla$. Therefore the element of orbital magnetic moment $d\mathbf{M}$ is

$$d\mathbf{M} = -\frac{e}{2\mu c}\,\mathrm{Re}\,[\Psi^*(\mathbf{r}, t)\mathbf{L}\Psi(\mathbf{r}, t)]\,dv. \tag{3.96}$$

When integrated over all space, this equation gives a total magnetic moment

$$\mathbf{M} = -\frac{e}{2\mu c}\langle\Psi(\mathbf{r}, t)\,|\,\mathbf{L}\,|\,\Psi(\mathbf{r}, t)\rangle, \tag{3.97}$$

where we have introduced the notation

$$\langle\Psi(\mathbf{r}, t)\,|\,\mathbf{L}\,|\,\Psi(\mathbf{r}, t)\rangle \equiv \int \Psi^*(\mathbf{r}, t)\mathbf{L}\Psi(\mathbf{r}, t)\,d\mathbf{r} \tag{3.98}$$

for the matrix element of the orbital angular momentum $\mathbf{L}$. This matrix element is the sum of three terms—$\hat{\mathbf{x}}\langle\Psi(\mathbf{r}, t)\,|\,L_x|\,\Psi(\mathbf{r}, t)\rangle$, and so on.

If we now define the operator for the orbital magnetic moment

$$\mathbf{M}_\ell \equiv -\frac{e}{2\mu c}\mathbf{L}, \tag{3.99}$$

then Eq. (3.97) states that the observable $\mathbf{M}$ is simply the expectation value of this operator. The orbital magnetic moment operator is more conventionally written[16]

$$\boxed{\mathbf{M}_\ell = -\frac{g_\ell \beta}{\hbar}\mathbf{L},} \quad \text{Orbital magnetic moment operator} \tag{3.100}$$

where β is the *Bohr magneton* defined as

$$\beta \equiv \frac{e\hbar}{2m_e c} = 0.927 \times 10^{-20}\ \text{erg/gauss} \tag{3.101}$$

and g_ℓ is the *orbital g factor*, which we introduce here so that Eq. (3.100) appears in the form of a more general result to be obtained later. For the

[16]Elsewhere you may encounter the symbol $\boldsymbol{\mu}_\ell$ for the orbital magnetic moment operator. We have chosen $\mathbf{M}_\ell$ instead to avoid confusion with the reduced mass. Similarly, the Bohr magneton is often represented by the symbol μ_B.

orbital magnetic moment, we have

$$g_\ell = 1. \tag{3.102}$$

For the stationary state $(n\ell m)$ of a one-electron atom, it follows from Eqs. (3.100), (3.79), and (3.80) that the square and z component of the orbital magnetic moment are *sharp*:

$$M_\ell^2 \Psi_{n\ell m}(\mathbf{r}, t) = \left(\frac{g_\ell \beta}{\hbar}\right)^2 \ell(\ell + 1)\hbar^2 \Psi_{n\ell m}(\mathbf{r}, t), \tag{3.103}$$

$$M_{\ell_z} \Psi_{n\ell m}(\mathbf{r}, t) = -\frac{g_\ell \beta}{\hbar}(m\hbar)\Psi_{n\ell m}(\mathbf{r}, t). \tag{3.104}$$

Consequently, for the state $(n\ell m)$, the magnitude of the square of the orbital magnetic moment is[17]

$$\langle M_\ell^2 \rangle = (g_\ell \beta)^2 \ell(\ell + 1), \tag{3.105}$$

and the value of its z component is

$$\langle M_{\ell_z} \rangle = -g_\ell \beta m. \tag{3.106}$$

Each of these results is a consequence of the fact that the functions $\Psi_{n\ell m}(\mathbf{r}, t)$ are eigenfunctions of L^2 and L_z.

This is about as far as we can develop the solution of the one-electron atom at this level of approximation of quantum theory. We have solved the Schroedinger equation for the hydrogenic atom, making only the assumption that the potential energy $V(\mathbf{r})$ was of the form $-Ze^2/r$. There is one major addition to be made to our theory of the one-electron atom: the introduction of electron spin. We shall discuss spin in Chapter 6 and then take another look at the hydrogenic atom in Chapter 7. But first we shall digress briefly and see what happens when our hydrogenic atom is exposed to an external field.

PROBLEMS

3.1 Solution of the Hydrogenic Radial Equation by Series Expansion (***)

In this problem we shall use the method of series expansion to solve Eq. (3.32).

(a) Expand the function $f_{n\ell}(\rho)$ in a power series in ρ,

$$f_{n\ell}(\rho) = \sum_{i=0}^{\infty} a_i \rho^i.$$

[17]These results can also be derived by discussing orbital magnetic moments in terms of the Bohr model of the hydrogenic atom. See Robert M. Eisberg, *Fundamentals of Modern Physics* (New York: Wiley, 1961), Chap. 10.

Substitute your expansion into Eq. (3.32) to obtain a recursion relation for the coefficients a_i. We shall choose $a_0 = 1$.

(b) Consider the function $R_{n\ell}(\rho)$ of Eq. (3.31) in the limit $\rho \longrightarrow \infty$, using the expansion of part (a). Show that in order to have $R_{n\ell}(\rho) \longrightarrow 0$ as $\rho \longrightarrow \infty$, the series must be truncated; that is, we must demand that all coefficients after the kth one (for any $k > 0$) be zero. [HINT: Consider the ratio of consecutive coefficients in the series at very large values of ρ.]

(c) Show that this truncation can be effected by setting

$$\gamma = \ell + k + 1$$

and that this condition leads to Eq. (3.37) for $R_{n\ell}(\rho)$.

3.2 The Deuteron (**)

The deuteron is a composite "particle" resulting from the binding together of a proton and neutron due to their mutual attraction. It has been empirically established that the interaction potential energy for this system may be represented by

$$V(r) = -Ae^{-r/a},$$

where r is the proton-neutron interparticle separation, A is a "strength" parameter of the order 32 MeV, and a is a "range" parameter of the order of 2.2 fermi (1 fermi $= 10^{-13}$ cm). Consider bound states of this system—thus $E < 0$.

(a) Write down the radial equation for $\chi(r) = rR(r)$ for the case $\ell = 0$. Perform a transformation to a new variable $\xi \equiv \exp(-r/2a)$ and obtain a differential equation in ξ. Verify that this is a form of Bessel's equation.[18] Determine the general solution before applying boundary conditions.

(b) Impose the boundary condition that $|R(r)| < \infty$ as $r \longrightarrow \infty$ and write the form of the solution that satisfies this condition. (You will need to employ the limiting properties of Bessel's functions.)

(c) Next, impose the additional boundary condition that $|R(r)| < \infty$ as $r \longrightarrow 0$ and show that it restricts the energy to certain discrete values. Describe how you would go about finding the energies.

(d) Use the values for A and a given above to calculate the ground-state binding energy $-E_{n\ell}$ of the deuteron.

3.3 Average Electrostatic Potential Field in the Vicinity of a One-Electron Atom (***)

Consider a one-electron atom (or ion) with nuclear charge Ze in its ground (1s) state. If classical physics were valid for this system, we should expect to be able to measure its *electrostatic potential field* φ. This quantity is valuable in the quantum

[18]See the Suggested Readings for Chapter 2. A graph of the solutions to Bessel's equation may be found in Eugene Jahnke and Fritz Emde, *Tables of Functions* (New York: Dover, 1945).

mechanical treatment as well. In this problem we consider the average field and point out the connection with real systems.

(a) Take the origin of coordinates at the nucleus and label the atomic electron with coordinates $\mathbf{r}_1$. Suppose that this electron is "frozen" at position $\mathbf{r}_1$. Derive an expression for the electrostatic potential field $\varphi(\mathbf{r}_1, \mathbf{r}_2)$ at an arbitrary point $\mathbf{r}_2$ due to this system.

(b) Calculate the average value of φ for the $1s$ ground state of the atom (or ion) and show that the result may be written

$$\bar{\varphi}(r_2) = \langle \varphi(\mathbf{r}_1, \mathbf{r}_2) \rangle_{1s} = \frac{e}{r_2}(Z-1) + \frac{e}{r_2}\left(1 + \frac{Zr_2}{a_0}\right)e^{-2Zr_2/a_0},$$

which is clearly spherically symmetric. [HINT: The evaluation of the integral is greatly simplified by using the following expansion for $|\mathbf{r}_1 - \mathbf{r}_2|^{-1}$:

$$\frac{1}{|\mathbf{r}_1 - \mathbf{r}_2|} = \begin{cases} \sum_{\lambda=0}^{\infty} \frac{r_1^\lambda}{r_2^{\lambda+1}} P_\lambda(\cos\theta_{12}) & \text{if } r_2 > r_1, \\ \sum_{\lambda=0}^{\infty} \frac{r_2^\lambda}{r_1^{\lambda+1}} P_\lambda(\cos\theta_{12}) & \text{if } r_2 < r_1, \end{cases}$$

where θ_{12} is the angle between $\mathbf{r}_1$ and $\mathbf{r}_2$. It is easiest to take the $\hat{z}$ axis along $\hat{\mathbf{r}}_2$ for the integration.]

(c) Examine the behavior of $\bar{\varphi}(r_2)$ in the limits $r_2 \longrightarrow \infty$ and $r_2 \longrightarrow 0$. Discuss your results in terms of "screening of the nucleus," taking into account the electron density distribution and the corresponding "electron charge density."

(d) Consider the He^+($Z = 2$) ion. Let us use the result of part (b) to obtain an approximate answer to a rather hard question: What is the binding energy (or energies) of a *second* electron added to the system? Suppose that we say that the total effect of the first electron is to "screen" the nucleus in the manner of parts (b) and (c). Then electron 2 interacts with He^+ via a potential energy

$$V(r_2) = -e\bar{\varphi}(r_2).$$

If the second term in $\bar{\varphi}(r_2)$ could be ignored, what would be the binding energies (i.e., $-E_{n\ell}$) for this He system? Should the second term raise or lower the binding energies? Why? Will the ℓ degeneracy be removed, and, if so, how will the energies vary with ℓ (qualitatively)?

(e) Think about the nature of the approximations considered in part (d). What is the main thing *wrong* with treating the He atom in this manner (aside from electron spin and the Pauli exclusion principle, which are discussed in Chapter 5 ff.)?

3.4 Penetration of the Centrifugal Barrier and Its Effects in Multielectron Atoms (**)

Wave functions corresponding to states of high angular momentum are distinctive in both their angular and radial dependence. In particular, the effect of the "centrifugal barrier" is to push out the radial density away from the region of small r. The purpose of this problem is to make this feature explicit and to point out some related physical effects.

(a) Consider the $n = 3$ states of hydrogen ($Z = 1$). Calculate radial probability densities of the $3s$, $3p$, and $3d$ states at $r = 0.5a_0$ and $1.0a_0$. Briefly discuss your results to make sure that they agree with your expectations. The $3s$ state is described as more "penetrating" than $3p$ and $3p$ as more penetrating than $3d$.

In multielectron atoms, the concept of penetration is important. You may recall from your chemistry background that the electrons in such an atom tend to form a shell structure. The physical explanation for the structure and the quantitative details will be dealt with later (Chapter 9). For the present, let us take the somewhat naive approach that, from the vantage point of the "outer" electron, all the other electrons serve mainly to screen the nuclear coulomb field. These other electrons, on the average, constitute a negative charge density that tends to weaken the effect of the attractive nuclear potential. Thus in a neutral atom—such as Na ($Z = 11$)—the ground state consists of a spherically symmetric core (nucleus + 10 electrons) surrounded by a single electron in a state similar to the $3s$ state of the hydrogen atom. We might then take as a model of Na an electron with potential energy

$$V(r) = -Z(r)\frac{e^2}{r},$$

where

$$Z(r) \longrightarrow \begin{cases} 11, & r \longrightarrow 0 \\ 1, & r \longrightarrow \infty. \end{cases}$$

(b) Explain why $V(r)$ suggested above is reasonable and sketch a qualitative plot consisting of curves for $-11e^2/r$, $-e^2/r$, and $V(r)$. [We shall see later that $Z(r) \sim 1$ for $r \gtrsim 3$ or $4a_0$, so that the screening occurs quite close in. You may use this fact in drawing your plots.]

(c) Using the results of part (a) and your ideas on penetration, determine which of the states of the outer electron (i.e., $3s$, $3p$, or $3d$) should be most affected and which least affected by the fact that the potential is not "pure coulomb" for all r. Explain your answer carefully. What is it that is being "penetrated," and in what way should this affect the wave functions and energies of the states?

(d) It has been determined spectroscopically that the binding energy (see Prob. 3.3) of the outer electron is 5.12 eV for the $3s$ state, 2.10 eV for the $3p$ state, and 1.50 eV for the $3d$ state. Are these results consistent with what we have determined about the relative penetration of these states? Explain. Use the pure one-electron atom result for the three $n = 3$ energies and determine an effective charge Z' for each (which is independent of r). What, in particular, is striking about the $3d$ state?

3.5 Probability Density Distributions for States of the One-Electron Atom (**)

(a) Consider the angular distributions associated with the electron probability density of a one-electron atom in an f ($\ell = 3$) state; we will consider all states $m = 0, \pm 1, \pm 2, \pm 3$. For each value of m, draw three-dimensional, spherical-polar graphs analogous to Fig. 3.7. Consider the probability current for each state and calculate $\langle M_\ell^2 \rangle$ and $\langle M_{\ell_z} \rangle$, the latter for each m value. (Leave in terms of β.) Draw a vector model diagram (see Fig. 3.9) for the f states.

(b) Consider radial distributions for the $n = 4$ states. Use Fig. 3.4 to guess the radial probability density functions for the $n = 4$ states, plotting curves for $4s$, $4p$, $4d$, and $4f$ radial distributions. Calculate $\langle r \rangle$ for these states and indicate this value on your plot.

(c) We can now combine information on the angular and radial distributions. Draw "boundary surface plots" (see Fig. 3.8) for the $4d$ and $4f$ states with $m = 0$. Indicate the spherical surface $r = \langle r \rangle$ and any nodal surfaces on your graphs.

3.6 Dipole Moments of States of the One-Electron Atom (**)

The *dipole moment operator* for a one-electron atom is simply

$$\mathbf{d} = -e\mathbf{r},$$

where $\mathbf{r}$ is the position vector for the electron relative to the nucleus.

(a) Show that the expectation value of $\mathbf{d}$ vanishes for all stationary states of a one-electron atom.

(b) If a one-electron atom is "irritated" by a time-dependent field, such as an electromagnetic field (i.e., light), then the atom is necessarily in a nonstationary state. (Actually, it is in this nonstationary state that absorption and emission of light occurs.) Suppose that the atom is found at time t in the state

$$\Psi(\mathbf{r}, t) = (1 + \gamma^2)^{-1/2}[\psi_{1s}(\mathbf{r})e^{-(i/\hbar)E_1 t} + \gamma\psi_{2p_0}(\mathbf{r})e^{-(i/\hbar)E_2 t}],$$

where γ is a fixed real constant. Verify that Ψ is normalized. Find the values of $\langle E \rangle$, $\langle L^2 \rangle$, and $\langle L_z \rangle$ for this state. Calculate $\langle \mathbf{d} \rangle$ for this state and show that, on the average, the atom behaves like an oscillating electric dipole. Calculate the frequency of oscillation. Classically, such a system would emit electromagnetic radiation of what frequency?

(c) In this example, Ψ is of the correct form for an atom interacting with an electromagnetic field to absorb or emit radiation. Calculate $\langle \mathbf{d} \rangle$ for a state in which ψ_{2p_0} is replaced by ψ_{2s} in Ψ above. Do the same for ψ_{2p_0} replaced by ψ_{3d_0}. What would you get for $\langle \mathbf{d} \rangle$ if the two functions in Ψ had different m values? What can you conclude about the ℓ and m values of the functions in Ψ necessary to give $\langle \mathbf{d} \rangle \neq 0$?

3.7 Inner-Shell Ionization of Heavy Atoms (*)

We shall see later on in our study of multielectron atoms that the two innermost electrons (occupying the so-called K shell) are fairly independent of one another and may be approximately described by $1s$ one-electron atom wave functions with appropriate values of Z.

(a) Using the preceding approximation, calculate the binding energy in thousands of electron volts (KeV) (see Prob. 3.3) and the value of $\langle r \rangle$ (in atomic units) for a $1s$ electron in the following atoms: Ca ($Z = 20$), Zr ($Z = 40$), Nd ($Z = 60$), Hg ($Z = 80$).

(b) Binding energies of inner-shell electrons may be measured experimentally by means of *X-ray absorption spectroscopy*. If light of a small enough wavelength λ (large enough energy) is incident on an atom, there is a finite probability that a K shell ($1s$) electron will be ejected. The maximum value of λ that will still result in K-shell ionization is called the *absorption edge*. Derive an expression for the K-shell binding energy (in KeV) in terms of the K-shell absorption edge (in Å). The observed edges for the atoms in part (a) in angstroms are Ca: 3.070, Zr: 0.6888, Nd: 0.2845, Hg: 0.1493. Calculate the "observed" binding energies for each of these atoms.

(c) Calculate the differences between each of the four approximate K-shell binding energies obtained in part (a) and the corresponding "observed" values from part (b). Most of the error in the results of part (a) is due to the fact that one of the $1s$ electrons partially screens the nucleus. A useful way to take this effect into account is simply to replace the nuclear charge Z by an *effective charge* Z', defined as

$$Z' \equiv Z - s,$$

where s represents the screening due to the second $1s$ electron. Calculate values of s for each of the four atoms in part (b).

3.8 Beta Decay of the Triton (**)

The nucleus ${}_1H^3$ is called the *triton*. It is composed of one proton and two neutrons and is known to be unstable with respect to beta decay, as shown below. This is a strictly nuclear process, which is outside the realm of atomic physics. The nuclear reaction for beta decay may be written

$${}_1H^3 \xrightarrow[\beta]{} {}_2He^3 + e.$$

Crudely speaking, this expression tells us that a neutron has been converted into a proton, which remains in the nucleus, forming He, and an electron, which is ejected.

Suppose that we have a hydrogen atom with a ${}_1H^3$ nucleus instead of ${}_1H^1$ (clearly, $Z = 1$ in both cases); this is called a *tritium atom*. (We shall ignore the mass difference between these two nuclei.) At time $t = 0$, beta decay occurs, and a fast electron is shot out of the nucleus, converting it to a ${}_2He^3$ nucleus with $Z = 2$. We may assume that the process is instantaneous and that the two electrons do not interact. Thus, from the point of view of the *atomic electron*, all that has happened is a sudden change of the nuclear charge from $Z = 1$ to $Z = 2$.

(a) Assume that the tritium atom is in the $1s$ state prior to beta decay. Calculate the probability that the He^+ ion will be found in the $1s$ state after beta decay.

(b) Repeat the calculation of part (a) for a $2s$ final state and then for any "non-s" final state (i.e., any final state for which $\ell \neq 0$, such as $2p$, $3p$, $3d$, . . .). What conclusions can you draw from your calculations? [HINT: The set of one-electron atom wave functions is complete for any choice of Z.]

4
The Wonderful World of Approximation Methods (Time Independent)

"The unforeseen does not exist," replied Phileas Fogg . . . "A well-used minimum suffices for everything."
Jules Verne, Around the World in Eighty Days *(N.Y.: Dell Publishing Co. Inc, Dell Laurel Library edition).*

The hydrogenic atom considered in Chapter 3 was assumed to be "isolated" —that is, free of all external influences. This assumption is clearly somewhat unrealistic; systems of interest in physics are almost invariably exposed to some small external influence or *perturbation*, the effects of which cannot be ignored. Perhaps the most familiar such perturbation is an external electromagnetic field, such as that due to an incident electron about to scatter off an atom. Hence finding a way to introduce perturbations into our theory is important; they give rise to additional potential energy terms in the Hamiltonian, and we would like to obtain eigenfunctions of this new Hamiltonian. Because of these additional terms, the exact solution of Schroedinger's equation is impossible and we must resort to *approximation methods*.

There is another reason for studying approximation methods. The hydrogen atom is one of the very few systems that can be solved exactly (another is the H_2^+ molecular ion). Since our knowledge of the universe should not be restricted to these few systems, we need ways to obtain approximate solutions to more complicated problems.

Many perturbations of interest give rise to potential energy terms in the Hamiltonian that are independent of time (e.g., a static external electric field). In this chapter several useful ways to handle such perturbations will be developed. Time-dependent perturbations will be considered in the next chapter.

Stationary-state eigenfunctions for a system perturbed by some external influence that gives rise to a potential energy term depending only on position coordinates are obtained by solution of the familiar time-independent Schroedinger equation,

$$\mathcal{H}\psi_n = E_n\psi_n, \tag{4.1}$$

where $\mathcal{H}$ now contains all potential energy terms plus, of course, the kinetic energy term. In Eq. (4.1) the subscript n represents a set of quantum numbers that uniquely identifies a state of the system.[1] The methods we shall study will be applicable to a wide variety of systems, so we shall not always specify the coordinate dependence of the wave functions. (These coordinates will collectively be denoted by the symbol τ.)

4.1 NONDEGENERATE TIME-INDEPENDENT PERTURBATION THEORY

In some problems the potential energy of the perturbed system may differ only slightly from that of a system that has already been solved. For example, the potential energy of a hydrogenic atom in an external static electric field E_z directed along the $\hat{z}$ axis is simply $V(r) = -Ze^2/r + ezE_z$, and we know the solutions for a potential energy $V(r) = -Ze^2/r$. If the additional term (or terms) in $V(\mathbf{r})$ is small, then we might expect the wave functions of the perturbed system to resemble the wave functions of the unperturbed system in many respects. Let us try to formulate a theory based on this idea.

Suppose that the Hamiltonian of the perturbed system can be written as

$$\mathcal{H} = \mathcal{H}^{(0)} + \mathcal{H}^{(1)}, \tag{4.2}$$

where $\mathcal{H}^{(0)}$ is a Hamiltonian, the eigenfunctions of which are known. $\mathcal{H}^{(1)}$ is called the *perturbation Hamiltonian*, $\mathcal{H}^{(0)}$ the *zeroth-order Hamiltonian*. Further suppose that the dominant term in $\mathcal{H}$ is $\mathcal{H}^{(0)}$; $\mathcal{H}^{(1)}$ is expected to

[1]Thus n is *not* the principal quantum number introduced in Chapter 3. For a hydrogenic atom, n would be the *set* of quantum numbers consisting of the principal quantum number, the orbital-angular-momentum quantum number, and the magnetic quantum number (and any others that may arise; see Chapter 6). We shall usually let $n = 1$ denote the ground state.

induce only small changes in the stationary-state wave functions and energies of the unperturbed system. A superscript (0) is used to denote the unperturbed problem,

$$\mathcal{H}^{(0)}\psi_n^{(0)} = E_n^{(0)}\psi_n^{(0)}; \tag{4.3}$$

it is assumed that we know (or can easily obtain) the unperturbed eigenfunctions $\{\psi_n^{(0)}\}$ and unperturbed energies $\{E_n^{(0)}\}$. In this section we shall deal only with nondegenerate systems; thus to each discrete eigenvalue $E_n^{(0)}$ there corresponds one and only one eigenfunction $\psi_n^{(0)}$.

For a simple illustration of these ideas, consider a system consisting of a particle in the one-dimensional potential shown in Fig. 4.1(a), a "modified" square well. Clearly, $\mathcal{H}$ may be separated as in Fig. 4.1(b). The strong interaction Hamiltonian $\mathcal{H}^{(0)}$ is that of the one-dimensional square well, a system for which the wave functions and energies are well known.[2] The perturbation Hamiltonian is the small "square hump." (The solution of this problem by perturbation theory is the subject of Prob. 4.6.)

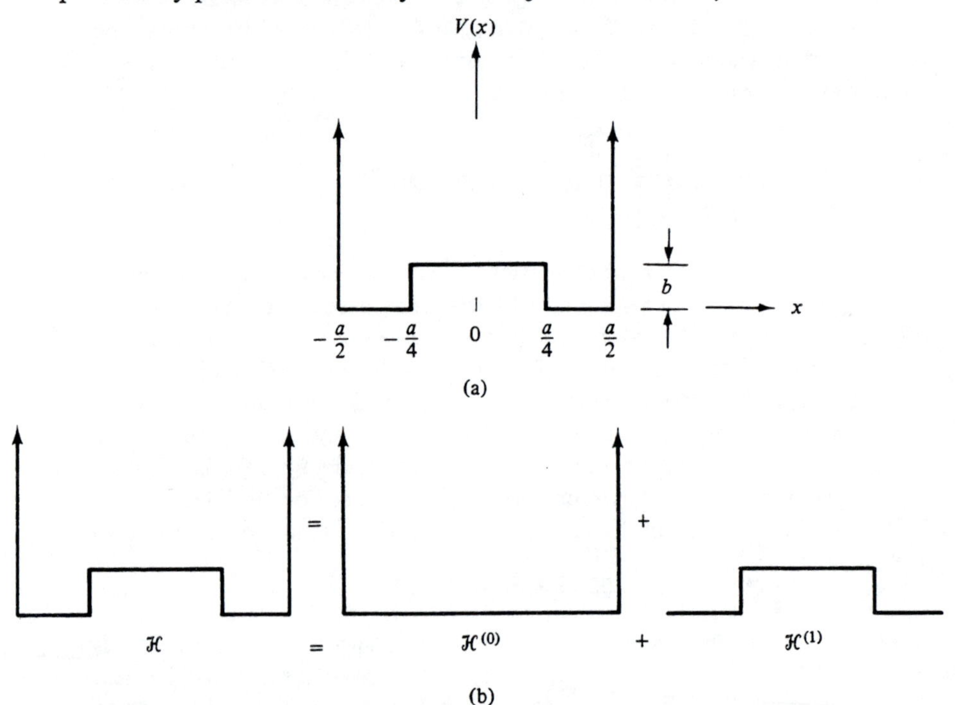

Figure 4.1 (a) A modified square-well potential; (b) the potential of (a) broken up into "strong" and "weak" interactions [$\mathcal{H}^{(0)}$ and $\mathcal{H}^{(1)}$, respectively] to illustrate the separation of the Hamiltonian.

[2]A discussion of the solution of the square well problem may be found in Robert M. Eisberg, *Fundamentals of Modern Physics* (New York: Wiley, 1961), Chap. 8. See also the references listed at the end of Chapter 1.

In order to simplify the mathematics of the derivations that are to be performed, let us write the perturbation Hamiltonian $\mathcal{H}^{(1)}$ in a more convenient form,

$$\mathcal{H}^{(1)} = \lambda \mathcal{H}', \tag{4.4}$$

where λ is a real constant. This constant is a measure of how small[3] the perturbation Hamiltonian is; we shall call it the *smallness parameter*. $\mathcal{H}'$ is a term of "normal magnitude." These definitions are illustrated for the modified square well in Fig. 4.2. The point to be emphasized is that λ is introduced solely for convenience; it will not appear in any of the final results. Since $\mathcal{H}^{(1)}$ is independent of time, so are λ and $\mathcal{H}'$.

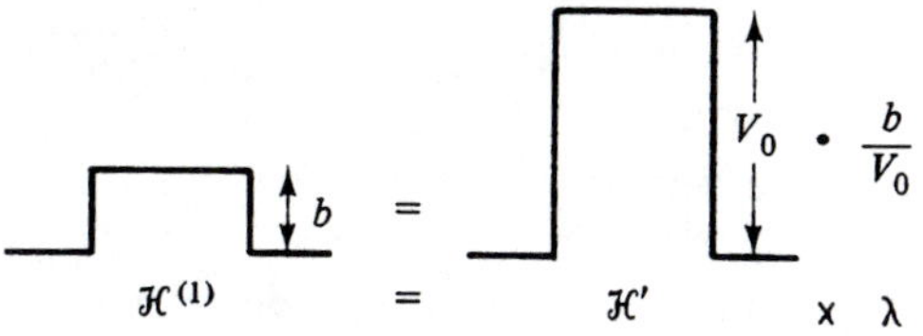

Figure 4.2 Application of the definition of the smallness parameter λ to the perturbation $\mathcal{H}^{(1)}$ of Fig. 4.1(b).

Setting Up the Solution

We shall now derive expressions for the perturbed wave functions ψ_n and energies E_n in terms of the unperturbed wave functions $\psi_n^{(0)}$ and $E_n^{(0)}$. Regardless of the magnitude of λ, we know that the set of functions $\{\psi_k^{(0)}\}$ is complete (since $\mathcal{H}^{(0)}$ is Hermitian). Therefore we could expand, say, the ground-state perturbed function ψ_1 in terms of this set:

$$\psi_1 = \sum_{k=1}^{\infty} c_k \psi_k^{(0)}. \tag{4.5}$$

We reasoned earlier that if λ is small—say $\lambda \ll 1$—we expect ψ_1 to be not too different from $\psi_1^{(0)}$, so the first term in this expansion will be the largest.

In particular, in the limit of vanishingly small λ, the full Hamiltonian $\mathcal{H}$ becomes the unperturbed Hamiltonian $\mathcal{H}^{(0)}$; that is,

$$\mathcal{H} \xrightarrow[(\lambda \to 0)]{} \mathcal{H}^{(0)}. \tag{4.6}$$

Therefore, as $\lambda \longrightarrow 0$, the perturbed wave functions and energies approach their unperturbed counterparts; that is,

$$\psi_n \xrightarrow[(\lambda \to 0)]{} \psi_n^{(0)}. \tag{4.7a}$$

$$E_n \xrightarrow[(\lambda \to 0)]{} E_n^{(0)}. \tag{4.7b}$$

[3]For a time our use of the word "small" will be a little vague. Eventually the various criteria for the applicability of perturbation theory will emerge naturally from our derivation.

Equation (4.7a) tells us that there is a one-to-one correspondence between the set of unperturbed wave functions $\{\psi_n^{(0)}\}$ and the set of perturbed wave functions $\{\psi_n\}$. Therefore we can expand ψ_n in powers of λ about the point $\lambda = 0$; we write the series

$$\psi_n = \varphi_n^{(0)} + \lambda\varphi_n^{(1)} + \lambda^2\varphi_n^{(2)} + \lambda^3\varphi_n^{(3)} + \cdots, \tag{4.8}$$

where the coefficients of the expansion are $\varphi_n^{(0)}, \varphi_n^{(1)}, \varphi_n^{(2)}, \ldots$ They are all functions of the same coordinates τ as appear in ψ_n. As yet we do not know the form of these functions;[4] our goal is to determine them.

Similarly, we can write

$$E_n = A_n^{(0)} + \lambda A_n^{(1)} + \lambda^2 A_n^{(2)} + \lambda^3 A_n^{(3)} + \cdots, \tag{4.9}$$

where the coefficients $A_n^{(0)}, A_n^{(1)}, \ldots$ are real numbers that we shall also determine.

Equations (4.8) and (4.9) are the starting point of our derivation of perturbation theory. We shall use what we know about ψ_n and E_n—namely, that they satisfy Eq. (4.1)—to obtain expressions for $\{\varphi_n^{(i)}\}$ and $\{A_n^{(i)}\}$ and hence for the perturbed wave functions and energies.

Since λ is small, the first term in each of these expansions will be the dominant one.[5] In fact, we see immediately from Eqs. (4.7) that

$$\varphi_n^{(0)} = \psi_n^{(0)} \quad \text{and} \quad A_n^{(0)} = E_n^{(0)}, \tag{4.10}$$

a result that should hardly be surprising.

If the perturbation Hamiltonian (and hence λ) is small enough, it is valid to drop all but the first two terms in Eqs. (4.8) and (4.9). This procedure is called *first-order perturbation theory*; if the first three terms are kept, we have *second-order perturbation theory*, and so on. The smallness parameter λ helps in keeping track of the order of approximation being considered.

To determine the other coefficients in the expansions of ψ_n and E_n, let us substitute these expansions into the time-independent Schroedinger equation for the perturbed system, Eq. (4.1). Grouping terms of the same order in λ

[4]Notice that the superscript labeling each coefficient is equal to the *order* (i.e., the power of λ) of the term it multiplies. Thus the coefficient that multiplies λ^0 is denoted $\varphi^{(0)}$, the coefficient that multiplies λ^1 is denoted $\varphi^{(1)}$, and so on.

[5]Of course, this does not ensure that the series converge. In most problems of interest to us, the problem of convergence can (and will) be avoided; if λ is not small enough so that the series converge, we will not use perturbation theory! See David Saxon, *Elementary Quantum Mechanics* (San Francisco: Holden-Day, 1968), pp. 194 ff, for a discussion of the problem of convergence.

(i.e., with like powers of λ), we obtain

$$\begin{aligned}\mathcal{H}^{(0)}\varphi_n^{(0)} &+ \lambda[\mathcal{H}^{(0)}\varphi_n^{(1)} + \mathcal{H}'\varphi_n^{(0)}] + \lambda^2[\mathcal{H}^{(0)}\varphi_n^{(2)} + \mathcal{H}'\varphi_n^{(1)}] + \cdots \\ &= A_n^{(0)}\varphi_n^{(0)} + \lambda[A_n^{(0)}\varphi_n^{(1)} + A_n^{(1)}\varphi_n^{(0)}] \\ &\quad + \lambda^2[A_n^{(0)}\varphi_n^{(2)} + A_n^{(1)}\varphi_n^{(1)} + A_n^{(2)}\varphi_n^{(0)}] + \cdots. \end{aligned} \tag{4.11}$$

By equating coefficients of $\lambda^0, \lambda^1, \lambda^2, \ldots$ on the left- and right-hand sides of Eq. (4.11), we obtain an infinite set of equations

$$\lambda^0 : \mathcal{H}^{(0)}\varphi_n^{(0)} = A_n^{(0)}\varphi_n^{(0)} \tag{4.12a}$$

$$\lambda^1 : \mathcal{H}^{(0)}\varphi_n^{(1)} + \mathcal{H}'\varphi_n^{(0)} = A_n^{(0)}\varphi_n^{(1)} + A_n^{(1)}\varphi_n^{(0)} \tag{4.12b}$$

$$\lambda^2 : \mathcal{H}^{(0)}\varphi_n^{(2)} + \mathcal{H}'\varphi_n^{(1)} = A_n^{(0)}\varphi_n^{(2)} + A_n^{(1)}\varphi_n^{(1)} + A_n^{(2)}\varphi_n^{(0)} \tag{4.12c}$$

$$\vdots \qquad\qquad\qquad \vdots$$

These equations can be solved for as many of the coefficients $\{\varphi_n^{(i)}\}$ and $\{A_n^{(i)}\}$ as needed. The first one, Eq. (4.12a), is simply the time-independent Schroedinger equation for the unperturbed system; it merely confirms Eq. (4.10), which we already deduced. [We say that $\psi_n^{(0)}$ and $E_n^{(0)}$ are *zeroth-order approximations* to ψ_n and E_n.] We can substitute Eq. (4.10) into (4.12b) and solve the resulting equation for $\varphi_n^{(1)}$ and $A_n^{(1)}$. Using these coefficients in the expansion of Eqs. (4.8) and (4.9) will provide expressions for the perturbed wave functions and energies good to *first order*. This process can be continued, using Eqs. (4.12c), (4.12d), and so on, to obtain results of any desired order.

First-Order Approximations

To illustrate, let us determine $\varphi_n^{(1)}$ and $A_n^{(1)}$, the first-order corrections to the unperturbed wave functions and energies, respectively; the first-order approximations to ψ_n and E_n are

$$\psi_n \simeq \psi_n^{(0)} + \lambda\varphi_n^{(1)}, \tag{4.13}$$

$$E_n \simeq E_n^{(0)} + \lambda A_n^{(1)}. \tag{4.14}$$

Equation (4.12b) can be solved by using the fact that any function of the coordinates τ can be expanded in terms of the complete set of unperturbed eigenfunctions $\{\psi_m^{(0)}\}$. In particular, the function $\varphi_n^{(1)}$ can be expanded:

$$\varphi_n^{(1)} = \sum_{m=1}^{\infty} a_{nm}^{(1)}\psi_m^{(0)}, \tag{4.15}$$

where the $a_{nm}^{(1)}$ are constants and are complex in general. As long as all the eigenfunctions $\{\psi_m^{(0)}\}$ are included in Eq. (4.15), it is exact. Of course, since the set of unperturbed wave functions is usually an infinite set, we shall eventually have to truncate this summation if we expect to use the method.[6]

To determine these coefficients, we merely substitute Eq. (4.15) into (4.12b):

$$\mathcal{H}^{(0)} \sum_m a_{nm}^{(1)} \psi_m^{(0)} + \mathcal{H}'\psi_n^{(0)} = E_n^{(0)} \sum_m a_{nm}^{(1)}\psi_m^{(0)} + A_n^{(1)}\psi_n^{(0)}. \tag{4.16}$$

Multiplying this equation on the left by $\psi_k^{(0)*}$, integrating over all coordinates, and using the orthonormality of the unperturbed wave functions, we isolate the coefficient $a_{nm}^{(1)}$. The result is

$$\sum_m a_{nm}^{(1)} E_m^{(0)} \delta_{km} + \langle \psi_k^{(0)} | \mathcal{H}' | \psi_n^{(0)} \rangle = E_n^{(0)} \sum_m a_{nm}^{(1)} \delta_{km} + A_n^{(1)} \delta_{kn}, \tag{4.17}$$

where we have used the Dirac bracket notation

$$\langle \psi_k^{(0)} | \mathcal{H}' | \psi_n^{(0)} \rangle \equiv \int \psi_k^{(0)*} \mathcal{H}' \psi_n^{(0)} \, d\tau \tag{4.18}$$

for the matrix element cooresponding to the kth row and the nth column of the "matrix of the perturbing Hamiltonian."[7] Carrying out the indicated summations in Eq. (4.17), we find

$$a_{nk}^{(1)} E_k^{(0)} + \langle \psi_k^{(0)} | \mathcal{H}' | \psi_n^{(0)} \rangle = E_n^0 a_{nk}^{(1)} + A_n^{(1)} \delta_{kn}. \tag{4.19}$$

We shall consider two cases. If $k = n$, Eq. (4.19) gives us an expression for $A_n^{(1)}$. If $n \neq k$, it gives us $a_{nk}^{(1)}$. In the former case, we have

$$A_n^{(1)} = \langle \psi_n^{(0)} | \mathcal{H}' | \psi_n^{(0)} \rangle. \tag{4.20}$$

From Eqs. (4.14) and (4.4), it follows that

$$\boxed{E_n \simeq E_n^{(0)} + \langle \psi_n^{(0)} | \mathcal{H}^{(1)} | \psi_n^{(0)} \rangle.} \quad \text{First-order perturbed energy} \tag{4.21}$$

If we consider Eq. (4.19) for the case $n \neq k$, we obtain

$$a_{nk}^{(1)} = \frac{\langle \psi_k^{(0)} | \mathcal{H}' | \psi_n^{(0)} \rangle}{E_n^{(0)} - E_k^{(0)}}, \qquad n \neq k. \tag{4.22}$$

[6]To be perfectly general, we should also include all continuum wave functions of the unperturbed system in the expansion of Eq. (4.15). The fact that we do not acknowledges that the problem is difficult enough keeping only discrete eigenfunctions. For most standard applications of perturbation theory the continuum is not important.

[7]This is a convenient notation, since it enables us to view the collection of integrals $\int \psi_k^{(0)*} \mathcal{H}' \psi_n^{(0)} d\tau$ as a matrix, the (kn) element of which is $\langle \psi_k^{(0)} | \mathcal{H}' | \psi_n^{(0)} \rangle$. Note that each integral is taken over all coordinates τ.

By invoking the requirement that the set of perturbed wave functions be normalized, we can show that $a_{nn}^{(1)} = 0$ (Prob. 4.1). Thus the nth perturbed wave function, correct to first order, is

$$\psi_n \simeq \psi_n^{(0)} + \sum_{k \neq n}^{\infty} \frac{\langle \psi_k^{(0)} | \mathcal{H}^{(1)} | \psi_n^{(0)} \rangle}{E_n^{(0)} - E_k^{(0)}} \psi_k^{(0)}. \qquad \text{First-order perturbed wave function} \tag{4.23}$$

What have we learned so far? We have seen that the perturbed wave function ψ_n can be viewed as a linear combination of unperturbed eigenfunctions $\{\psi_k^{(0)}\}$; the coefficient $a_{nk}^{(1)}$ tells us "how much" of the kth unperturbed function contributes to the nth perturbed function. We say that "the unperturbed functions $\psi_k^{(0)}$ for a particular $k \neq n$ *mix* with $\psi_n^{(0)}$ to form the perturbed function ψ_n." Moreover, Eq. (4.21) indicates that the energy of the nth perturbed state of the system is given by the energy of the corresponding unperturbed state plus the expectation value of the perturbation Hamiltonian, $\mathcal{H}^{(1)}$, with respect to $\psi_n^{(0)}$. This term can be of either sign and thus can raise or lower the energy from the unperturbed value. As promised, λ has disappeared from all our results.

Since $a_{nk}^{(1)}$ is inversely proportional to the energy separation between the nth and kth unperturbed states, we expect states that are closest together in energy to mix most strongly with $\psi_n^{(0)}$. This fact provides a clue on how to avoid calculating the infinity of terms in Eq. (4.23): depending on the degree of accuracy desired, we need only evaluate terms for the states closest in energy to the nth state. It is important to note that if any of these coefficients is large, then perturbation theory might break down and the problem probably must be handled in another way.

Finally, it is apparent from Eq. (4.23) that *this result is restricted to nondegenerate states only*, for if some unperturbed function, say $\psi_m^{(0)}$, is degenerate with $\psi_n^{(0)}$, then $E_m^{(0)} = E_n^{(0)}$ and the mth term in the summation of Eq. (4.23) is infinite. [This situation would not occur if $\langle \psi_m^{(0)} | \mathcal{H}^{(1)} | \psi_n^{(0)} \rangle$ happened to be zero, but such will not generally be the case.] Thus all degenerate states must be treated differently (see Sec. 4.6).

Example 4.1
The Anharmonic Oscillator

One way to handle recalcitrant one-dimensional potentials is to expand them in powers of x—that is,

$$V(x) = c_0 + c_1 x + c_2 x^2 + c_3 x^3 + c_4 x^4 + \cdots. \tag{4.24}$$

Some potentials (e.g., molecular potentials like the Morse potential) are especially important at small values of x (measured from a well-chosen origin) and have c_0 and c_1 equal to zero. The Hamiltonian for a system consisting of a particle of mass m in such a potential is

$$\mathcal{H} = -\frac{\hbar^2}{2m}\frac{d^2}{dx^2} + \frac{1}{2}m\omega^2 x^2 + \alpha x^3, \tag{4.25}$$

where we have set $c_2 = \frac{1}{2}m\omega^2$ and $c_3 = \alpha$ and dropped all higher terms. If α is small, then the problem can be handled by using perturbation theory. The unperturbed system is simply the one-dimensional harmonic oscillator; therefore

$$\mathcal{H}^{(0)} = -\frac{\hbar^2}{2m}\frac{d^2}{dx^2} + \frac{1}{2}m\omega^2 x^2. \tag{4.26}$$

The unperturbed energies and eigenfunctions are[8]

$$E_n^{(0)} = \hbar\omega\left(n + \frac{1}{2}\right), \qquad n = 0, 1, 2, \ldots \tag{4.27}$$

and

$$\psi_n^{(0)}(x) = \left(\frac{1}{2^n n!}\sqrt{\frac{\lambda}{\pi}}\right)^{1/2} H_n(\sqrt{\lambda}\, x)e^{-(1/2)\lambda x^2}, \tag{4.28}$$

where we have defined

$$\lambda \equiv \frac{m\omega}{\hbar}. \tag{4.29}$$

The functions $H_n(\sqrt{\lambda}\, x)$ are the *Hermite polynomials*. The perturbation Hamiltonian is obviously

$$\mathcal{H}^{(1)} = \alpha x^3. \tag{4.30}$$

Suppose that we wish to determine the perturbed energies to first order. By Eq. (4.21), E_n is given by

$$E_n \simeq \hbar\omega\left(n + \frac{1}{2}\right) + \langle\psi_n^{(0)}\,|\,\alpha x^3\,|\,\psi_n^{(0)}\rangle. \tag{4.31}$$

The unperturbed wave functions $\psi_n^{(0)}$ of Eq. (4.28) are even functions of x for $n = 0, 2, 4, \ldots$ [$\psi_n^{(0)}(x) = +\psi_n^{(0)}(-x)$] and odd functions of x for $n = 1, 3, 5, \ldots$ [$\psi_n^{(0)}(x) = -\psi_n^{(0)}(-x)$]. Hence the product $\psi_n^{(0)*}(x)\,\psi_n^{(0)}(x)$ is an even function of x for all values of n. Since αx^3 is odd, the integrand in the matrix element $\langle\psi_n^{(0)}\,|\,\alpha x^3\,|\,\psi_n^{(0)}\rangle$ is zero regardless of the value of n. Therefore

[8]See almost any introductory quantum mechanics book for a discussion of the simple harmonic oscillator problem in one dimension—for example, Linus Pauling and E. B. Wilson, *Introduction to Quantum Mechanics* (New York: McGraw-Hill, 1935), Chap. 11. Notice that, according to convention, we have denoted the ground state of the harmonic to oscillator by $n = 0$, not by $n = 1$.

the perturbed energy to first order is simply equal to the unperturbed energy; that is,

$$E_n \simeq E_n^{(0)} = \hbar\omega\left(n + \frac{1}{2}\right), \qquad n = 0, 1, 2, \ldots . \tag{4.32}$$

What went wrong? Nothing, actually. Perturbation theory will work for this problem, but because of the symmetry of this anharmonic oscillator, we must calculate higher-order correction terms in order to determine the effect of the perturbation on the energies.[9]

This situation occurs because the functions used in calculating first-order corrections to $E_n^{(0)}$ for any problem are unperturbed; they do not take into consideration the effect of the perturbation on the states of the system. In higher-order calculations of the energy, the functions used will take into account distortion due to the perturbation and give rise to nonzero corrections to $E_n^{(0)}$. Of course, in general, the first-order correction to the energy is nonzero (see Prob. 4.2).

Exercise 4.1 (a) Calculate a first-order expression for $\psi_0(x)$, mixing in only the two largest nonzero terms.

(b) Make a rough sketch of $\psi_0(x)$ and comment on the deformations induced by the perturbing Hamiltonian.

Higher-Order Corrections to the Energy

Example 4.1 revealed that it is sometimes necessary to calculate $A_n^{(2)}$ so that perturbed energies correct to second order can be obtained. The derivation is a straightforward mathematical exercise that closely parallels work already done. One merely expands both $\varphi_n^{(1)}$ and $\varphi_n^{(2)}$ in terms of $\{\psi_m^{(0)}\}$ and substitutes into Eq. (4.12c). The result is

$$E_n \simeq E_n^{(0)} + \langle\psi_n^{(0)}|\mathcal{H}^{(1)}|\psi_n^{(0)}\rangle + \sum_{k\neq n}^{\infty} \frac{|\langle\psi_k^{(0)}|\mathcal{H}^{(1)}|\psi_n^{(0)}\rangle|^2}{E_n^{(0)} - E_k^{(0)}}.$$

Second-order perturbed energy (4.33)

Here again is an expression that involves an infinite number of terms, and all but the largest terms must be discarded. These terms involve the same matrix elements of $\mathcal{H}^{(1)}$ that appeared in the first-order correction to the wave function [see Eq. (4.23)]. They take into account, at least to first order,

[9]Note carefully that this result *does not imply* that $E_n = E_n^{(0)}$ or that we can stop at first order. Indeed, we are compelled to go on! It does, however, suggest that the effect of the perturbation on the energies will be small.

deformations due to the perturbation. The summation term in Eq. (4.33) represents a sort of "average" over these distorted states. If the distortion is significant, the matrix elements will be of substantial magnitude.

Although Eq. (4.33) may look rather formidable, it entails few new calculations. The hard work lies in the evaluation of the matrix elements appearing in this expression, a problem that we already faced when calculating the wave functions to first order in Eq. (4.23).[10]

Exercise 4.2 Although the first-order correction to the energy can either raise or lower the energy from its unperturbed value, Eq. (4.33) reveals that second-order corrections can only lower the energy of the ground state. Provide a justification for this fact.

We could continue calculating corrections of as high an order as our patience allows; but it should be clear by now that, as the order increases, the calculations rapidly become extremely cumbersome. Normally, perturbation theory is not employed unless the perturbation series converge rapidly enough that terms of higher than second order can be neglected.

Exercise 4.3 Calculate the second-order correction to the unperturbed energy for the anharmonic oscillator of Example 4.1, keeping only the terms $k = 1$ and $k = 3$ in Eq. (4.33).

4.2 THE VARIATIONAL PRINCIPLE

Considering all the special conditions that must be satisfied in order for perturbation theory to be applicable, this technique may seem of dubious value. Indeed, although the method is relatively easy to apply, it is often difficult to separate out a small part of the Hamiltonian in such a way that the corresponding unperturbed problem is soluble. Therefore an alternate method that will enable us to handle systems that cannot be treated by perturbation theory is needed.

It might be possible to use our physical intuition to guess a reasonable wave function for the system and use it to calculate energies, but such a guess would be of little value unless we have some way to assess the error and systematically improve the results. The variational principle provides a systematic method of approximation and becomes the basis for a whole class of simple but powerful methods. In fact, in Sec. 4.5 we shall show that the perturbation method of the previous section is a special case of this variational theory.

[10]This, by the way, exemplifies a general result: knowledge of the perturbed wave function correct to order j is sufficient to permit calculation of the perturbed energy to order $j + 1$.

Background

We wish to solve the time-independent Schroedinger equation $\mathcal{H}\psi_n = E_n\psi_n$ for a perfectly general system, which, for example, might consist of several particles in three dimensions and an arbitrary potential energy function. If this equation is multiplied on the left by ψ_n^* and integrated over all coordinates, the result is

$$E_n\langle\psi_n|\psi_n\rangle = \langle\psi_n|\mathcal{H}|\psi_n\rangle. \tag{4.34}$$

(Notice that we have not assumed that the ψ_n are normalized.) From Eq. (4.34) we immediately recover the familiar statement (see Sec. 1.5) that the energy of the nth state is given by the expectation value of the Hamiltonian with respect to normalized exact eigenfunctions; that is,

$$E_n = \frac{\langle\psi_n|\mathcal{H}|\psi_n\rangle}{\langle\psi_n|\psi_n\rangle}. \tag{4.35}$$

Next, suppose that instead of knowing the exact eigenfunction ψ_n, we have only a rough idea, based in part on intuition derived from experience, of its properties and behavior. We propose to use this rather vague knowledge to guess an analytic form for a *trial wave function*, φ; we shall then use φ to calculate the quantity $E'[\varphi]$, defined by[11]

$$E'[\varphi] \equiv \frac{\langle\varphi|\mathcal{H}|\varphi\rangle}{\langle\varphi|\varphi\rangle}. \tag{4.36}$$

Clearly, $E'[\varphi]$ will not, in general, be equal to one of the exact energies E_n unless we are uncommonly good guessers.

In Sec. 4.3 we shall examine E' and show that, regardless of the choice of φ, it is an *upper bound* on the exact ground-state energy of the system—that is, that $E' \geq E_1$. Moreover, the more closely ψ_1 is approximated by φ, the closer E' will be to E_1.

Derivation of the Variational Principle

Before verifying these assertions, we must lay some groundwork. In particular, we shall now prove a variational principle. Suppose, for the moment, that we know the exact wave function ψ_n. Let us construct a trial function φ that is related to ψ_n by

$$\varphi = \psi_n + \alpha\chi, \tag{4.37}$$

[11]We use the notation $E'[\varphi]$ when we wish to recall explicitly that the number E' depends on the function φ. Such a "function of a function" is called a *functional* by mathematicians. (Do not confuse φ, the trial function, with the expansion coefficients $\varphi_n^{(i)}$ of Sec. 4.1. They are unrelated.)

where α is a constant and χ is some function that depends on the same set of coordinates as ψ_n and satisfies the same boundary conditions. Clearly, an unlimited number of such trial functions is available to us. By introducing $\alpha\chi$, we are allowing an arbitrary *variation* of ψ_n about its "true value."

We can use φ in Eq. (4.37) to evaluate $E'[\varphi]$, defined above. To see how close $E'[\varphi]$ is to E_n, the "true energy," we use the fact that $\mathcal{H} - E_n$ annihilates ψ_n [i.e., $(\mathcal{H} - E_n)\varphi = (\mathcal{H} - E_n)\alpha\chi$, since $(\mathcal{H} - E_n)\psi_n = 0$] to evaluate the difference:

$$E'[\varphi] - E_n = \frac{\langle\varphi\,|\,\mathcal{H} - E_n\,|\,\alpha\chi\rangle}{\langle\varphi\,|\,\varphi\rangle}. \tag{4.38}$$

Recalling the definition of the expectation value and the fact that $\mathcal{H}$ is Hermitian, we write Eq. (4.38) as

$$\begin{aligned} E'[\varphi] - E_n &= \frac{\langle(\mathcal{H} - E_n)\varphi\,|\,\alpha\chi\rangle}{\langle\varphi\,|\,\varphi\rangle} \\ &= \frac{\langle(\mathcal{H} - E_n)\alpha\chi\,|\,\alpha\chi\rangle}{\langle\varphi\,|\,\varphi\rangle} \\ &= |\alpha|^2\frac{\langle\chi\,|\,\mathcal{H} - E_n\,|\,\chi\rangle}{\langle\varphi\,|\,\varphi\rangle}. \end{aligned} \tag{4.39}$$

Equation (4.39) reveals that $E'[\varphi]$ differs from E_n by a quantity that is of second order in the variation $\alpha\chi$; even if the wave function is allowed to vary by a first-order variation, the resulting approximate energy will vary only by a second-order term. This means that if α is small, so φ is "close" to ψ_n, then E' will be "even closer" to E_n. It also follows from Eq. (4.39) that $E' = E_n$ if and only if $\varphi = \psi_n$. (The important point is that χ is arbitrary.)

A formal statement of this result is as follows: *The energy is stationary with respect to small arbitrary variations in the eigenfunction about its true value.* This statement defines for us the term *stationary* and provides a verbal statement of the *variational principle.*[12] This important principle is entirely equivalent to the Schroedinger equation (see Prob. 4.3).

4.3 THE VARIATIONAL METHOD

The variational principle assures us that *if* we can find a trial wave function φ that is reasonably close to the true eigenfunction ψ_n, then the energy $E'[\varphi]$ will be a rather good approximation to the true energy E_n. It is not apparent, however, that this principle can help us solve a problem, for unless we know the function ψ_n, we cannot construct $\varphi = \psi_n + \alpha\chi$. After a few comments on

[12] See David Saxon, *Elementary Quantum Mechanics* (San Francisco: Holden-Day, 1968), Chap. 7, for an alternate derivation of the variational principle.

other ways to choose trial wave functions, we shall develop a method based on the variational principle that can be used in problem solving.

Variational Wave Functions

Suppose that we guess a "reasonable" trial function φ that depends on all the coordinates of the system and satisfies the correct boundary conditions. In order to introduce some flexibility in φ, included in it will be several parameters that we are free to choose at will. Called *variational parameters*, they will be denoted by Greek letters (α, β, γ, . . .). The trial function can contain as many variational parameters as we wish.

Example 4.2
The One-Dimensional Simple Harmonic Oscillator: Variational Wave Functions

Suppose that we do not know how to obtain an exact solution to the standard problem of a particle of mass m in a one-dimensional simple harmonic oscillator potential. The Hamiltonian of the system is

$$\mathcal{H} = -\frac{\hbar^2}{2m}\frac{d^2}{dx^2} + \frac{1}{2}m\omega^2 x^2, \tag{4.40}$$

where we have used the notation of Example 4.1. As a trial wave function for this system, we might consider

$$\varphi(x) = \begin{cases} (\alpha^2 - x^2), & |x| < \alpha \\ 0, & |x| > \alpha. \end{cases} \tag{4.41a}$$

This is clearly not an eigenfunction of $\mathcal{H}$, but it is "reasonable". It satisfies the boundary conditions by vanishing as $|x| \longrightarrow \infty$, it is symmetrical about $x = 0$, as is $V(x)$, and it contains a variational parameter, α.

Alternatively, we could try a trial function $\bar{\varphi}$ of the form

$$\bar{\varphi}(x) = Ce^{-\gamma x^2}, \tag{4.41b}$$

where C is a normalization constant and γ is the variational parameter. This is an even better choice than φ; it is smooth, symmetrical about $x = 0$, and behaves properly in the limit. This so-called *gaussian function* is especially convenient because numerous integrals of Gaussians have been tabulated, and these tables can save us considerable unpleasant labor.

We shall continue with this problem after noting how these variational wave functions are used.

The Variational Method for Ground States

It is particularly easy to use variational theory to obtain approximate ground-state energies, since (as we shall show) $E'[\varphi]$, calculated via Eq. (4.36), satisfies a minimum principle to the effect that E' is an upper bound on E_1; that is,

$$E'[\varphi] \geq E_1, \tag{4.42}$$

where E_1 is the exact ground-state energy of the system. As the free variational parameters in φ are changed, $E'[\varphi]$ takes on different values. Since $E'[\varphi]$ is an upper bound on the true energy, the best choice of parameters is the one which yields the smallest value of E'. Thus the problem reduces to one of minimizing E' with respect to all the variational parameters; that is,

$$\boxed{\min\{E'(\alpha, \beta, \gamma, \ldots)\} \geq E_1.} \quad \text{Variational method} \tag{4.43}$$

Of course, if we include enough variational parameters to allow a completely arbitrary and independent variation of the real and imaginary parts of φ, the equality in (4.43) will hold according to the variational principle of Sec. 4.2.[13]

Proof of the Minimum Principle

The minimum principle as expressed in Eq. (4.42) is not hard to prove. We know that the set of eigenfunctions $\{\psi_k\}$ that solve the time-independent Schroedinger equation is *complete*. Therefore we can expand the trial function φ as

$$\varphi = \sum_{k=1}^{\infty} a_k \psi_k.$$

(To be completely general, integration over any continuum functions would have to be included in this expansion.) Substituting this function into the definition of $E'[\varphi]$, Eq. (4.36), we obtain

$$E'[\varphi] = \frac{\sum_{k=1}^{\infty} |a_k|^2 E_k}{\sum_{k=1}^{\infty} |a_k|^2}. \tag{4.44}$$

[13]There also exists a method of obtaining lower bounds on the energy [see Hendrik F. Hameka, *Introduction to Quantum Theory* (New York: Harper and Row, 1967), Chap. 11]. Using a combination of the two, one can obtain a very good approximation to E_1. However, the calculations are often very involved, for they require taking matrix elements of the square of the Hamiltonian.

Since E_1 is the exact ground-state energy, $E_k \geq E_1$ for all values of k, and we have

$$E' \geq \frac{\sum_{k=1}^{\infty} |a_k|^2 E_1}{\sum_{k=1}^{\infty} |a_k|^2} = E_1, \tag{4.45}$$

which is just Eq. (4.42); this completes the proof.

One further point should be emphasized. It is possible to show from Eq. (4.44) that even if E' is very close to E_1, it does not necessarily follow that the trial function φ is very close to the exact ground-state wave function ψ_1. In other words, the energy is not necessarily a good criterion by which to choose the best of several possible trial wave functions.

To see this point, notice that, in order for φ to be close to ψ_1, we would need $a_1 \simeq 1$ and $a_k \simeq 0$ for $k = 2, 3, \ldots$. Now, a fairly large coefficient for one of the wave functions other than ψ_1, say $a_3 = 0.1$, contributes less than 1% to the energy expectation value. Yet such a large value of a_3 could badly distort the wave function from its true value. In other words, we must always exercise caution when using a variational wave function to calculate quantities other than the energy—for example, expectation values of operators that depend critically on a range of values of the coordinates not heavily weighted in our energy calculations.

Example 4.3
The Variational Method Applied to the One-Dimensional Oscillator

Consider again the problem of Example 4.2. We know [see Eq. (4.27) and footnote 8] that the exact ground-state energy is

$$E_0 = \frac{1}{2}\hbar\omega, \tag{4.46}$$

where we again use $n = 0$ rather than $n = 1$ for the ground state. First trying the trial function φ of Eq. (4.41a), we find, after some simple integration, that

$$E'(\alpha) = \frac{5}{4}\frac{\hbar^2}{m}\alpha^{-2} + \frac{1}{14}m\omega^2\alpha^2. \tag{4.47}$$

According to the minimum principle, the best approximation to E_0 is obtained by minimizing E' with respect to α—that is, by setting

$$\frac{dE'}{d\alpha} = 0. \tag{4.48}$$

We find that the minimum value of E' is obtained for

$$\alpha^2 = \sqrt{\frac{35}{2}} \frac{\hbar}{m\omega}. \tag{4.49}$$

Substituting this result into Eq. (4.47), we obtain the approximate energy for this trial wave function:

$$E' = (0.6)\hbar\omega. \tag{4.50}$$

Not too bad. E' is about 20% too large.

Next, let us try our other variational wave function, $\bar{\varphi}$ of (4.41b). We find, again after some integration, that

$$E'(\gamma) = \frac{m\omega^2}{8}\gamma^{-1} + \frac{\hbar^2}{2m}\gamma. \tag{4.51}$$

This expression for E' attains its minimum value for

$$\gamma = \frac{1}{2}\frac{m\omega}{\hbar}, \tag{4.52}$$

which corresponds to an approximate energy of

$$E' = \frac{1}{2}\hbar\omega. \tag{4.53}$$

But this is the exact ground-state energy! Therefore Eq. (4.41b), with γ as given in (4.52), is the exact ground-state wave function [see Eq. (4.27)].

The point of this example is that we should always take a few minutes to think about the choice of the trial function φ before beginning to calculate integrals. We want φ to resemble as closely as possible the wave function we expect on physical grounds. In Example 4.3 the discontinuity in the slope of φ at $x = \alpha$ has no physical justification; consequently, we did not obtain the best possible approximation to E_0 when we used this function. (Incidentally, only in a very few cases is it possible to obtain the exact wave function by means of the variational method.)

Generalization to Excited States

The minimum principle $E'[\varphi] \geq E_1$ can be generalized to excited states. If a trial function φ is chosen that is orthogonal to all the exact eigenfunctions ψ_k

for $k = 1, 2, 3, \ldots, n-1$; that is, if

$$\langle\varphi|\psi_1\rangle = \langle\varphi|\psi_2\rangle = \cdots = \langle\varphi|\psi_{n-1}\rangle = 0, \tag{4.54}$$

then $E'[\varphi]$ is an upper bound on the energy of the nth excited state:

$$E' \geq E_n. \tag{4.55}$$

(A proof of this assertion is developed in Prob. 4.5.)

Pondering this result for a moment, we find it hard to imagine a realistic problem to which it could be applied, for we must know $\psi_1, \psi_2, \psi_3, \ldots, \psi_{n-1}$ in order to be able to enforce the orthogonality condition, Eq. (4.54). However, sometimes it is possible to use symmetry arguments to choose a φ that satisfies a few of these orthogonality conditions. For example, the ground-state eigenfunction of the one-dimensional simple harmonic oscillator Hamiltonian is an *even* function of x. If we choose φ to be an *odd* function of x, it will automatically be orthogonal to the true ground-state function, and we can use it to obtain an upper bound to the energy of the first excited state.

4.4 THE LINEAR VARIATIONAL METHOD

The key result of Sec. 4.3 is Eq. (4.43); it gives us a procedure for obtaining an approximation to the energy of a system that we cannot solve exactly. This approach can be used, for example, to handle problems to which perturbation theory does not apply.

The variational parameters in the trial function could appear anywhere—in exponents, multiplying functions, as additive constants, and so on. The choice is up to us. A useful and systematic method can be developed if we restrict the class of functions from which φ is selected. In particular, we shall consider in this section only trial functions that depend linearly on the variational parameters. The resulting linear variational method (or Rayleigh-Ritz variational method) is often used in research applications, for it is especially suited to computer adaptation.

We construct a *linear variational wave function* by forming a linear combination of N functions $\{u_i\}$ ($i = 1, 2, \ldots, N$), each of which depends on all the coordinates in $\mathcal{H}$ and satisfies the appropriate boundary conditions. Thus we write

$$\varphi = \sum_{i=1}^{N} c_i u_i. \tag{4.56}$$

The only variational parameters in the function φ are the coefficients c_i, $i = 1, 2, \ldots, N$. We call these coefficients *linear variational parameters* and the functions u_i *basis functions*. (The number and type of basis functions to be included in this trial function depend on the properties of the particular system under consideration and on the degree of accuracy required. We shall see how to go about making these decisions in the examples and problems below.)

Calculation of Energies and Wave Functions

Suppose that we have somehow selected the set of basis functions $\{u_i\}$. Then the function φ formed from this basis set, according to Eq. (4.56), can be used to calculate an approximate energy, $E'[\varphi]$. Substituting φ into the definition $E' = \langle\varphi|\, \mathcal{H}\,|\varphi\rangle/\langle\varphi\,|\,\varphi\rangle$ we obtain

$$E' \sum_{j=1}^{N} \sum_{i=1}^{N} c_j^* c_i \langle u_j\,|\,u_i\rangle = \sum_{j=1}^{N} \sum_{i=1}^{N} c_j^* c_i \langle u_j\,|\,\mathcal{H}\,|\,u_i\rangle, \tag{4.57}$$

where we have multiplied by $\langle\varphi\,|\,\varphi\rangle$ for convenience. It will also be convenient to define two new quantities

$$S_{ij} \equiv \langle u_i\,|\,u_j\rangle = \int u_i^*(\tau) u_j(\tau)\, d\tau \tag{4.58}$$

and

$$H_{ij} \equiv \langle u_i\,|\,\mathcal{H}\,|\,u_j\rangle = \int u_i^*(\tau) \mathcal{H} u_j(\tau)\, d\tau. \tag{4.59}$$

Since i and j both take on all integral values from 1 to N, there are N^2 values of S_{ij} and N^2 values of H_{ij}. We can arrange these quantities in two $N \times N$ square matrices, S and H. We call S the *overlap matrix*, since the element S_{ij} is a measure of the overlap of the basis functions φ_i and φ_j. Similarly, H is called the *Hamiltonian matrix*. Introducing these definitions into (4.57), we obtain

$$\sum_{j=1}^{N} \sum_{i=1}^{N} c_j^* c_i (H_{ji} - E' S_{ji}) = 0. \tag{4.60}$$

In order to calculate an upper bound on the ground-state energy, we must minimize E' with respect to the parameters $c_j\ (j = 1, 2, 3, \ldots, N)$. If we differentiate (4.60) with respect to each c_i and demand that

$$\frac{\partial E'}{\partial c_i} = 0, \qquad i = 1, \ldots, N \tag{4.61}$$

we thereby acquire a set of N *linear homogeneous equations* in the unknown parameters c_j:[14]

$$\boxed{\sum_{j=1}^{N} c_j (H_{ij} - E' S_{ij}) = 0, \qquad i = 1, \ldots, N.} \quad \text{Secular equations} \tag{4.62}$$

[14]In going from Eq. (4.60) to (4.62), we have interchanged the indices i and j and used the fact that $\mathcal{H}$ is Hermitian, so $H_{ij} = H_{ji}^*$.

These are called the *secular equations*. This set of equations possesses non-trivial solutions for the variational parameters if and only if the determinant formed from the square matrix $H - E'S$ is zero, that is, if

$$|H_{ij} - E'S_{ij}| = 0. \qquad \text{Determinant equation} \tag{4.63}$$

By explicitly writing out Eq. (4.63) and expanding the determinant,[15] we obtain an algebraic equation of order N in the unknown E'. The solution of this equation yields N roots, which we denote E'_n, $n = 1, 2, \ldots, N$. Let us arrange these roots in ascending order so that $E'_1 \leq E'_2 \leq \cdots \leq E'_N$. Then, by the minimum principle, we know that the *lowest root* E'_1 *is an upper bound on the ground-state energy*; that is, $E'_1 \geq E_1$.

In fact, all the roots are bounds on exact energies of the system:

$$E'_1 \geq E_1, \quad E'_2 \geq E_2, \quad E'_3 \geq E_3, \ldots, E'_N \geq E_N. \tag{4.64}$$

The proof of this result, which is not obvious, is beyond the scope of this text.[16] Clearly, the linear variational method does not give us an upper bound on E_{N+1}.

An expression for the optimum linear variational wave function φ_n corresponding to the nth root is obtained by substituting E'_n into the secular equations,

$$\sum_{j=1}^{N} c_j^{(n)}(H_{ij} - E'_n S_{ij}) = 0, \qquad i = 1, \ldots, N, \tag{4.65}$$

and solving for the coefficients $c_j^{(n)}$. (The superscript on c_j reminds us that it is the jth coefficient for the nth root.) This yields all but one of the coefficients $c_j^{(n)}$, $j = 1, 2, \ldots, N$. The remaining coefficient may be determined by requiring that φ be normalized, $\langle \varphi | \varphi \rangle = 1$.

Matrix Notation

The principal results of this section are Eqs. (4.62) and (4.63); together they constitute the *linear variational method*. Before turning to an example of the use of this method, let us look briefly at the matrices S and H. The secular equations (4.65) for the nth root can be written in matrix form as

$$(H - E'_n S)c^{(n)} = 0, \tag{4.66}$$

[15]For a review of matrix algebra and the properties of determinants, see Henry Margenau and George M. Murphy, *The Mathematics of Physics and Chemistry*, 2nd ed. (London: D. Van Nostrand Co. Ltd., 1956), Chap. 10.

[16]See E. A. Hylleraas and B. Undheim, *Z. Physik* **65**, 759 (1930).

where $c^{(n)}$ is a column vector (i.e., an $N \times 1$ matrix) whose elements are the N linear variational parameters $c_j^{(n)}$, $j = 1, 2, \ldots, N$, which appear in the trial wave function φ corresponding to the nth energy. There will be one set of equations like (4.66) for each root: N sets altogether.

The matrix H *represents* the Hamiltonian of the system. Of course, this representation of $\mathcal{H}$ is directly related to the particular set of basis functions used in constructing φ, since $H_{ij} = \langle u_i | \mathcal{H} | u_j \rangle$. This matrix is only an approximate representation of $\mathcal{H}$, which is why we get approximate energies when using this technique. However, suppose that we let $N \longrightarrow \infty$ in the definition of φ, Eq. (4.56), and choose the basis functions so that the set $\{u_i\}$ is complete. If we could solve the resulting secular and determinant equations for E'_n $(n = 1, \ldots, \infty)$ and then arrange the roots in order of ascending magnitude, these roots would be equal to the exact energy eigenvalues.[17] In this case, the Hamiltonian eigenvalue equation would be completely represented by a matrix equation like Eq. (4.66). Moreover, the infinite expansions of the trial wave functions corresponding to various values of n would exactly equal the true wave functions of the system.

The obvious problem is that if we let $N \longrightarrow \infty$, we must deal with $\infty \times \infty$ matrices. This being impractical, we must truncate the basis set and accept approximate results.

Example 4.4
Return of the Anharmonic One-Dimensional Oscillator

Consider once more a particle of mass m in the anharmonic oscillator potential. Suppose that this time the Hamiltonian is

$$\mathcal{H} = \mathcal{H}^{(0)} + \beta x^4, \tag{4.67}$$

where $\mathcal{H}^{(0)}$ is given by Eq. (4.26), and let us assume that β is so large that perturbation theory is not valid. (For example, suppose that βx^4 is of the same order of magnitude as $\frac{1}{2}m\omega^2 x^2$ for values of x where the unperturbed ground state is large.)

In applying the linear variational method to this problem, we must first select a set of basis functions that are physically reasonable. Such a set is conveniently available—namely, the set of eigenfunctions of the unperturbed simple harmonic oscillator. These functions $\{\psi_n^{(0)}\}$ are given in Eq. (4.28).

Let us construct a trial wave function from two of these functions, say

$$u_1 \equiv \psi_0^{(0)}(x) \quad \text{and} \quad u_2 \equiv \psi_2^{(0)}(x). \tag{4.68}$$

[17]This assertion is proved in B. L. Moiseiwitsch, *Variational Principles* (New York: Wiley-Interscience, 1966), p. 166.

[We know that we should not include $\psi_1^{(0)}(x)$, since it is an odd function of x and the ground-state eigenfunction of the Hamiltonian (4.67) is expected to be even.] The linear variational wave function is then

$$\varphi = c_1\psi_0^{(0)}(x) + c_2\psi_2^{(0)}(x). \tag{4.69}$$

To obtain the approximate energies, we must solve the determinant equation for $N = 2$:

$$\begin{vmatrix} H_{11} - E' & H_{12} \\ H_{21} & H_{22} - E' \end{vmatrix} = 0. \tag{4.70}$$

(The overlap matrix is equal to the unit matrix here.) The matrix elements are found to be[18]

$$\begin{aligned} H_{11} &= E_0^{(0)} + \beta\langle\psi_0^{(0)}|x^4|\psi_0^{(0)}\rangle \\ &= \frac{1}{2}\hbar\omega + \frac{3}{4}\beta\lambda^{-2}; \end{aligned} \tag{4.71a}$$

$$\begin{aligned} H_{12} &= \beta\langle\psi_0^{(0)}|x^4|\psi_2^{(0)}\rangle \\ &= \frac{3}{\sqrt{2}}\beta\lambda^{-2}; \end{aligned} \tag{4.71b}$$

$$\begin{aligned} H_{22} &= E_2^{(0)} + \beta\langle\psi_2^{(0)}|x^4|\psi_2^{(0)}\rangle \\ &= \frac{5}{2}\hbar\omega + \frac{39}{4}\beta\lambda^{-2}. \end{aligned} \tag{4.71c}$$

Exercise 4.4 Derive Eq. (4.71a). [Equations (4.71b) and (4.71c) follow from similar integrations.]

We now merely substitute Eqs. (4.71) into the determinant in (4.70), expand this determinant, and solve the resulting quadratic equation for E'. We find that

$$E'_{1,2} = \frac{3}{2}\hbar\omega + \frac{21}{4}\beta\lambda^{-2} \pm \sqrt{\hbar^2\omega^2 + 9\beta\hbar\omega\lambda^{-2} + \frac{99}{4}\beta^2\lambda^{-4}}. \tag{4.72}$$

[18]Evaluation of these integrals is facilitated by use of two useful properties of the Hermite polynomials $H_n(\sqrt{\lambda}\,x)$:

$$H_{n+1}(\sqrt{\lambda}\,x) + 2nH_{n-1}(\sqrt{\lambda}\,x) - 2\sqrt{\lambda}\,xH_n(\sqrt{\lambda}\,x) = 0,$$

where λ is given by Eq. (4.29), and

$$\int_{-\infty}^{\infty} e^{-\xi^2}H_n(\xi)H_{n'}(\xi)\,d\xi = 2^n n!\sqrt{\pi}\,\delta_{nn'}.$$

See the references on special functions listed at the end of Chapter 2 for proofs of these properties.

We get two roots, E'_1 and E'_2, corresponding, respectively, to the $-$ and $+$ signs in (4.72). There are also two linear variational wave functions, one for each root.

To obtain these wave functions, we substitute the two roots E'_1 and E'_2 into the secular equations

$$\begin{aligned} c_1^{(n)}(H_{11} - E'_n) + c_2^{(n)}H_{12} &= 0, \\ c_1^{(n)}H_{21} + c_2^{(n)}(H_{22} - E'_n) &= 0, \end{aligned} \qquad n = 1, 2 \tag{4.73}$$

and solve the resultant pair of simultaneous linear equations for the ratios $c_1^{(n)}/c_2^{(n)}$, $n = 1$, 2. We have only to normalize the two variational wave functions and we are through.

The prospect of actually carrying out the steps outlined in the previous paragraph is unattractive. Instead let us consider the energy (4.72) in the limit of small β; remember that it is in this limit that we expect to be able to use the perturbation theory of Sec. 4.1. In particular, suppose that $\beta \ll \lambda^2\hbar\omega$. Expanding the square root in (4.72) and keeping terms to first order in β, we obtain the approximate energies

$$E'_1 \simeq \frac{1}{2}\hbar\omega + \frac{3}{4}\beta\lambda^{-2}, \tag{4.74a}$$

$$E'_2 \simeq \frac{5}{2}\hbar\omega + \frac{39}{4}\beta\lambda^{-2}. \tag{4.74b}$$

Now, for comparison, the perturbation theory result for the approximate ground-state energy in first order is [see Eq. (4.21)]

$$\begin{aligned} E_1 &= E^{(0)} + \langle\psi_0^{(0)}\,|\,\beta x^4\,|\,\psi_0^{(0)}\rangle \\ &= \frac{1}{2}\hbar\omega + \frac{3}{4}\beta\lambda^{-2}, \end{aligned} \tag{4.75}$$

where we have used Eq. (4.71a) in the last step. This is precisely Eq. (4.74a)! [We can similarly verify Eq. (4.74b).] Thus, in the limit of small β, the results of the linear variational method and perturbation theory agree.

4.5 NONDEGENERATE PERTURBATION THEORY—REVISITED

Example 4.4 suggests a link between the linear variational method and time-independent perturbation theory. Indeed, a connection does exist, and it is worth noting because it unites these seemingly disparate approximation methods. In fact, the familiar perturbation theory equations of Sec. 4.1

result when the full Hamiltonian $\mathcal{H}$ separates into a "large" and a "small" part; that is,

$$\mathcal{H} = \mathcal{H}^{(0)} + \mathcal{H}^{(1)}. \tag{4.76}$$

Let us suppose that this is the case and construct the linear variational function φ, using eigenfunctions of the unperturbed Hamiltonian $\mathcal{H}^{(0)}$. The basis functions are then

$$u_i = \psi_i^{(0)}, \qquad i = 1, 2, \ldots, N, \tag{4.77}$$

and the trial function is

$$\varphi = \sum_{i=1}^{N} c_i \psi_i^{(0)}. \tag{4.78}$$

The linear variational calculation yields approximate energies E_n' from the determinant equation (4.63) and coefficients $c_j^{(n)}$ from the secular equations (4.62). Since the basis functions are orthogonal and normalized, the overlap matrix is

$$S_{ij} = \delta_{ij}, \tag{4.79}$$

and the elements of the Hamiltonian matrix are

$$H_{ij} = E_i^{(0)} \delta_{ij} + H_{ij}^{(1)}, \tag{4.80}$$

where $H_{ij}^{(1)}$ is the (ij) element of the matrix of the perturbation Hamiltonian,

$$H_{ij}^{(1)} = \langle \psi_i^{(0)} | \mathcal{H}^{(1)} | \psi_j^{(0)} \rangle. \tag{4.81}$$

If we write $\mathcal{H}^{(1)} = \lambda \mathcal{H}'$, λ being the smallness parameter of Sec. 4.1, the matrix element $H_{ij}^{(1)}$ becomes

$$H_{ij}^{(1)} = \lambda \langle \psi_i^{(0)} | \mathcal{H}' | \psi_j^{(0)} \rangle, \tag{4.82}$$

or

$$H_{ij}^{(1)} = \lambda H'_{ij}. \tag{4.83}$$

Clearly, if λ is small, as it must be if perturbation theory is to be valid, then the "off-diagonal matrix elements" H_{ij} $(i \neq j)$ are small.

Suppose that the unperturbed level of interest is nondegenerate. If we substitute S_{ij} and H_{ij} into the determinant equation, expand the determinant, and neglect all terms of higher order than the first in λ, the result is an algebraic equation that yields roots E_n',

$$E_n' \simeq E_n^{(0)} + \langle \psi_n^{(0)} | \mathcal{H}^{(1)} | \psi_n^{(0)} \rangle, \qquad n = 1, 2, \ldots, N. \tag{4.84}$$

This is precisely the nth first-order perturbed energy of Eq. (4.21). Discarding terms of higher order than the second in λ, we obtain

$$E'_n \simeq E_n^{(0)} + \langle \psi_n^{(0)} | \mathcal{H}^{(1)} | \psi_n^{(0)} \rangle + \sum_{j \neq n}^{N} \frac{|\langle \psi_j^{(0)} | \mathcal{H}^{(1)} | \psi_n^{(0)} \rangle|^2}{E_n^{(0)} - E_j^{(0)}} .$$

If $N \longrightarrow \infty$, so that φ contains all unperturbed wave functions, then E'_n becomes identical to the nth second-order perturbed energy of Eq. (4.33).

The first-order perturbed wave function is obtained by solving the secular equations and discarding terms of higher order than the first in λ. Specifically, we expand the coefficients c_j in the smallness parameter λ,

$$c_j = a_j^{(0)} + \lambda a_j^{(1)} + \lambda^2 a_j^{(2)} + \cdots,$$

substitute the result into the secular equations, and then solve for the $a_j^{(0)}$ and $a_j^{(1)}$, which result when only zeroth- and first-order terms are kept. After a little algebra, we find that the approximate eigenfunction of $\mathcal{H}$ is

$$\psi'_n \simeq \psi_n^{(0)} + \sum_{j \neq n}^{N} \frac{\langle \psi_j^{(0)} | \mathcal{H}^{(1)} | \psi_n^{(0)} \rangle}{E_n^{(0)} - E_j^{(0)}} \psi_j^{(0)}, \qquad n = 1, 2, \ldots, N.$$

If $N \longrightarrow \infty$ and all the zeroth-order functions are included in the basis set, this equation becomes identical to the perturbation theory result of Eq. (4.23).

The point is that by selecting a basis set consisting of unperturbed wave functions and dropping terms of order $n + 1$ and higher in the secular and determinant equations, we can obtain nth-order perturbation theory results. Thus perturbation theory can be viewed as a special case of linear variational theory.

4.6 THE DISAPPEARANCE OF DIABOLICAL DEGENERACIES

In Sec. 4.1 it was noted that the perturbation theory results derived there would not work for degenerate states. In particular, the expression for the first-order perturbed wave function,

$$\psi_n \simeq \psi_n^{(0)} + \sum_{k \neq n} \frac{\langle \psi_k^{(0)} | \mathcal{H}^{(1)} | \psi_n^{(0)} \rangle}{E_n^{(0)} - E_k^{(0)}} \psi_k^{(0)}, \tag{4.85}$$

contains terms that diverge if any of the $\psi_k^{(0)}$, for $k \neq n$, are degenerate with $\psi_n^{(0)}$. (Similar problems arise in the calculation of the perturbed energies to second order, although the first-order expression,

$$E_n \simeq E_n^{(0)} + \langle \psi_n^{(0)} | \mathcal{H}^{(1)} | \psi_n^{(0)} \rangle,$$

is still finite.) Consequently, although Eq. (4.85) can be used to calculate the corrections to $\psi_n^{(0)}$ due to nondegenerate states, we must look elsewhere in order to handle degenerate ones. Clearly, we will wish to do so, since nearly all systems of interest to us (e.g., the hydrogenic atom) have some degree of degeneracy.

Therefore let us consider a hypothetical system consisting of N bound states, the energy levels of which are shown in Fig. 4.3(a). Levels 1 and 2 are

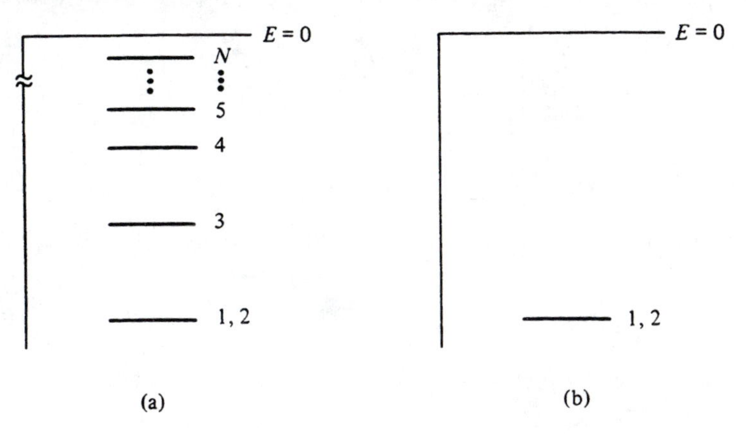

Figure 4.3 (a) Energy level diagram for a hypothetical N-level system with states 1 and 2 degenerate; (b) hypothetical two-level system obtained by ignoring states 3, 4, 5, . . . , N in the system of Fig. 4.3(a).

degenerate; that is, $\psi_1^{(0)}$ and $\psi_2^{(0)}$ have the same energy $E_1^{(0)} = E_2^{(0)}$, but no other function $\psi_n^{(0)}$, $n \neq 1, 2$, has this energy. [Thus the energy separation in zeroth order, $\Delta E^{(0)} \equiv E_1^{(0)} - E_2^{(0)}$, is zero.] We would like to see what happens to the energies when the system is exposed to some external perturbation. In particular, what will the ground-state energy be in this event? Will this level still be degenerate?

To answer these questions, let us focus on those two degenerate states. For the moment, we shall assume that all other levels are so far from the ground level in energy that the corrections due to them are negligible. So we are now dealing with the system whose zeroth-order energy levels are shown in Fig. 4.3(b); the degenerate functions $\psi_1^{(0)}$ and $\psi_2^{(0)}$ corresponding to these energies are eigenfunctions of the unperturbed Hamiltonian $\mathcal{H}^{(0)}$. We seek wave functions and energies of the perturbed system.

The divergence trouble arises if we attempt to use perturbation theory; consequently, we shall ignore it for the present. At no point in the derivation of the linear variational method in Sec. 4.4 did we require that the system be nondegenerate; let us try to use this method to solve for the desired wave functions and energies.

As usual in variational calculations, we begin by constructing functions φ

that we think will approximate the true eigenfunctions of $\mathcal{H}$. A logical set of basis functions is the collection of eigenfunctions of the unperturbed system, $\{\psi_1^{(0)}, \psi_2^{(0)}\}$. So we shall try the function

$$\varphi = c_1\psi_1^{(0)} + c_2\psi_2^{(0)}, \tag{4.86}$$

where c_1 and c_2 are the linear variational parameters.

Proceeding as in Sec. 4.4, we calculate variational energies E'_1 and E'_2 from the determinant equation (4.63)

$$\begin{vmatrix} H_{11} - E' & H_{12} \\ H_{21} & H_{22} - E' \end{vmatrix} = 0, \tag{4.87a}$$

where
$$H_{ij} = E_i^{(0)}\,\delta_{ij} + \langle\psi_i^{(0)}\,|\,\mathcal{H}^{(1)}\,|\,\psi_j^{(0)}\rangle. \tag{4.87b}$$

Expanding the determinant, we obtain a quadratic equation with solutions

$$E'_{1,2} = \frac{1}{2}(H_{11} + H_{22}) \pm \frac{1}{2}[(H_{11} - H_{22})^2 + 4H_{12}H_{21}]^{1/2}, \tag{4.88}$$

where the subscripts 1 and 2 on E' refer to the lower (— sign) and upper (+ sign) roots, respectively.

Exercise 4.5 Carry out the algebra leading to Eq. (4.88).

There are two sets of secular equations, one for each root E'_n,

$$\begin{aligned} c_1^{(n)}(H_{11} - E'_n) + c_2^{(n)}H_{12} &= 0, \\ c_1^{(n)}H_{21} + c_2^{(n)}(H_{22} - E'_n) &= 0, \end{aligned} \qquad n = 1, 2. \tag{4.89}$$

The solution of these equations yields two sets of coefficients; each set is associated with one of the roots E'_1 and E'_2. Substituting these coefficients into the trial function, we obtain two variational wave functions, which we shall call ψ'_1 and ψ'_2:

$$\psi'_1 = N_1\left[\psi_1^{(0)} + \frac{H_{21}}{E'_1 - H_{22}}\psi_2^{(0)}\right] \qquad \text{(lower root)}, \tag{4.90a}$$

$$\psi'_2 = N_2\left[\psi_2^{(0)} + \frac{H_{12}}{E'_2 - H_{11}}\psi_1^{(0)}\right] \qquad \text{(upper root)}. \tag{4.90b}$$

Here N_1 and N_2 are constants that may be determined by normalizing ψ'_1 and ψ'_2. The significance of these new wave functions will be discussed shortly.

Thus we have derived approximate energies and wave functions for the perturbed system. The energies are E'_1 and E'_2 of Eq. (4.88). Notice that

$E'_1 \neq E'_2$; the energy levels of the perturbed system are not degenerate. We say that the unperturbed degenerate levels 1 and 2 are "split" by the perturbation and illustrate this fact schematically in Fig. 4.4.

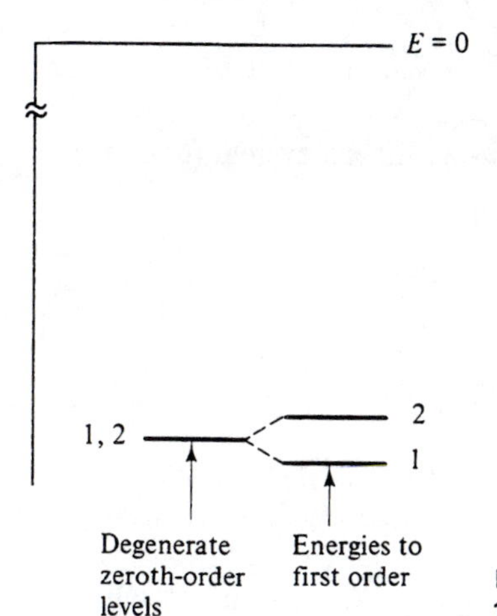

Figure 4.4 Splitting of degenerate energy levels 1 and 2 due to the effects of a perturbation $\mathcal{H}^{(1)}$.

Exercise 4.6 Consider the special case of the "almost degenerate system" for which the zeroth-order energy separation, $\Delta E^{(0)} = E_1^{(0)} - E_2^{(0)}$, is small but not zero. If the perturbation Hamiltonian $\mathcal{H}^{(1)}$ is small, then we can make the approximation that $|H_{12}/(H_{11} - H_{22})| \ll 1$.

(a) Use this approximation and the binomial expansion $(1 \pm x)^n \simeq 1 \mp nx$ in Eq. (4.88) to obtain expressions for E'_1 and E'_2.

(b) Use Eq. (4.87b) for the matrix elements H_{ij} with $\mathcal{H}^{(1)} = \lambda\mathcal{H}'$ to show that, for small λ, your results of part (a) reduce to the time-independent perturbation theory expression of Sec. 4.1.

The Variational Wave Functions

In addition to the energies E'_1 and E'_2 just discussed, we have obtained expressions for ψ'_1 and ψ'_2. Equations (4.90) show that each of these functions is a linear combination of the degenerate zeroth-order functions $\psi_1^{(0)}$ and $\psi_2^{(0)}$. Therefore ψ'_1 and ψ'_2 may be considered *new degenerate zeroth-order wave functions.*[19] They are "better" zeroth-order wave functions than the old ones, $\psi_1^{(0)}$ and $\psi_2^{(0)}$, in the sense that they have the desirable property

$$\langle \psi'_1 | \mathcal{H}^{(1)} | \psi'_2 \rangle = 0. \tag{4.91}$$

Exercise 4.7 (a) Verify Eq. (4.91).

(b) Determine N_1 and N_2 and show that ψ'_1 and ψ'_2 are orthogonal.

[19] This point is not hard to see: H_{21} and $E'_1 - H_{22}$ are each of order 1 in λ, so the factor $H_{21}/(E'_1 - H_{22})$ is of order 0.

Nondegenerate Levels

Having successfully conquered the simple two-level system of Fig. 4.3(b), let us return to the N-level system of Fig. 4.3(a) and calculate the corrections to the degenerate zeroth-order energies $E_1^{(0)}$ and $E_2^{(0)}$ due to the states with energies $E_j^{(0)}, j \neq 1, 2$. *Since none of these levels is degenerate with the ground state in zeroth order, we can use nondegenerate perturbation theory.* For example, the energies of levels 1 and 2 of the perturbed system, correct to second order, are [see Eq. (4.33)]

$$E_1 \simeq E'_1 + \sum_{j \neq 1,2}^{N} \frac{|\langle \psi_j^{(0)} | \mathcal{H}^{(1)} | \psi'_1 \rangle|^2}{E_1^{(0)} - E_j^{(0)}}, \tag{4.92a}$$

$$E_2 \simeq E'_2 + \sum_{j \neq 1,2}^{N} \frac{|\langle \psi_j^{(0)} | \mathcal{H}^{(1)} | \psi'_2 \rangle|^2}{E_2^{(0)} - E_j^{(0)}}, \tag{4.92b}$$

where E'_1 and E'_2 are the variational energies of Eq. (4.88). [Notice that we use ψ'_1 and ψ'_2 in the matrix elements of (4.92) because these are the functions that correspond to E'_1 and E'_2.] No "infinities" appear in these summations, since $E_1^{(0)} - E_j^{(0)}$ and $E_2^{(0)} - E_j^{(0)}$ are nonzero for $j \neq 1, 2$—that is, for all the terms in the sum.

Summarizing, we now have a procedure for calculating approximate energies and wave functions for a system that consists of two degenerate levels. We should emphasize that no new tools have been introduced; our procedure is a blend of variational theory for the corrections due to the degenerate states and nondegenerate perturbation theory for the corrections due to the other states. We expect that the latter correction terms will be smaller than the former.

Of course, many systems have more than two degenerate unperturbed levels, and the techniques of this section can easily be extended to such systems. If, for example, the first three unperturbed eigenfunctions are degenerate, we shall have to solve a cubic equation for E'_1, E'_2, and E'_3. The secular equation will yield three new degenerate zeroth-order wave functions; they can then be used to obtain second-order energies E_1, E_2, and E_3.

4.7 THE STARK EFFECT: AN EXAMPLE

In order to illustrate some of the approximation methods of this chapter, we shall now solve the problem of a hydrogen atom ($Z = 1$) in an external, static, homogeneous electric field $\mathbf{E}$. This field is constant in space and time. If the $\hat{\mathbf{z}}$ axis is chosen to be directed along $\mathbf{E}$, we can write[20]

$$\mathbf{E} = E_z \hat{\mathbf{z}}. \tag{4.93}$$

[20]Do not confuse the magnitude of the electric field E_z with the energies E_n.

The interaction of the hydrogen atom with this field gives rise to a term $+ezE_z$ in the potential energy. This term can conveniently be described by introducing the electric dipole moment operator,

$$\mathbf{d} \equiv -e\mathbf{r}, \tag{4.94}$$

where $\mathbf{r}$ is the position vector of the electron relative to the nucleus. Then the potential energy term $+ezE_z$ can be written

$$\mathcal{H}^{(1)} = -\mathbf{d} \cdot \mathbf{E}; \tag{4.95}$$

we have chosen this term as the perturbation Hamiltonian.

The full Hamiltonian for a hydrogen atom ($Z = 1$) in the field $\mathbf{E} = E_z\hat{\mathbf{z}}$ is

$$\mathcal{H} = -\frac{\hbar^2}{2\mu}\nabla^2 - \frac{e^2}{r} + ezE_z; \tag{4.96}$$

the unperturbed Hamiltonian is therefore

$$\mathcal{H}^{(0)} = -\frac{\hbar^2}{2\mu}\nabla^2 - \frac{e^2}{r}. \tag{4.97}$$

We know the solutions of the time-independent Schroedinger equation for the Hamiltonian $\mathcal{H}^{(0)}$; they are the hydrogenic wave functions $\psi_{n\ell m}(\mathbf{r})$ of Chapter 3 (see Table 3.3). The energies are simply the hydrogenic energies [see Eq. (3.61)]

$$E_n = -\frac{\mu e^4}{2\hbar^2 n^2}, \qquad n = 1, 2, \ldots. \tag{4.98}$$

Recall that n in Eq. (4.98) is the principal quantum number and that there are n^2 degenerate functions $\psi_{n\ell m}(\mathbf{r})$ with energy E_n. We shall append a superscript (0) to these unperturbed functions and energies—$\psi_{n\ell m}^{(0)}(\mathbf{r})$ and $E_n^{(0)}$.

Let us consider only small external fields (say $E_z \lesssim 10^3$ V/cm) and use perturbation theory to calculate energies and wave functions of a hydrogen atom in this field. The effect of $\mathbf{E}$ on the atomic states is called the *Stark effect.*

Ground State

The ground state, $n = 1, \ell = 0, m = 0$, of the unperturbed hydrogen atom is nondegenerate, so we can use simple nondegenerate perturbation theory to calculate the approximate ground-state energy. To first order, the perturbed energy is [see Eq. (4.21)]

$$E_1 \simeq E_1^{(0)} + \langle \psi_{1s}^{(0)}(\mathbf{r}) | ezE_z | \psi_{1s}^{(0)}(\mathbf{r}) \rangle, \tag{4.99}$$

where

$$\psi_{1s}^{(0)}(\mathbf{r}) = \frac{1}{\sqrt{\pi a_0^3}} e^{-r/a_0}. \tag{4.100}$$

Since e and E_z are constant, the matrix element in Eq. (4.99) is

$$eE_z\langle\psi_{1s}^{(0)}(\mathbf{r})|z|\psi_{1s}^{(0)}(\mathbf{r})\rangle = eE_z\int \psi_{1s}^*(\mathbf{r})z\psi_{1s}(\mathbf{r})\,d\mathbf{r}. \tag{4.101}$$

We could use $z = r\cos\theta$ and Eq. (4.100) for $\psi_{1s}(\mathbf{r})$ to evaluate this integral. However, much effort can be saved by pausing a moment to consider *symmetry*. Notice that $\psi_{1s}(\mathbf{r})$ has even parity.[21] Since z has odd parity, the integral in this matrix element is zero. Therefore, to first order, the perturbed energy is equal to the unperturbed energy; that is,

$$E_1 \simeq E_1^{(0)} = -\frac{e^2}{2a_0}. \tag{4.102}$$

We know from Example 4.1 that this means that we must go to a higher order to obtain the correction to $E_1^{(0)}$.

Second-Order Corrections

To calculate the second-order correction to the energy of the ground state, we insert $\mathcal{H}^{(1)} = ezE_z$ into the perturbation theory result (4.33) to obtain

$$E_1 \simeq E_1^{(0)} + e^2E_z^2 \sum_{(n\ell m)\neq(1,0,0)} \frac{|\langle\psi_{n\ell m}^{(0)}|z|\psi_{1s}^{(0)}\rangle|^2}{E_1^{(0)} - E_n^{(0)}}, \tag{4.103}$$

where we have summed over the set of quantum numbers $(n\ell m) \neq (1, 0, 0)$. Clearly, the second-order correction will not be zero, since the matrix element $\langle\psi_{n\ell m}^{(0)}|z|\psi_{1s}^{(0)}\rangle$ is nonzero for states of odd symmetry with respect to z. Moreover, this correction is proportional to the square of E_z; hence this effect is called the *quadratic Stark effect*.

We show in Prob. 4.8 how to evaluate the right-hand side of Eq. (4.103); all that is required is the solution of a differential equation by power series methods and the evaluation of an integral. The ground-state energy, correct to second order, is found to be

$$\boxed{E_1 \simeq E_1^{(0)} - \tfrac{9}{4}a_0^3E_z^2.} \quad \text{Quadratic Stark effect} \tag{4.104}$$

Atomic Polarizability

Physically, the correction term $-\frac{9}{4}a_0^3E_z^2$ is due to the *deformation* of the 1s unperturbed wave function by the electric field $\mathbf{E}$. Thus the charge distribution of the electron is distorted by the external electric field; we say that the atom has been *polarized* by the field.

[21]An *even parity* function satisfies $f(-\mathbf{r}) = f(\mathbf{r})$; an odd parity function satisfies $f(-\mathbf{r}) = -f(\mathbf{r})$.

A measure of the extent of this polarization is the *polarizability* α, defined as the ratio of the induced dipole moment to the external electric field. For the electric field $\mathbf{E} = E_z\hat{\mathbf{z}}$, we have

$$\mathbf{d} = \alpha E_z \hat{\mathbf{z}}, \tag{4.105}$$

where α is the polarizability of the hydrogen atom. But the shift in the energy levels is (see Prob. 4.8)

$$E_1 - E_1^{(0)} = -\tfrac{1}{2}\alpha E_z^2. \tag{4.106}$$

From this result and (4.104), we find that the polarizability[22] for a hydrogen atom in the ground state is

$$\alpha = \tfrac{9}{2}a_0^3. \tag{4.107}$$

First Excited State

In order to determine the effect of the electric field $\mathbf{E}$ on the first excited state ($n = 2$), we must use the method of Sec. 4.6, since this level is fourfold degenerate (see Table 3.5). Thus we form a linear variational wave function φ, using as basis functions the unperturbed eigenfunctions of $\mathcal{H}^{(0)}$ for $n = 2$. That is,

$$\varphi = \sum_{i=1}^{4} c_i \psi_i^{(0)}, \tag{4.108}$$

where

$$\psi_1^{(0)} \equiv \psi_{200}^{(0)} = \psi_{2s}^{(0)}, \tag{4.109a}$$

$$\psi_2^{(0)} \equiv \psi_{210}^{(0)} = \psi_{2p_0}^{(0)}, \tag{4.109b}$$

$$\psi_3^{(0)} \equiv \psi_{211}^{(0)} = \psi_{2p_1}^{(0)}, \tag{4.109c}$$

$$\psi_4^{(0)} \equiv \psi_{21-1}^{(0)} = \psi_{2p_{-1}}^{(0)}. \tag{4.109d}$$

We obtain first-order approximate energies by solving the formidable-looking equation

$$\begin{vmatrix} H_{11} - E' & H_{12} & H_{13} & H_{14} \\ H_{21} & H_{22} - E' & H_{23} & H_{24} \\ H_{31} & H_{32} & H_{33} - E' & H_{34} \\ H_{41} & H_{42} & H_{43} & H_{44} - E' \end{vmatrix} = 0, \tag{4.110}$$

where

$$H_{ij} = E_i^{(0)}\,\delta_{ij} + \langle \psi_{n\ell m}^{(0)} | ezE_z | \psi_{n'\ell'm'}^{(0)} \rangle. \tag{4.111}$$

[22]Other quantities used to describe the effect of an external electric field on an atom are the *electrical susceptibility* χ and the *dielectric constant* ϵ. For a gas of atoms that are, on the average, far apart, the electrical susceptibility is defined to be $\chi \equiv N\alpha$, where N is the number of atoms per cubic centimeter. The dielectric constant is then defined to be $\epsilon \equiv 1 + 4\pi\chi$.

We hope that many of these matrix elements are zero. To see if they are, let us use two helpful facts from our study of central potentials in Chapter 2.

1. The matrix element in Eq. (4.111) is zero unless $m' = m$, since the functions $\Phi_m(\varphi)$ are orthogonal [see Eq. (2.45)].
2. The wave functions $\psi^{(0)}_{n\ell m}$ have even or odd parity, depending on whether ℓ is even or odd. Since z has odd parity, it follows that if ℓ and ℓ' are of the same parity, the matrix element will vanish. Thus of all the "off-diagonal" matrix elements in the determinant in (4.110), only H_{21} and H_{12} are nonzero. From (4.111) we see that $H_{21} = H^{(1)}_{21}$ and $H_{12} = H^{(1)}_{12}$. The remaining unknown matrix element is easily evaluated as

$$\begin{aligned} H^{(1)}_{12} &= +eE_z\langle\psi^{(0)}_{2s}|z|\psi^{(0)}_{2p_0}\rangle \\ &= +\frac{eE_z}{16a_0^4}\int_0^\infty r^4\left(2-\frac{r}{a_0}\right)e^{-r/a_0}\,dr\int_0^\pi \sin\theta\cos^2\theta\,d\theta, \end{aligned} \tag{4.112}$$

or

$$H^{(1)}_{12} = 3a_0eE_z. \tag{4.113}$$

Equation (4.110) therefore reduces to a far more appealing form:

$$\begin{vmatrix} E^{(0)}-E' & 3a_0eE_z & 0 & 0 \\ 3a_0eE_z & E^{(0)}-E' & 0 & 0 \\ 0 & 0 & E^{(0)}-E' & 0 \\ 0 & 0 & 0 & E^{(0)}-E' \end{vmatrix} = 0. \tag{4.114}$$

Much has been learned, although we have yet to solve a single equation. We have determined, for example, that the first-order Stark effect mixes only states with different orbital angular momentum quantum numbers but the same magnetic quantum number. Moreover, Eq. (4.114) reveals that two of the roots, E'_3 and E'_4, are simply $E^{(0)}$, so, to first order, the $2p_1$ and $2p_{-1}$ levels are unaffected by the perturbation.

The other two roots are obtained from the factored equation

$$\begin{vmatrix} E^{(0)}-E' & 3a_0eE_z \\ 3a_0eE_z & E^{(0)}-E' \end{vmatrix} = 0, \tag{4.115}$$

which immediately yields

$$E' = E^{(0)} \pm 3a_0eE_z. \tag{4.116}$$

Therefore the energies perturbed to first order are

$$E'_1 = E^{(0)}_2 - 3a_0eE_z = -\frac{e^2}{8a_0} - 3a_0eE_z, \tag{4.117a}$$

$$E'_2 = E^{(0)}_2 + 3a_0eE_z = -\frac{e^2}{8a_0} + 3a_0eE_z. \tag{4.117b}$$

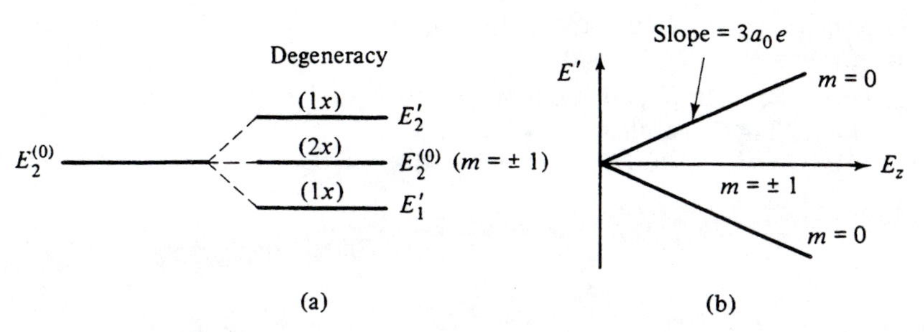

Figure 4.5 Linear Stark effect. (a) Splitting of the fourfold degenerate $n = 2$ unperturbed level of the hydrogen atom due to a constant external electric field E_z; (b) the variation of the perturbed energies with the magnitude of the electric field **E**.

These energies are shown in Fig. 4.5(a), where we have also indicated the degeneracies of the perturbed levels. Notice that the degeneracy is not entirely lifted by the first-order corrections. Two of the levels (with $m = 0$) are split symmetrically about the unperturbed energy $E_2^{(0)}$. We see from (4.117) that E' depends linearly on the magnitude of the electric field; hence this effect is called the *linear Stark effect*. The variation of E' with E_z is shown in Fig. 4.5(b). The new degenerate zeroth-order wave functions are now easily obtained by using the energies E_1' and E_2' in Eqs. (4.90):

$$\psi_1' = \frac{1}{\sqrt{2}}[\psi_{2s}^{(0)} - \psi_{2p_0}^{(0)}] \qquad \text{(lower root),} \tag{4.118a}$$

$$\psi_2' = \frac{1}{\sqrt{2}}[\psi_{2s}^{(0)} + \psi_{2p_0}^{(0)}] \qquad \text{(upper root).} \tag{4.118b}$$

Of course, the quadratic Stark effect of (4.104) is not restricted to the ground state. For the $n = 2$ level, we could calculate second-order corrections to the energies just obtained by employing the new zeroth-order functions of (4.118). The results will clearly be proportional to E_z^2 and thus constitute a quadratic Stark effect. However, at low field strengths the linear terms will dominate.

We have restricted this calculation to the case of a constant uniform electric field, since that is obviously the easiest example. In the next chapter we shall turn to a different type of perturbation, a time-dependent electric field.

PROBLEMS

4.1 Coefficients in the First-Order Wave Function (**)

In deriving expressions for the coefficients in the first-order correction to the unperturbed wave function [see Eq. (4.23)], we claimed that $a_{nn}^{(1)} = 0$.

(a) Requiring that the first-order perturbed eigenfunctions ψ_n be normalized to first order, show that $\text{Re}\,[a_{nn}^{(1)}] = 0$.

(b) Given the result of part (a), show that, with no loss of generality, we may choose $\text{Im}\,[a_{nn}^{(1)}] = 0$, so that $a_{nn}^{(1)} = 0$.
[HINT: It is often useful to recall that for $x \ll 1$,

$$(1 \pm x)^{-1} \simeq 1 \mp x$$

and

$$\exp(\pm x) \simeq 1 \pm x.]$$

4.2 Anharmonic Oscillator in Perturbation Theory (***)

Consider a particle of mass m in a one-dimensional anharmonic oscillator potential with potential energy

$$V(x) = \tfrac{1}{2}m\omega^2 x^2 + \alpha x^3 + \beta x^4.$$

It was noted in Example 4.1 that the x^3 term does not contribute a first-order correction to the energy. We shall now show that this term does give a nonzero first-order correction to the wave function and that the x^4 term contributes first-order corrections to both.

(a) Calculate the first-order correction to the energy of the nth unperturbed state and write down an expression for the energy perturbed to first order.

(b) Evaluate all the required matrix elements of x^3 and x^4 and write the wave function of the nth state perturbed to first order.

4.3 Equivalence of the Variational Principle and the Schroedinger Equation (**)

An alternate way of expressing the variational principle of Sec 4.2 is

$$\delta\langle\psi|\mathcal{H} - E|\psi\rangle = 0,$$

where ψ is the exact wave function, E the corresponding exact energy, and δ symbolizes a small but completely arbitrary variation of the wave function, and hence of E, from their true values. Of course, we have that

$$E = \frac{\langle\psi|\mathcal{H}|\psi\rangle}{\langle\psi|\psi\rangle}.$$

(a) Show that if we assume that ψ satisfies the time-independent Schroedinger equation, then the variational principle holds. [HINT: Treat δ operationally just like an ordinary differential.]

(b) Show that if the variational principle holds, then the time-independent Schroedinger equation results. [HINT: Use the fact that $\mathcal{H}$ is Hermitian.] Taking the results of parts (a) and (b) together, we conclude that the variational principle is necessary and sufficient for the time-independent Schroedinger equation; that is, they are completely equivalent!

4.4 A Variational Calculation of the Deuteron Ground-State Energy (***)

The deuteron was discussed in Prob. 3.2. We wish to use the "empirical" potential energy $V(r)$ given there, with $A = 32.7$ MeV and $a = 2.18 \times 10^{-13}$ cm, to obtain a variational approximation to the energy of the ground state ($\ell = 0$.)

(a) Try a simple variational function of the form

$$\varphi(r) = e^{-\alpha r/2a},$$

where α is the variational parameter to be determined. Calculate E' in terms of α and minimize it. Give your result for α and for E' [in millions of electron volts (MeV)]. The experimental value is -2.23 MeV. Is your answer above this? [HINT: Do not forget about "reduced mass" in this problem.]

(b) Draw a graph of your normalized wave function (corresponding to the minimum energy) versus r from 0 to $3a$. On the same graph, plot the normalized exact wave function

$$\psi(r) = \frac{N}{r} J_q(ce^{-r/2a}),$$

where $q = 1$, $c = 3.832$, $a = 2.18$ fermi, and N is the normalization constant, which is equal to 0.216. (J_q is a Bessel function.) In the range $0 < r < 3a$, where is the trial function best and where is it worst? Explain why it works out this way. (Recall our previous warning that the variational method does not guarantee a good wave function everywhere.)

4.5 The Variational Method Applied to Excited States (**)

(a) In Sec. 4.3 we claimed that it is possible to apply the variational method to *excited states* of physical systems provided that certain properties are satisfied by the trial wave function φ. Suppose that we are interested in the nth state, whose true energy is E_n. Show that if φ is orthogonal to all exact lower-lying states (i.e., states with energy lower than E_n), then the minimum principle applies:

$$E' \geq E_n.$$

(b) To illustrate this result, consider again the one-dimensional harmonic oscillator of Example 4.3. Consider the trial function

$$\varphi = \begin{cases} x(x^2 - c^2)^2, & |x| \leq c \\ 0, & |x| > c. \end{cases}$$

Find the energy of the first excited state of this system, using φ with c^2 as the variational parameter. Does the minimum principle apply? Why or why not?

4.6 Perturbed Square Well (**)

Consider a one-dimensional system consisting of a particle of mass m and charge e in the potential well $V(x)$ shown (Fig. 4.6).

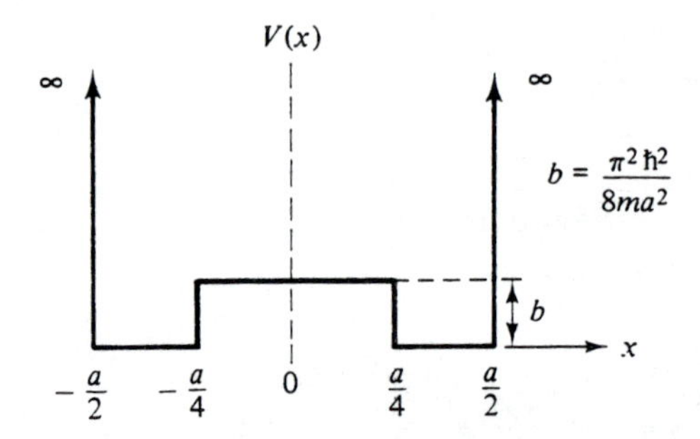

Figure 4.6

The potential energy function is

$$V(x) = \begin{cases} \infty, \; x < -\dfrac{a}{2}, \; x > \dfrac{a}{2} \\ 0, \; -\dfrac{a}{2} < x < -\dfrac{a}{4}, \; \dfrac{a}{2} > x > \dfrac{a}{4} \\ b, \; -\dfrac{a}{4} < x < \dfrac{a}{4}. \end{cases}$$

Suppose that b is sufficiently small that we can treat this problem by means of perturbation theory.

(a) Write expressions for the unperturbed eigenfunctions and energy eigenvalues.

(b) Obtain an expression for the first-order perturbed energy of the ground state of the system, expressing your result in terms of b and fundamental constants.

(c) Write an expression for the first-order perturbed wave function of the ground state in terms of the complete set of unperturbed functions of part (a). (You need not evaluate matrix elements in your expression.)

(d) What is the time-dependent stationary-state wave function for the ground state of the unperturbed system? For the ground state of the perturbed system?

(e) Would you expect any of the unperturbed eigenfunctions not to contribute to $\psi_1(x)$? If so, which ones? Explain your answer, using physical arguments. Which unperturbed wave function will mix most strongly with $\psi_1^{(0)}$? Which next most strongly?

(f) Evaluate the appropriate matrix elements for the perturbed wave function of part (e) for states that mix strongly.

(g) Suppose that we apply a weak uniform electric field in the $\hat{\mathbf{x}}$ direction. Describe quantitatively how you would expect $\mathcal{H}$ to be affected. Would your answer to part (e) be affected by the addition of this new term to the Hamiltonian of the system? If so, tell how.

4.7 Finite Spatial Extent of the Nucleus (***)

In all we have done so far, the nucleus has been treated as a positively charged point particle. In fact, the nucleus does possess a finite size with a radius given approximately by the empirical formula

$$R \simeq r_0 A^{1/3},$$

where $r_0 = 1.2 \times 10^{-13}$ cm (i.e., 1.2 fermi) and A is the atomic weight (essentially, the total number of protons and neutrons in the nucleus). A reasonable assumption is to take the total nuclear charge $+Ze$ as being uniformly distributed over the nuclear volume.

(a) Derive the following expression for the electrostatic potential energy of an electron in the field of the "finite" nucleus of charge $+Ze$:

$$V(r) = \begin{cases} -\dfrac{Ze^2}{r}, & r > R \\ \dfrac{Ze^2}{R}\left(\dfrac{r^2}{2R^2} - \dfrac{3}{2}\right), & r < R. \end{cases}$$

Draw a graph comparing this potential energy and the point nucleus potential energy.

(b) Since you know the solution of the point nucleus problem, choose this as $\mathcal{H}^{(0)}$ and construct a perturbation $\mathcal{H}^{(1)}$ such that the total Hamiltonian contains the $V(r)$ derived above; write an expression for $\mathcal{H}^{(1)}$.

(c) Calculate the first-order perturbed energy for a $1s$ state, obtaining an expression in terms of Z and fundamental constants. Do not assume a particular value for the electron mass.

(d) Use your result from part (c) to compare the effect of this finite nuclear-size correction on the ground-state energies of a hydrogen atom and a one-electron carbon ion ($Z = 6$). Compare the percentage correction to the energies due to the perturbation. Explain your results.

(e) Repeat part (d) for a muonic hydrogen atom and a muonic carbon ion, where a *muonic atom* results when the electron is replaced by a *muon* (a particle of charge $= -e$ and mass $= 207m_e$). Compare your results here with those of part (d) and discuss your findings. [HINT: It might help to consider the values of $\langle r \rangle$ for all the cases; just use the zeroth-order result.]

(f) For the general muonic atom, determine for what values of Z you would expect perturbation theory to fail. What approach would you use in such circumstances?

4.8 Quadratic Stark Effect and Polarizability of the Ground State of Atomic Hydrogen (***)

Consider a ground-state hydrogen atom in a constant uniform electric field $\mathbf{E} = E_z\hat{\mathbf{z}}$. It is easily shown that the energy is unperturbed to first order (see Sec. 4.7). The purpose of this problem is to demonstrate that the ground-state energy perturbed to second order is

$$E_1 = E_1^{(0)} - \tfrac{1}{2}\alpha E_z^2,$$

where $E_1^{(0)}$ is the unperturbed ground state energy of H and α is a constant, the electric dipole polarizability of the H atom in its ground state. We will also show that α is given exactly by

$$\alpha = \tfrac{9}{2}a_0^3.$$

(a) Our first thought is simply to write down the summation that makes up the second-order correction to the energy. In this particular example, we can avoid this step by going directly to the differential equation for $\varphi_1^{(1)}(\mathbf{r})$ in terms of the perturbation Hamiltonian $\mathcal{H}^{(1)}$ and zeroth-order function $\psi_1^{(0)}(\mathbf{r})$. Write out this equation, putting in $\mathcal{H}^{(0)}$, $\mathcal{H}^{(1)}$, and $\psi_1^{(0)}$ explicitly [see Eq. (4.8)].

(b) Show that the angular dependence of $\varphi_1^{(1)}$ is particularly simple, so that we can write

$$\varphi_1^{(1)}(\mathbf{r}) = f(r)\cos\theta.$$

Obtain the radial equation satisfied by $f(r)$. [HINT: Expand $\varphi_1^{(1)}(\mathbf{r})$ in a series of spherical harmonics, substitute into the differential equation, and voila!]

(c) The form of the differential equation for $f(r)$ should suggest a solution of the form

$$f(r) = g(r)e^{-r/a_0}.$$

Obtain a differential equation for $g(r)$ and solve it by expansion in a power series of r. In this way, show that the wave function, correct through first order, is

$$\psi_1 \simeq (\pi a_0^3)^{-1/2}e^{-r/a_0}\left[1 - \frac{E_z}{e}\left(a_0 r + \frac{1}{2}r^2\right)\cos\theta\right].$$

(d) Use your result for $\varphi_1^{(1)}(\mathbf{r})$ to derive the second-order correction to the energy and the exact electric dipole polarizability α for hydrogen.

4.9 Ground-State Hydrogen Atom Perturbed by a Proton: Short-Range Effects (****)

Consider a hydrogen atom in the $1s$ ground state located at a distance R from a proton. It is convenient to take the origin of coordinates at the nucleus of the H atom (proton a) and to locate the second proton (proton b) a distance R along the positive $\hat{z}$ axis. (See Fig. 4.7.) For now, we will consider the distance R to be fixed.

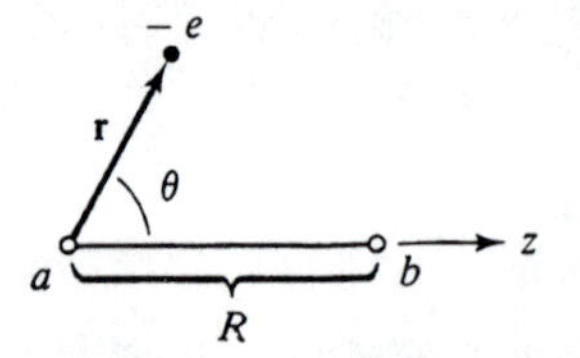

Figure 4.7

(a) Using perturbation theory, calculate the first-order correction to the energy for an arbitrary proton separation R, obtaining

$$E_1^{(1)}(R) = \frac{e^2}{R}e^{-2R/a_0}\left(1 + \frac{R}{a_0}\right),$$

where R may be thought of as a "parametric" argument of $E_1^{(1)}(R)$, since, for different choices of R, we will clearly obtain different values of $E_1^{(1)}(R)$. Discuss the behavior of $E_1^{(1)}(R)$ in the limits $R \longrightarrow 0$ and $R \longrightarrow \infty$ and draw a rough plot of $E_1^{(1)}(R)$ versus R. [HINT: Expand $|\mathbf{r} - \mathbf{R}|^{-1}$ in a series of Legendre polynomials as in Prob. 3.3]

It would seem that aside from higher-order corrections of perturbation theory, we have done the problem correctly. The result seems to indicate that the nearer the proton is to the H atom, the higher is the energy. Thus we would never expect to find a stable molecule H_2^+, since this system is simply a proton "bound" to an H atom. But stable H_2^+ (hydrogen molecule ion) does exist in nature; in fact, it takes $\sim$ 2.0 eV to dissociate it. Let us see if we can resolve this apparent conflict.

Should we look at the second-order correction? Doing so is quite reasonable, for we know that the second-order correction always lowers the ground-state energy. We shall see in Prob. 4.10 that this effect is actually most pronounced at large separations R, and, in fact, is not the "molecular binding" mechanism.

What did we do wrong? We forgot that this is a degenerate perturbation problem! Once we establish that the new system consists of one electron in the field of two protons, it becomes clear that two zeroth-order states are possible: $\psi_{1s_a}^{(0)}$ refers to the electron bound to proton a in a $1s$ state; and $\psi_{1s_b}^{(0)}$ refers to the electron bound to proton b in a $1s$ state. In particular, we have

$$\psi_{1s_a}^{(0)}(\mathbf{r}) = 2a_0^{-3/2}e^{-r_a/a_0}Y_{00},$$

$$\psi_{1s_b}^{(0)}(\mathbf{r}) = 2a_0^{-3/2}e^{-r_b/a_0}Y_{00},$$

where r_a and r_b are defined in Fig. 4.8. Both functions have the same zeroth-order

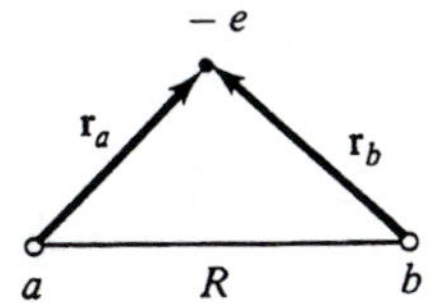

Figure 4.8

energy, $E_1^{(0)} = -e^2/2a_0$, and hence are degenerate. They are also normalized but are not orthogonal.

(b) Apply degenerate perturbation theory to this two-fold degenerate zeroth-order level by first showing that the two "new" zeroth-order functions

$$\psi_1' \equiv (2 + 2S)^{-1/2}[\psi_{1s_a}^{(0)} + \psi_{1s_b}^{(0)}]$$

$$\psi_2' \equiv (2 - 2S)^{-1/2}[\psi_{1s_a}^{(0)} - \psi_{1s_b}^{(0)}]$$

are normalized and orthogonal, where the "overlap integral" is

$$S = \langle 1s_a | 1s_b \rangle = \int \psi_{1s_a}^{(0)}(\mathbf{r})\psi_{1s_b}^{(0)}(\mathbf{r})\, d\mathbf{r}.$$

Using ψ'_1 and ψ'_2 as basis functions, set up the 2×2 secular equation and solve for the two roots, obtaining (do not evaluate the integrals)

$$E_1 = \frac{H_{aa} + H_{ab}}{1 + S} \quad \text{and} \quad E_2 = \frac{H_{aa} - H_{ab}}{1 - S},$$

where

$$H_{aa} = \langle 1s_a | \mathcal{H} | 1s_a \rangle = \int \psi_{1s_a}^{(0)}(\mathbf{r}) \mathcal{H} \psi_{1s_a}^{(0)}(\mathbf{r})\, d\mathbf{r},$$

$$H_{ab} = \langle 1s_a | \mathcal{H} | 1s_b \rangle = \int \psi_{1s_a}^{(0)}(\mathbf{r}) \mathcal{H} \psi_{1s_b}^{(0)}(\mathbf{r})\, d\mathbf{r}.$$

You should show that $H_{bb} = H_{aa}$ and $H_{ba} = H_{ab}$. Also show that $H_{ab} \longrightarrow 0$ and $S \longrightarrow 0$ as $R \longrightarrow \infty$. (Do not evaluate these integrals!)

(c) Next, consider the matrix element H_{aa}. Show that

$$H_{aa} = E_{1s}^{(0)} + E_1^{(1)}(R),$$

where $E_1^{(1)}(R)$ is simply the first-order correction to the energy we obtained in part (a) by ignoring the presence of degeneracy—that is, by not allowing the electron to be "shared" by proton b in the zeroth-order wave function. We already know that this term increases monotonically with decreasing R. We could go through the calculation of H_{ab} and S, but these so-called two-center integrals are much harder to evaluate and we will not attempt it here. It turns out that H_{ab} is negative and decreases monotonically with decreasing R. A rough plot of the two

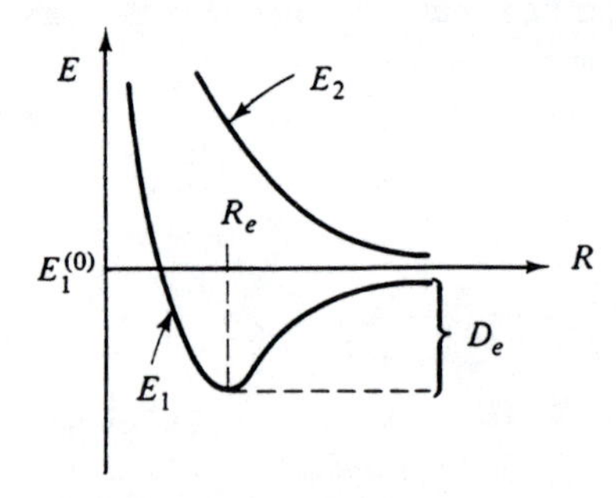

Figure 4.9

energies is given in Figure 4.9. Starting with part (a), discuss what may be learned by this problem. Include comments on

(i) the symmetry of the functions ψ'_1 and ψ'_2; draw a rough plot of both functions.

(ii) why you would expect ψ'_1 to correspond to the lowest energy just because of its symmetry.

(iii) the nature of the electron probability density for the two states and the physical implications with regard to molecular binding.

(iv) the reasons for the discrepancy between experimental values for the equilibrium separation, $R_e = 1.06$ Å, and the dissociation energy, $D_e =$

2.79eV, as compared with what we would obtain by evaluating all the expressions in our approximate treatment—$r_e = 1.32$ Å and $D_e = 1.76$ eV.

4.10 Long-Range Interaction of a Proton with a Ground-State Hydrogen Atom—The Polarization Potential (***)

[This problem requires that you have done Prob. 4.9.]

Consider the same system as discussed in Prob. 4.9—a ground-state hydrogen atom in the field of a proton. (Refer to the diagrams in Prob. 4.9 for details.) We now wish to study the energy of the system at large separations $R \gg a_0$, where essentially all the interesting "bonding" features of Prob. 4.9 have disappeared.

(a) Show that, for $R \gg r$, the perturbation Hamiltonian may be written

$$\mathcal{H}^{(1)} \simeq -\frac{e^2}{R^2}(r \cos \theta) = -\frac{e^2 z}{R^2},$$

where all coordinates are defined in Prob. 4.9. What is the first-order correction to the "long-range" energy based on this approximation to $\mathcal{H}^{(1)}$? Explain how this is consistent with the first-order results in Prob. 4.9.

(b) For the second-order correction to the energy, we can use the results of Prob. 4.8. Show that to second-order,

$$E \simeq E_1^{(0)} - \frac{\alpha e^2}{2R^4},$$

where $\alpha = 4.5a_0^3$. Explain your reasoning. Explain to what extent this result might be generalized to other atoms in the field of a proton. In other words, does anything really tie this result to an H atom?

5
More Approximation Methods (Time Dependent)

Science is the captain, practice the soldiers.

Leonardo da Vinci

The principal means available to probe the structure of atoms and molecules is through observations of the absorption and emission of electromagnetic energy by these systems as they change from one state to another. Such experiments constitute the huge and important field of *spectroscopy*.

The early spectral observations of atomic hydrogen were crucial to quantum mechanics in its infancy. Correct calculation of wavelengths for transitions between energy levels of hydrogen was the first real test of Schroedinger's wave theory. We know that atoms are observed to have only certain allowed energies. Therefore, in transitions between states of atoms, only certain frequencies will be observed. It is now necessary to establish how the atom interacts with the electromagnetic radiation fields so that energy can be absorbed or emitted.

5.1 THE CARE AND TREATMENT OF TIME-DEPENDENT PERTURBATIONS

Except in unusually simple cases (like the one studied in the last chapter), the external electromagnetic field that we shall use to "perturb" the atom will depend on the time as well as on spatial coordinates. Thus we cannot solve this problem by the approximation methods of Chapter 4, all of which assume that the full Hamiltonian $\mathcal{H}$ is independent of time.

We must therefore develop a way to solve the quantum mechanical problem of a system with a time-dependent Hamiltonian. This new problem will require a substantial revision in our thinking. In cases where $\mathcal{H}$ did not explicitly depend on t, we concentrated on stationary states; these states arose from a separation in space and time of the wave function into the form

$$\Psi_E(\mathbf{r}, t) = e^{-i(E/\hbar)t}\psi_E(\mathbf{r}). \tag{5.1}$$

But this separation could be carried out only because the potential energy was independent of t. Clearly, stationary states do not exist for a system where the Hamiltonian does depend on time. We cannot even write down a meaningful equation of the form $\mathcal{H}\psi_E = E\psi_E$ for such a system because states do not exist in which the energy is sharp.

Instead we must return to the more fundamental time-dependent Schroedinger equation

$$\mathcal{H}\Psi(\mathbf{r}, t) = i\hbar\frac{\partial}{\partial t}\Psi(\mathbf{r}, t). \tag{5.2}$$

The solution of Eq. (5.2) is no trivial matter. In this section we shall develop a method of solution that is somewhat similar to the perturbation theory of Sec. 4.1. In particular, suppose that the full Hamiltonian $\mathcal{H}$ can be split up into two parts:

$$\mathcal{H} = \mathcal{H}^{(0)} + \mathcal{H}^{(1)}(t). \tag{5.3}$$

The zeroth-order Hamiltonian $\mathcal{H}^{(0)}$ is time independent; all the time dependence is in the perturbation Hamiltonian $\mathcal{H}^{(1)}(t)$. [Of course, both $\mathcal{H}^{(0)}$ and $\mathcal{H}^{(1)}$ depend on $\mathbf{r}$ in general.]

Since $\mathcal{H}^{(0)}$ is independent of t, there do exist stationary-state wave functions for the zeroth-order Hamiltonian $\mathcal{H}^{(0)}$. These functions satisfy the time-dependent Schroedinger equation for $\mathcal{H}^{(0)}$ and can be separated as in Eq. (5.1). Thus we can write a time-independent equation for $\mathcal{H}^{(0)}$,

$$\mathcal{H}^{(0)}\psi_E(\mathbf{r}) = E\psi_E(\mathbf{r}), \tag{5.4}$$

where $\psi_E(\mathbf{r})$ is an eigenfunction of the zeroth-order Hamiltonian $\mathcal{H}^{(0)}$ corresponding to eigenvalue E.[1] Suppose further that we can solve Eq. (5.4) for the zeroth-order wave functions and energies.

Of course, this does not answer the question of how to solve the time-dependent equation (5.2) for the full wave functions $\Psi(\mathbf{r}, t)$. We shall proceed in a manner that should be familiar by now. We expand the unknown function $\Psi(\mathbf{r}, t)$ in a complete set of known functions of $\mathbf{r}$ and t, substitute the expansion into the equation satisfied by $\Psi(\mathbf{r}, t)$, and try to obtain expressions for the expansion coefficients.

To see how this process goes, we must first choose a complete set in which to expand $\Psi(\mathbf{r}, t)$. The set of eigenfunctions of $\mathcal{H}^{(0)}$ is complete in $(\mathbf{r})$ and is particularly convenient for a perturbation approach. By introducing coefficients $a_j(t)$, which depend on the time t, we can write the expansion

$$\Psi(\mathbf{r}, t) = \sum_{j=1}^{\infty} a_j(t)\psi_j(\mathbf{r})e^{-i(E_j/\hbar)t}, \tag{5.5}$$

where the subscript j represents a set of quantum numbers that uniquely identifies a state of the unperturbed system. The nonstationary character of $\Psi(\mathbf{r}, t)$ is reflected in this expansion. The state $\Psi(\mathbf{r}, t)$ can be viewed as a mixture of stationary states; however the "mixing coefficients" $a_j(t)$ depend on time.

Transitions

The next step is to substitute this expansion into the time-dependent Schroedinger equation and solve for the coefficients. Before doing so, let us look more closely at $a_j(t)$ and try to discover its physical significance. Suppose that we consider a whole ensemble of identical systems (see footnote 9 of Chapter 1), each of which is known to be in the kth stationary state at $t = 0$ (before the perturbation is "turned on"). This means that, in the expansion of Eq. (5.5), at $t = 0$ the kth coefficient $a_k(0)$ is one and all the others $a_j(0)$, $j \neq k$, are zero; that is,

$$a_j(0) = \delta_{jk}. \tag{5.6}$$

For $t > 0$, the perturbation is "turned on" and, at a later time, "turned off" again. At some time t_1 after the perturbation has been turned off, the state of each system will be a complicated linear combination expressed by

[1]We shall not bother to write superscript (0) on the zeroth-order eigenfunctions and energies in this chapter. [Remember that once we introduce the time-dependent perturbation $\mathcal{H}^{(1)}(t)$, we no longer even talk about eigenfunctions and eigenvalues of the Hamiltonian of the system.] Note also that we shall use $\mathbf{r}$ for the spatial dependence of the wave functions. The techniques of this section can be generalized to systems of more than one particle, in which case the spatial coordinate dependence is more complicated.

Eq. (5.5). Thus, in general, many of the coefficients $a_j(t_1)$ are nonzero. The probability that a system will be found in the qth stationary state at time t_1 is

$$P_{k\to q}(t_1) = \left| \int \psi_q^*(\mathbf{r}) e^{i(E_q/\hbar)t_1} \Psi(\mathbf{r}, t_1)\, d\mathbf{r} \right|^2. \tag{5.7}$$

From Eq. (5.5) and the orthogonality of the eigenfunctions $\psi_q(\mathbf{r})$, we see that

$$P_{k\to q}(t_1) = |a_q(t_1)|^2. \tag{5.8}$$

$P_{k\to q}(t_1)$ is the fraction of systems in the ensemble that would be found in state Ψ_q at time t_1; thus the probability that a transition has taken place from initial state k to final state q in time t_1 is $|a_q(t_1)|^2$. Since (in general) many of the coefficients $a_j(t_1)$ are nonzero, there will be nonzero probabilities of transition from the kth state to several other final states.

Solution of the Time-Dependent Equation

Returning to the time-dependent Schroedinger equation (5.2), we determine the coefficients by substituting the expansion (5.5) for $\Psi(\mathbf{r}, t)$; we obtain

$$[\mathcal{H}^{(0)} + \mathcal{H}^{(1)}(t)] \sum_{j=1}^{\infty} a_j(t) e^{-i\omega_j t} \psi_j(\mathbf{r}) = i\hbar \frac{\partial}{\partial t} \sum_{j=1}^{\infty} a_j(t) e^{-i\omega_j t} \psi_j(\mathbf{r}), \tag{5.9}$$

where we have defined an angular frequency ω_j by

$$\omega_j = \frac{E_j}{\hbar}. \tag{5.10}$$

We now multiply Eq. (5.9) by $\psi_q^*(\mathbf{r}) e^{+i\omega_q t}$, integrate over $d\mathbf{r}$, and use the time-independent zeroth-order Schroedinger equation (5.4) to derive an expression for the change in the qth coefficient $a_q(t)$ with time:

$$\frac{d}{dt} a_q(t) = \frac{1}{i\hbar} \sum_{j=1}^{\infty} a_j(t) e^{-i(\omega_j - \omega_q)t} H_{qj}^{(1)}(t), \qquad q = 1, 2, \ldots, \tag{5.11}$$

where $H_{qj}^{(1)}(t)$ is the matrix element of the perturbation Hamiltonian taken between unperturbed states q and j,

$$H_{qj}^{(1)}(t) \equiv \langle \psi_q(\mathbf{r}) | \mathcal{H}^{(1)}(\mathbf{r}, t) | \psi_j(\mathbf{r}) \rangle, \tag{5.12}$$

or

$$H_{qj}^{(1)}(t) = \int \psi_q^*(\mathbf{r}) \mathcal{H}^{(1)}(\mathbf{r}, t) \psi_j(\mathbf{r})\, d\mathbf{r}. \tag{5.13}$$

There are an infinite number of equations in the set (5.11), one for each $q = 1, 2, 3, \ldots$. Thus we have an infinite set of coupled, first-order differential equations that must be solved simultaneously for the coefficients $a_q(t)$, $q = 1, 2, 3, \ldots$; notice that the matrix elements $H_{qj}^{(1)}$ couple the qth and jth equations. How can we possibly solve such equations?

Exercise 5.1 Derive Eq. (5.11) from (5.9).

A first step is to convert Eq. (5.11) to a set of coupled integral equations. This may be accomplished simply by integrating Eq. (5.11) over time from 0 to t:

$$a_q(t) = -\frac{i}{\hbar}\sum_{j=1}^{\infty}\left\{\int_0^t a_j(t)e^{-i(\omega_j-\omega_q)t}H_{qj}^{(1)}(t)\,dt\right\} + c_q, \quad q = 1, 2, \ldots. \tag{5.14}$$

In Eq. (5.14) we have introduced a constant of integration, c_q, which can be determined by the initial conditions of each particular problem to which we apply these results. This set of equations is an exact formal solution of the time-dependent Schroedinger equation—no approximations yet. However, it has several drawbacks. First, there are an infinite number of equations in the set, each of which contains the sum of an infinite number of integrals. Clearly, solving these equations will not be easy unless they can be simplified in some way. Second, to solve for the qth coefficient $a_q(t)$, we must already know all the coefficients, including $a_q(t)$, since they appear in the sum on the right-hand side of (5.14).

Small Perturbations

To resolve this dilemma, we approximate. In particular, suppose that the perturbation $\mathcal{H}^{(1)}(t)$ is small. Then the rate of change of each coefficient with time will also be small [see Eq. (5.11)]. If the system is initially in the kth state, at $t = 0$, then $a_k(0) = 1$ and $a_q(0) = 0$ for all $q \neq k$. At some t a short time later, none of the coefficients will have changed much, so $a_k(t) \simeq 1$ will be the largest, and the superposition of stationary states making up $\Psi(\mathbf{r}, t)$ [Eq. (5.5)] will be dominated by the kth state. The simplest approximation we could make would be to set

$$a_j(t) \simeq \delta_{jk} \qquad \text{for small } t > 0,$$

so that

$$\Psi(\mathbf{r}, t) \simeq \psi_k(\mathbf{r})e^{-i(E_k/\hbar)t} \qquad \text{for small } t > 0.$$

This zeroth-order approximation is actually very crude; in effect, it ignores the perturbation. A more reasonable set of coefficients can be obtained by substituting these zeroth-order coefficients into the right-hand side of each of Eqs. (5.14) and evaluating the integrals. In this way, we obtain the first-order coefficients satisfying the initial conditions of Eq. (5.6):

$$a_k(t) \simeq 1 - \frac{i}{\hbar}\int_0^t H_{kk}^{(1)}(t)\,dt, \qquad t \geq 0. \tag{5.15a}$$

$$a_q(t) \simeq -\frac{i}{\hbar}\int_0^t e^{-i(\omega_k-\omega_q)t}H_{qk}^{(1)}(t)dt, \qquad q \neq k, \tag{5.15b}$$

Given a particular perturbation $\mathcal{H}^{(1)}(t)$, we can evaluate these coefficients and substitute them into the expansion (5.5) to obtain the first-order perturbed wave function

$$\Psi(\mathbf{r}, t) = \sum_{j=1}^{\infty} a_j(t)\psi_j(\mathbf{r})e^{-i(E_j/\hbar)t}.$$

Higher-order approximations for $\Psi(\mathbf{r}, t)$ can be derived if absolutely necessary by continuing this iterative process.

Equations (5.15) are the principal results of time-dependent perturbation theory. To see how to apply this method to an actual problem, we shall now return to the hydrogen atom and discover what effect a time-dependent electromagnetic field will have on it.

5.2 THE ELECTROMAGNETIC RADIATION FIELD

A highly accurate and precise treatment of the problem of a hydrogen atom in a time-dependent electromagnetic field can be carried out quantum mechanically by quantizing the field itself; this process calls for the beautiful but awesome formalism of *quantum electrodynamics*.[2] In most cases, however, the errors made in not quantizing the field are exceedingly small. Moreover, a better physical understanding of the processes involved may be obtained by treating the field classically, as we shall do here. The atom, of course, will be treated quantum mechanically.

Thus we shall first write down the perturbation Hamiltonian $\mathcal{H}^{(1)}(t)$ due to an electromagnetic field interacting with an atom and then use time-dependent perturbation theory to expand $\Psi(\mathbf{r}, t)$ in the stationary states of an isolated hydrogenic atom,[3] $\psi_{n\ell m}(\mathbf{r})e^{-i(E_n/\hbar)t}$, and to obtain first-order expressions for the expansion coefficients.

[2]See, for example, R. P. Feynman, *Quantum Electrodynamics* (New York: W. A. Benjamin, 1962).

[3]Since the zeroth-order Hamiltonian will be chosen to be that of the isolated hydrogen atom, the eigenfunctions of $\mathcal{H}^{(0)}$ will be labeled with the hydrogenic quantum numbers. Thus in this section n is the principal quantum number, ℓ is the orbital angular momentum quantum number, and m is the magnetic quantum number.

Far from its source, a classical electromagnetic field consists of perpendicular field components **E** and **B** and propagates as a plane wave through space in a direction **k** perpendicular to both **E** and **B** (see the Suggested Readings at the end of this chapter). Moreover, the effects due to the magnetic field components of the radiation field are usually much smaller than those of the electric field components and can be neglected.

Thus let us take as our field a time-dependent monochromatic[4] electric field propagating in the direction **k**. Then the field at any point in space **R** can be written

$$\mathbf{E}(\mathbf{R}, t) = E_x(\mathbf{R}, t)\hat{\mathbf{x}} + E_y(\mathbf{R}, t)\hat{\mathbf{y}} + E_z(\mathbf{R}, t)\hat{\mathbf{z}}, \tag{5.16a}$$

where each component is of the form

$$E_x(\mathbf{R}, t) = E_x^0 e^{i\mathbf{k}\cdot\mathbf{R}} e^{i\omega t} + E_x^0 e^{-i\mathbf{k}\cdot\mathbf{R}} e^{-i\omega t}, \tag{5.16b}$$

where we take E_x^0 to be real, and so on for $E_y(\mathbf{R}, t)$ and $E_z(\mathbf{R}, t)$. The (peak to zero) amplitude of oscillation is $2E_x^0$, the frequency of the radiation is ω, and the *wave number* k is related to the wavelength of the radiation by

$$k = \frac{2\pi}{\lambda}. \tag{5.17}$$

Since $E_x = 2E_x^0 \cos(\mathbf{k}\cdot\mathbf{R} + \omega t)$, the field oscillates at frequency ω. We refer to $\mathbf{E}(\mathbf{R}, t)$ as a harmonic perturbation (see Sec. 5.3). The point **R** is defined with respect to the source of the radiation.

An Atom in the Field

Now consider a hydrogen atom located a fixed distance R_0 from the source, as shown in Fig. 5.1. The interaction of this atom with the field $\mathbf{E}(\mathbf{R}, t)$ will

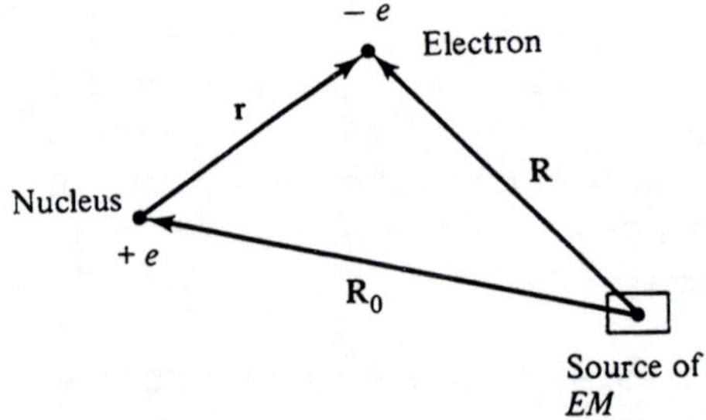

Figure 5.1 Symbolic sketch of a hydrogen atom in the presence of an external electromagnetic (*EM*) field.

[4]A *monochromatic wave* is one that oscillates at a single frequency. See also Sec. 5.4.

give rise to an additional term in the Hamiltonian of the form [see Eq. (4.95)]

$$\mathcal{H}^{(1)}(t) = -\mathbf{d}\cdot\mathbf{E}(\mathbf{R}, t), \tag{5.18}$$

where $\mathbf{d}$ is the electric dipole moment operator

$$\mathbf{d} \equiv -e\mathbf{r}. \tag{5.19}$$

We must be careful about coordinates here; there are three vectors of interest: $\mathbf{r}$, the position of the electron relative to the nucleus, $\mathbf{R}_0$, the position of the nucleus relative to the source of the radiation field, and $\mathbf{R}$, the position of the electron relative to the source. These vectors are drawn in Fig. 5.1. Notice that

$$\mathbf{R} = \mathbf{R}_0 + \mathbf{r}. \tag{5.20}$$

The full Hamiltonian of the system is

$$\mathcal{H} = \mathcal{H}^{(0)} + \mathcal{H}^{(1)}(t), \tag{5.21}$$

or

$$\mathcal{H} = -\frac{\hbar^2}{2\mu}\nabla^2 - \frac{e^2}{r} + e\mathbf{r}\cdot\mathbf{E}(\mathbf{R}, t). \tag{5.22}$$

The Hamiltonian of the isolated atom, $\mathcal{H}^{(0)}$, is independent of time; but since $\mathcal{H}^{(1)}$ explicitly contains t, we shall have to solve the time-dependent Schroedinger equation (5.2) for wave functions of the system.

Since we know the solutions of the time-independent equation for $\mathcal{H}^{(0)}$,

$$\mathcal{H}^{(0)}\psi_{n\ell m}(\mathbf{r}) = E_n\psi_{n\ell m}(\mathbf{r}), \tag{5.23}$$

from Chapter 3, we can use them in a time-dependent perturbation theory calculation of $\Psi(\mathbf{r}, t)$ provided that $\mathcal{H}^{(1)}$ is small. Let us suppose that such is the case.[5] Thus we write the expansion[6]

$$\Psi(\mathbf{r}, t) = \sum_{n'\ell'm'} a_{n'\ell'm'}(t)\psi_{n'\ell'm'}(\mathbf{r})e^{-i(E_{n'}/\hbar)t}, \tag{5.24}$$

[5]For all cases of conventional atomic spectroscopy, the magnitude of $\mathcal{H}^{(1)}$ is small and the use of perturbation theory is justified. With the advent of lasers, however, spectroscopy using high-intensity light beams has become possible. Such beams are composed of electromagnetic waves having large electric fields, in which case perturbation theory may fail. We shall not be concerned with the laser problem at present. See Richard H. Pantell and Harold E. Puthoff, *Fundamentals of Quantum Electronics* (New York: Wiley, 1969), Chap. 4.

[6]Usually we shall denote the initial state of the atom by $n\ell m$. For example, if the atom is in the ground state before the perturbation is "turned on," then $n = 1$, $\ell = 0$, $m = 0$. Other states will be denoted by $n'\ell'm'$; for example, the $2p_1$ state has $n' = 2$, $\ell' = 1$, $m' = +1$. Of course, the sum in Eq. (5.24) includes the initial state as well as all the excited states.

where the sum is taken over all stationary states[7] of the hydrogen atom $(n'\ell'm')$. We shall obtain first-order expressions for the coefficients $a_{n'\ell'm'}(t)$ by using (5.15). Let us assume that the atom is initially in state $(n\ell m)$; then we have

$$a_{n\ell m}(t) \simeq 1 - \frac{i}{\hbar}\int_0^t H^{(1)}_{n\ell m,n\ell m}(t)\,dt, \tag{5.25a}$$

and

$$a_{n'\ell'm'}(t) \simeq -\frac{i}{\hbar}\int_0^t e^{-i(\omega_n-\omega_{n'})t}H^{(1)}_{n'\ell'm',n\ell m}(t)\,dt, \qquad (n'\ell'm') \neq (n\ell m), \tag{5.25b}$$

where $\omega_{n'} = E_{n'}/\hbar$ and $\omega_n = E_n/\hbar$. Clearly, we cannot proceed further until we have expressions for the matrix elements of $\mathcal{H}^{(1)}$ that appear in these equations—namely,

$$H^{(1)}_{n'\ell'm',n\ell m}(t) = \langle\psi_{n'\ell'm'}(\mathbf{r})|\,\mathcal{H}^{(1)}(\mathbf{r},t)\,|\psi_{n\ell m}(\mathbf{r})\rangle.$$

Using Eqs. (5.18) and (5.19), we can write $H^{(1)}_{n'\ell'm',n\ell m}$ as

$$H^{(1)}_{n'\ell'm',n\ell m} = \int \psi^*_{n'\ell'm'}(\mathbf{r})[e\mathbf{r}\cdot\mathbf{E}(\mathbf{R},t)]\psi_{n\ell m}(\mathbf{r})\,d\mathbf{r}. \tag{5.26}$$

Since

$$e\mathbf{r}\cdot\mathbf{E}(\mathbf{R},t) = exE_x(\mathbf{R},t) + eyE_y(\mathbf{R},t) + ezE_z(\mathbf{R},t), \tag{5.27}$$

the matrix element will consist of three terms, one for each term in Eq. (5.27). Consider the x term,

$$H^{(1)}_x = eE^0_x\int \psi^*_{n'\ell'm'}(\mathbf{r})[xe^{i\mathbf{k}\cdot\mathbf{R}}e^{i\omega t} + xe^{-i\mathbf{k}\cdot\mathbf{R}}e^{-i\omega t}]\psi_{n\ell m}(\mathbf{r})\,d\mathbf{r}. \tag{5.28}$$

A glance at Fig. 5.1 reminds us that $\mathbf{R} = \mathbf{R}_0 + \mathbf{r}$; hence the factors $e^{\pm i\mathbf{k}\cdot\mathbf{R}}$ cannot be removed from under the integral.

Spatial Variation of the Field

Digressing briefly, we shall take advantage of one further simplification in order to evaluate these matrix elements. Let us concentrate on the first term in Eq. (5.28), in which we can write $e^{i\mathbf{k}\cdot\mathbf{R}}$ as

$$e^{i\mathbf{k}\cdot\mathbf{R}} = e^{i\mathbf{k}\cdot\mathbf{R}_0}e^{i\mathbf{k}\cdot\mathbf{r}}. \tag{5.29}$$

In the matrix elements, this term is multiplied by the bound-state wave functions of the hydrogen atom $\psi^*_{n'\ell'm'}(\mathbf{r})$ and $\psi_{n\ell m}(\mathbf{r})$. One of the properties

[7]This sum should contain integration over continuum states as well. However, we shall restrict ourselves explicitly to bound states here.

of these functions is that they die away exponentially to zero for large r (see Fig. 3.2). Hence the contribution to these integrals from large r, say $r \gg \langle r \rangle_{n\ell m}$, will be quite small. We shall use the expectation value $\langle r \rangle_{n\ell m}$ to characterize roughly the dimensions of the atom [see Eq. (3.67)].

The point of making this argument is that if the wavelength of the incident radiation λ $(= 2\pi/k)$ is much larger than the dimensions of the atom; that is, if

$$\lambda \gg \langle r \rangle_{n\ell m}, \tag{5.30}$$

then, to a good approximation, we can neglect all but the first term in an expansion[8] of $e^{i\mathbf{k}\cdot\mathbf{r}}$,

$$e^{i\mathbf{k}\cdot\mathbf{r}} = 1 + i\mathbf{k}\cdot\mathbf{r} + \cdots, \tag{5.31}$$

or

$$e^{i\mathbf{k}\cdot\mathbf{r}} \simeq 1. \tag{5.32}$$

Let us see how good this approximation is by using as a reasonable guess $\langle r \rangle_{n\ell m} = 10a_0 = 5.29$ Å. Then Eq. (5.30) demands that $\lambda \gg 30$ Å. But the wavelengths normally used for spectroscopy are in the ultraviolet, visible, and infrared regions of the electromagnetic spectrum, usually of the order $\lambda \gtrsim 1000$ Å, a value that satisfies this condition.

Using this simplification in Eq. (5.28), the x term in the matrix element becomes

$$H_x^{(1)} = eE_x^0 \int \psi^*_{n'\ell'm'}(\mathbf{r})(xe^{i\mathbf{k}\cdot\mathbf{R}_0}e^{i\omega t} + xe^{-i\mathbf{k}\cdot\mathbf{R}_0}e^{-i\omega t})\psi_{n\ell m}(\mathbf{r})\, d\mathbf{r}, \tag{5.33}$$

or

$$\begin{aligned} H_x^{(1)} = eE_x^0\Big(&e^{i\mathbf{k}\cdot\mathbf{R}_0}e^{i\omega t}\int \psi^*_{n'\ell'm'}(\mathbf{r})x\psi_{n\ell m}(\mathbf{r})\, d\mathbf{r} \\ &+ e^{-i\mathbf{k}\cdot\mathbf{R}_0}e^{-i\omega t}\int \psi^*_{n'\ell'm'}(\mathbf{r})x\psi_{n\ell m}(\mathbf{r})\, d\mathbf{r}\Big). \end{aligned} \tag{5.34}$$

By making precisely the same arguments for the y and z terms in Eq. (5.26), we can write the matrix element $H^{(1)}_{n'\ell'm',n\ell m}(t)$ as

$$H^{(1)}_{n'\ell'm',n\ell m}(t) = -\mathbf{d}_{n'\ell'm',n\ell m}\cdot[\mathbf{E}^+e^{i\omega t} + \mathbf{E}^-e^{-i\omega t}], \tag{5.35}$$

where we have introduced the vector $\mathbf{E}^\pm$, defined as

$$\mathbf{E}^\pm \equiv e^{\pm i\mathbf{k}\cdot\mathbf{R}_0}[E_x^0\hat{\mathbf{x}} + E_y^0\hat{\mathbf{y}} + E_z^0\hat{\mathbf{z}}], \tag{5.36}$$

[8]For large λ, the wave number k will be small; so the higher terms—$i\mathbf{k}\cdot\mathbf{r}$, etc.—in the expansion will be small except for large r. But at large r, these terms are multiplied by zero—that is, by the function $\psi^*_{n'\ell'm'}(\mathbf{r})\psi_{n\ell m}(\mathbf{r})$.

and the matrix element of the dipole moment vector

$$\mathbf{d}_{n'\ell'm',n\ell m} \equiv \langle \psi_{n'\ell'm'}(\mathbf{r}) | \mathbf{d} | \psi_{n\ell m}(\mathbf{r}) \rangle \tag{5.37}$$

with $\mathbf{d} = -e\mathbf{r}$ as usual.

Exercise 5.2 Carry out the algebra leading to Eq. (5.35).

Let us summarize briefly what has been done so far. In order to obtain an approximate expression for the wave function $\Psi(\mathbf{r}, t)$ of a hydrogen atom in a small, time-varying, electromagnetic radiation field, we have had to evaluate matrix elements of the perturbation Hamiltonian $\mathcal{H}^{(1)}(t)$. By assuming that the wavelength of the incident radiation was much larger than the characteristic dimension of the atom, we were able to derive a fairly simple expression for the matrix elements. The remaining integral, $\mathbf{d}_{n'\ell'm',n\ell m}$, can easily be evaluated and the resulting matrix elements $H^{(1)}_{n'\ell'm',n\ell m}(t)$ substituted into Eqs. (5.25) to obtain the expansion coefficients $a_{n'\ell'm'}(t)$.

Selection Rules

We are especially interested in the transitions (between unperturbed states) caused by the external radiation field. In particular, we can use our results for the expansion coefficients $a_{n'\ell'm'}(t)$ to calculate the probability that a transition from initial state $(n\ell m)$ to final state $(n'\ell'm')$ will occur in time t—according to Eq. (5.8),

$$P_{n\ell m \to n'\ell'm'}(t) = |a_{n'\ell'm'}(t)|^2. \tag{5.38}$$

If the dipole matrix element $\mathbf{d}_{n'\ell'm',n\ell m}$ is zero for any final state $(n'\ell'm')$, then $H^{(1)}_{n'\ell'm',n\ell m}(t)$ and $a_{n'\ell'm'}(t)$ will be zero for any t, and the transition probability $P_{n\ell m \to n'\ell'm'}(t)$ will be zero. Consequently, the transition $(n\ell m) \longrightarrow (n'\ell'm')$ can never take place; it is a *forbidden transition*. In fact, unless the initial and final quantum numbers satisfy some very specific conditions, the transition $(n\ell m) \longrightarrow (n'\ell'm')$ will be forbidden. Such a restriction on initial and final quantum numbers is called a *selection rule*.

To illustrate, let us explicitly write out the matrix element $\mathbf{d}_{n'\ell'm',n\ell m}$:

$$\mathbf{d}_{n'\ell'm',n\ell m} = -e \int_0^\infty \int_0^{2\pi} \int_0^\pi \psi^*_{n'\ell'm'}(\mathbf{r})\mathbf{r}\psi_{n\ell m}(\mathbf{r}) \sin\theta \, d\theta \, d\varphi \, r^2 \, dr. \tag{5.39}$$

It is convenient at this point to make some rather obvious definitions:

$$(d_x)_{n'\ell'm',n\ell m} \equiv -e \int_0^\infty \int_0^{2\pi} \int_0^\pi \psi^*_{n'\ell'm'}(\mathbf{r}) x \psi_{n\ell m}(\mathbf{r}) \sin\theta \, d\theta \, d\varphi \, r^2 \, dr, \tag{5.40}$$

and so on for $(d_y)_{n'\ell'm',n\ell m}$ and $(d_z)_{n'\ell'm',n\ell m}$. We can then use the properties of the spherical harmonics (e.g., orthogonality), familiar from Chapter 2, to derive the following selection rules[9] for dipole transitions:

$$\left.\begin{aligned}(d_x)_{n'\ell'm',n\ell m} &= 0\\ (d_y)_{n'\ell'm',n\ell m} &= 0\end{aligned}\right\}\ \text{unless}\ \begin{cases}\Delta m = \pm 1\\ \quad\text{and}\\ \Delta\ell = \pm 1,\end{cases}$$

$$(d_z)_{n'\ell'm',n\ell m} = 0\ \ \text{unless}\ \begin{cases}\Delta m = 0\\ \quad\text{and}\\ \Delta\ell = \pm 1,\end{cases} \tag{5.41}$$

where

$$\Delta m \equiv m' - m \quad \text{and} \quad \Delta\ell \equiv \ell' - \ell. \tag{5.42}$$

We therefore conclude that the transition probability is zero unless

$$\boxed{\Delta m = 0, \pm 1 \quad \text{and} \quad \Delta\ell = \pm 1.} \quad \text{Electric dipole selection rules} \tag{5.43}$$

Exercise 5.3 Derive Eqs. (5.41).

The point to be remembered is that for any set of quantum numbers $(n\ell m)$ and $(n'\ell'm')$ violating these rules, the coefficient $a_{n'\ell'm'}(t)$ is zero. Then the state $(n'\ell'm')$ will not contribute to the superposition of states making up the perturbed state with wave function $\Psi(\mathbf{r}, t)$. Moreover, a transition $(n\ell m) \longrightarrow (n'\ell'm')$ cannot take place, since $P_{n\ell m\to n'\ell'm'} = 0$. Thus, for example, transitions $(n \ 0 \ 0) \longrightarrow (n' \ 1 \ 0)$ and $(n \ 2 \ 1) \longrightarrow (n' \ 3 \ 2)$ are allowed, but $(n \ 1 \ 0) \longrightarrow (n' \ 3 \ 0)$ is forbidden.[10] Transitions that are allowed by the selection rules of Eq. (5.43) are called *electric dipole transitions.*

Selection rules have important consequences for atomic spectroscopy. Suppose that a transition $(n\ell m) \longrightarrow (n'\ell'm')$ is allowed and takes place.

[9]You will also need the convenient identities

$$\sin\theta P_\ell^{m-1}(\cos\theta) = \frac{1}{2\ell+1}[P_{\ell+1}^m(\cos\theta) - P_{\ell-1}^m(\cos\theta)], \qquad m \geq 1,$$

$$\cos\theta P_\ell^m(\cos\theta) = \frac{\ell-m-1}{2\ell+1}P_{\ell+1}^m(\cos\theta) + \frac{\ell+m}{2\ell+1}P_{\ell-1}^m(\cos\theta), \qquad m \geq 0.$$

See E. U. Condon and G. H. Shortley, *The Theory of Atomic Spectra* (New York: Cambridge University Press, 1951), Chap. 3.

[10]In some cases, the electric field may only have a component in the z direction; that is, $E_x^0 = 0$ and $E_y^0 = 0$. (We say that the field is *polarized* along the $\hat{z}$ axis.) If so, then only the z component of $\mathbf{d}_{n'\ell'm',n\ell m}$ contributes to $a_{n'\ell'm'}(t)$ and we need only consider selection rules for this component. Thus if the radiation field is polarized along the $\hat{z}$ axis, the transitions from a state with magnetic quantum number m to one with quantum number $m' = m \pm 1$ are forbidden.

Moreover, suppose that $E_n > E_{n'}$, so that the atom ends up in a state of lower energy than its initial state. Then the energy lost by the atom is *emitted* into the radiation field with frequency (angular) $\omega_{nn'} = (1/\hbar)(E_n - E_{n'})$; the emitted radiation will show up in an experimental measurement. However, if the quantum numbers $(n\ell m)$ and $(n'\ell' m')$ violate Eqs. (5.43), we know that the transition cannot occur and that there is no need to look for evidence of it.

These remarks do not mean that we never observe the so-called *forbidden transitions*. If some of the restrictions under which we obtained Eq. (5.43) are relaxed—for example, if we allow for a slight r-dependence in $\mathbf{E}^{\pm}$—the result is a small probability for a transition of the type $\Delta\ell = \pm 2$, a so-called *electric quadrupole transition* (see Prob. 5.2). There also exists a very weak interaction between the atom and the magnetic field of the incident electromagnetic wave, and selection rules can be derived for *magnetic dipole transitions*, and so on. However, the probabilities for such transitions are usually very small.

Although the results obtained are important, they do not help us actually to evaluate the nonzero coefficients $a_{n'\ell'm'}(t)$. Let us now return to that task.

5.3 HARMONIC PERTURBATIONS

This is a convenient place to remark that the perturbation being considered, $\mathcal{H}^{(1)} = -\mathbf{d}\cdot\mathbf{E}(\mathbf{r}, t)$, is one of a class of perturbations called *harmonic perturbations*. A harmonic perturbation has the general form

$$\mathcal{H}^{(1)}(\mathbf{r}, t) = \mathcal{H}^{+}(\mathbf{r})e^{i\omega t} + \mathcal{H}^{-}(\mathbf{r})e^{-i\omega t}; \tag{5.44}$$

that is, it is composed of terms that oscillate at frequency ω. In the example of this chapter, we have

$$\mathcal{H}^{\pm}(\mathbf{r}) = -\mathbf{d}\cdot\mathbf{E}^{\pm}, \tag{5.45}$$

where $\mathbf{E}^{\pm} = e^{\pm i\mathbf{k}\cdot\mathbf{R}_0}\,[E_x^0\hat{\mathbf{x}} + E_y^0\hat{\mathbf{y}} + E_z^0\hat{\mathbf{z}}]$ as in Eq. (5.36).

Let us evaluate the probability of a transition $(n\ell m) \longrightarrow (n'\ell'm')$ having occurred by time t due to a general harmonic perturbation. This quantity is simply $|a_{n'\ell'm'}(t)|^2$, where, from Eq. (5.25b), the coefficient, to first order, is

$$a_{n'\ell'm'}(t) \simeq -\frac{i}{\hbar}\int_0^t [H^{+}_{n'\ell'm',n\ell m}e^{-i(\omega_n-\omega_{n'}-\omega)t} + H^{-}_{n'\ell'm',n\ell m}e^{-i(\omega_n-\omega_{n'}+\omega)t}]\,dt. \tag{5.46}$$

Remember that $\omega_n = E_n/\hbar$, $\omega_{n'} = E_{n'}/\hbar$, and ω is the frequency of the incident radiation. The matrix elements, of course, are

$$H^{\pm}_{n'\ell'm',n\ell m} = \langle\psi_{n'\ell'm'}(\mathbf{r})|\,\mathcal{H}^{\pm}(\mathbf{r})\,|\psi_{n\ell m}(\mathbf{r})\rangle.$$

Since the $H^{\pm}_{n'\ell'm',n\ell m}$ are independent of time, we can remove them from under the integral. Carrying out the remaining simple integration, we obtain

$$a_{n'\ell'm'}(t) \simeq \frac{H^{+}_{n'\ell'm',n\ell m}}{E_n - E_{n'} - \hbar\omega}[e^{-i(\omega_n - \omega_{n'} - \omega)t} - 1] + \frac{H^{-}_{n'\ell'm',n\ell m}}{E_n - E_{n'} + \hbar\omega}[e^{-i(\omega_n - \omega_{n'} + \omega)t} - 1]. \tag{5.47}$$

Since $a_{n'\ell'm'}(t)$ depends on the frequency of the incident electromagnetic radiation ω, the transition probability $P_{n\ell m \to n'\ell'm'}$ also depends on ω. It attains its maximum value at the frequency $\omega = \omega_{nn'}$, where

$$\omega_{nn'} \equiv \frac{1}{\hbar}|E_n - E_{n'}|. \tag{5.48}$$

Thus a transition between an initial state $(n\ell m)$ and any final state $(n'\ell'm')$ will be most likely to occur if the frequency of the radiation field (times $\hbar$) is equal to the energy separation of the two states.

Notice that if $E_n > E_{n'}$ so that the initial state is higher in energy than the final state, the first term in Eq. (5.47) dominates; then $a_{n'\ell'm'}(t)$ is the probability amplitude for a de-excitation process accompanied by the *emission* of electromagnetic radiation. If $E_n < E_{n'}$ the second term dominates; then $a_{n'\ell'm'}(t)$ describes an excitation process accompanied by the *absorption* of energy from the electromagnetic field. In other words, transitions are preferentially induced in which a quantum of energy $\hbar\omega$ is either emitted from the radiation field or absorbed into the field. The first case is called *stimulated emission* and the latter is called *absorption*. The probabilities for these two particularly likely transitions are

$$P_{n\ell m \to n'\ell'm'}(t) \simeq \frac{|H^{\mp}_{n'\ell'm',n\ell m}|^2 \sin^2[\frac{1}{2}(\omega_n - \omega_{n'} \pm \omega)t]}{\hbar^2[\frac{1}{2}(\omega_n - \omega_{n'} \pm \omega)]^2}, \tag{5.49}$$

where the upper signs refer to the case $E_n < E_{n'}$ (absorption) and the lower signs to $E_n > E_{n'}$ (stimulated emission).

Exercise 5.4 (a) Derive Eqs. (5.47) and (5.49).
(b) Show that $P_{n\ell m \to n'\ell'm'}$ attains its maximum for $\omega = \omega_{nn'}$.

Substitution of $\mathcal{H}^{\pm} = -\mathbf{d} \cdot \mathbf{E}^{\pm}$ from Eq. (5.45) into (5.47) yields the desired expression for the expansion coefficients $a_{n'\ell'm'}(t)$ for a hydrogen atom perturbed by a time-varying electromagnetic field. Remember that this step must be done only if the selection rules (5.43) are satisfied; otherwise $a_{n'\ell'm'} = 0$. The transition probability, in this case, is

$$P_{n\ell m \to n'\ell'm'}(t) = \frac{|\mathbf{d}_{n'\ell'm',n\ell m} \cdot \mathbf{E}^0|^2}{\hbar^2} \frac{\sin^2[\frac{1}{2}(\omega_n - \omega_{n'} \pm \omega)t]}{[\frac{1}{2}(\omega_n - \omega_{n'} \pm \omega)]^2}, \tag{5.50}$$

where, as before, the upper sign refers to absorption and the lower sign to emission. The dipole matrix element $\mathbf{d}_{n'\ell'm',n\ell m}$ was defined in Eq. (5.39), and $2\mathbf{E}_0$ is simply the vector amplitude of the wave, where

$$\mathbf{E}^0 \equiv E_x^0\hat{\mathbf{x}} + E_y^0\hat{\mathbf{y}} + E_z^0\hat{\mathbf{z}}. \tag{5.51}$$

Thus, by application of time-dependent perturbation theory to the problem of a hydrogen atom in an external electromagnetic field, we have been able to derive expressions for transition probabilities between unperturbed (stationary) states of the atom. We have seen that these probabilities depend on the frequency ω of the incident radiation. Given a frequency and a set of levels, we could compute the probabilities and compare them to the results of experimental observations.

*5.4 FREQUENCY BANDS

The electromagnetic wave $\mathbf{E}(\mathbf{R}, t)$ considered in the last two sections was monochromatic; that is, it oscillated at a single frequency. In most experiments, however, the incident electromagnetic radiation used to perturb atoms includes waves that cover a range or *band* of frequencies. (The spread of frequencies in the radiation, usually specified by the *width* of the band, depends on the nature of the source.) Thus it is worth taking a moment to see how to modify the results of Sec. 5.3 so as to take into account frequency bands of polychromatic (as opposed to monochromatic) radiation.

Consider the case of *incoherent* electromagnetic radiation; that is, suppose that waves of different frequency have no particular phase relationship with one another and thus do not interfere. Then we can assume that each wave interacts with the hydrogen atom independently of all the other waves; we shall apply the theory developed in the last section to each wave separately and add up the resulting probabilities at the end.

We can still write the electric field vector as in Eq. (5.16) if we let the amplitude $\mathbf{E}^0$ be a function of frequency ω. Since we are no longer dealing with a single frequency, we must reinterpret the notion of a transition probability. For a system perturbed by a collection of incoherent polychromatic waves, we obtain the probability for a transition from state $(n\ell m)$ to $(n'\ell'm')$ by summing over all frequencies present in the collection of waves,

$$P_{n\ell m\to n'\ell'm'}(t) = \sum_{\omega} |a_{n'\ell'm'}(t)|^2, \tag{5.52}$$

where

$$|a_{n'\ell'm'}(t)|^2 = \frac{|\mathbf{d}_{n'\ell'm',n\ell m}\cdot\hat{\mathbf{e}}|^2\,|E^0(\omega)|^2}{\hbar^2}\,\frac{\sin^2[\frac{1}{2}(\omega_n - \omega_{n'} \pm \omega)t]}{[\frac{1}{2}(\omega_n - \omega_{n'} \pm \omega)]^2}. \tag{5.53}$$

Here $\hat{\mathbf{e}}$ is a unit vector along the direction of the amplitude $\mathbf{E}^0$; it is called the *polarization unit vector*. For plane-wave radiation, this vector is defined as

$$\hat{\mathbf{e}} \equiv \frac{\mathbf{E}^0}{|\mathbf{E}^0|}. \tag{5.54}$$

The *intensity* of the electromagnetic radiation in the frequency range ω to $\omega + d\omega$ may be written in terms of the electric field amplitude as

$$I(\omega)\,d\omega = \frac{c}{2\pi}|E^0(\omega)|^2, \tag{5.55}$$

and assuming a continuous distribution of frequencies to be present, the sum in Eq. (5.52) becomes a standard integral. Straightforward evaluation results in

$$P_{n\ell m \to n'\ell'm'}(t) = \frac{4\pi^2 t}{\hbar^2 c} I(\omega_{n'n})\,|\mathbf{d}_{n'\ell'm',n\ell m}\cdot\hat{\mathbf{e}}|^2, \tag{5.56}$$

where $\omega_{n'n} \equiv \omega_{n'} - \omega_n$ is the angular frequency corresponding to the energy separation of the two energy levels. Thus, according to Eq. (5.56), the transition probability, summed over a band of frequencies, depends linearly on the time. A constant *transition rate* can then be defined as

$$R_{n\ell m \to n'\ell'm'} = \frac{4\pi^2}{\hbar^2 c} I(\omega_{n'n})\,|\mathbf{d}_{n'\ell'm',n\ell m}\cdot\hat{\mathbf{e}}|^2. \tag{5.57}$$

Alternatively, it can be written in terms of the energy density of the electromagnetic waves by substituting $\rho(\omega) = I(\omega)/c$, and obtaining

$$R_{n\ell m \to n'\ell'm'} = \frac{4\pi^2}{\hbar^2} \rho(\omega_{n'n})\,|\mathbf{d}_{n'\ell'm',n\ell m}\cdot\hat{\mathbf{e}}|^2. \tag{5.58}$$

Note that $\rho(\omega_{n'n})$ is the *energy density* (in erg/cm^3/sec).

Equation (5.58) gives correctly the "rate" (i.e., probability per second) of "dipole" transitions $n\ell m \longrightarrow n'\ell'm'$ of an atom absorbing an amount of energy $|\hbar\omega_{n'n}|$ from (if $\omega_{n'} > \omega_n$) or emitting this amount of energy into (if $\omega_{n'} < \omega_n$) the electromagnetic field. The latter case is called *stimulated emission.* The rate depends on both the intensity of the electromagnetic waves at frequency $\omega_{n'n}$ and the magnitude of the dipole matrix element coupling the two atomic states. If the dipole matrix element vanishes, then the rate given by Eq. (5.58) is zero. However, as discussed in Sec. 5.2, there are subtle corrections to the theory we have described, which, when properly included, "allow" weak transitions even when they are "dipole forbidden."

One very important phenomenon completely missing in the semiclassical theory of radiation is that of *spontaneous emission*—that is, the emission of an electromagnetic wave by an excited atom in empty space (even in the absence of any other source of electromagnetic radiation). In order to account for it properly, the quantum field theory of electrodynamics, which is not discussed in this book, must be used. However, this phenomenon can be studied somewhat indirectly by using thermodynamics (see Prob. 5.5).

The principal reasons for treating the time-dependent electromagnetic field were to illustrate the use of time-dependent perturbation theory and to satisfy our curiosity about the behavior of a hydrogen atom in such a field. The point to be retained is that this behavior is very different from the Stark effect of Sec. 4.7; the concept of stationary states loses its meaning when the full Hamiltonian depends on time. Instead we speak of transitions occurring between stationary states of the unperturbed system. Other examples of time-dependent perturbations can be found in the problems for this chapter.

SUGGESTED READINGS

Almost any undergraduate book on classical electricity and magnetism will contain the rudimentary aspects of the theory used in this chapter. Two useful references are

LORRAIN, P., and D. R. CORSON, *Electromagnetic Fields and Waves*, 2nd ed. San Francisco: W. H. Freeman, 1970.

PARIS, DEMETRIUS T., and F. KENNETH HURD, *Basic Electromagnetic Theory*. New York: McGraw-Hill, 1969.

At a somewhat more advanced level is the well-written text

EYGES, LEONARD, *The Classical Electromagnetic Field*. Reading, Mass: Addison-Wesley, 1972.

PROBLEMS

5.1 Switched-on Perturbation Theory (**)

Consider a small perturbation that is "switched on" at time $t = 0$ and then "switched off" at a later time t_1. Otherwise the perturbation Hamiltonian $\mathcal{H}^{(1)}$ is independent of time; that is,

$$\mathcal{H}^{(1)}(\mathbf{r}, t) = \begin{cases} 0, & t < 0 \\ \mathcal{H}^{(1)}(\mathbf{r}), & 0 \leq t \leq t_1 \\ 0, & t \geq t_1. \end{cases}$$

Suppose that the system is initially in the state labeled by quantum numbers k.

(a) Write an expression for the full wave function of the system (valid to first order) at a time $t > t_1$.

(b) Derive an analytic formula for the transition probability to a final state $n \neq k$.

(c) Consider now two perturbations of this form, one of which lasts for a time t_1, the other for a time $t_2 = 2t_1$. Sketch $P_{k\to n}(t)$ versus energy separation $E_n - E_k$ for the two cases. To which states are transitions most likely?

(d) Consider transitions from state k to a large number of closely spaced levels. If there are dN_ℓ states in a small energy interval dE about E_ℓ, then the total probability of a transition to all states having energy in an interval ΔE about E_ℓ is,

$$P_{k\to\ell}(t) = \int_{(\Delta E)} |a_\ell(t)|^2 \, dN_\ell.$$

Write an expression for $P_{k\to\ell}$ in terms of the *density of states* $\rho_\ell(E)$, where $\rho_\ell(E) \equiv dN_\ell/dE$. [HINT: The limits of integration can be extended to $-\infty$ and ∞. Why?] Use your sketch of part (c) to help you evaluate the integral and thus show that

$$P_{k\to\ell}(t_1) \simeq \frac{2\pi}{\hbar} |H^{(1)}_{\ell k}|^2 \rho_\ell(E) t_1,$$

and that the transition rate is

$$R_{k\to\ell} \simeq \frac{2\pi}{\hbar} |H^{(1)}_{\ell k}|^2 \rho_\ell(E).$$

This is a form of the *Golden Rule*.

5.2 The Electric Quadrupole Transitions (**)

In deriving expressions for the matrix elements of the perturbation due to the interaction of an electromagnetic field with a hydrogen atom, we made the approximation

$$e^{i\mathbf{k}\cdot\mathbf{r}} \simeq 1,$$

which is valid for radiation of moderately long wavelengths. This approximation led ultimately to the dipole selection rules that describe so-called allowed transitions [see Eq. (5.41)].

If the first correction term $i\mathbf{k}\cdot\mathbf{r}$ is kept in the expansion of the foregoing exponential, then the so-called *quadrupole transitions* are also seen to occur, but with much smaller probabilities. Consider the particular case of an electromagnetic wave propagating along the $\hat{x}$ axis and polarized along the $\hat{z}$ axis. Derive selection rules for a hydrogen atom transition ($n\ell m \longrightarrow n'\ell'm'$) "induced" by this correction term. Explain why the term "quadrupole transition" is appropriate.

5.3 A Square Well Perturbed by an Electric Field (**)

At time $t = 0$, an electron is known to be in the $n = 1$ eigenstate of a one-dimensional infinite square well potential

$$V(x) = \begin{cases} \infty, & |x| > \dfrac{a}{2} \\ 0, & -\dfrac{a}{2} < x < \dfrac{a}{2}. \end{cases}$$

At $t = 0$, a uniform electric field of magnitude E is applied in the direction of increasing x. This electric field is left on for a short time τ and then removed. Use time-dependent perturbation theory to calculate the probability that the electron will be in the $n = 2, 3$, or 4 eigenstates at some time $t > \tau$. [HINT: You will encounter integrals involving $\sin(n\pi x/a)$ and/or $\cos(n\pi x/a)$, where n is an integer. Integration by parts becomes easy if the following substitutions are made:

$$z = \frac{\pi x}{a}$$

$$\sin nz = \frac{1}{2i}(e^{inz} - e^{-inz})$$

$$\cos nz = \frac{1}{2}(e^{inz} + e^{-inz}).$$

5.4 Adiabatic Approximation (****)

Adiabatic behavior can perhaps best be described by considering a cat moving around in a room full of mice. As long as the cat moves at a velocity much smaller than that of the mice, the latter will keep out of the way. The "state" of the mice and their "spatial distribution" will change as the cat moves around; however, provided that the cat moves slowly enough, nothing really exciting will occur! We would say that the system of mice "responds adiabatically" to the cat's motion. At velocities comparable to the mice, the cat can cause a great deal of "excitement" and can even cause the "population number" to change—clearly nonadiabatic behavior.

In quantum physics, a convenient way to study adiabatic behavior is through time-dependent perturbation theory. We will derive a special approximation called the *adiabatic approximation*.

(a) Consider an arbitrary system with Hamiltonian $\mathcal{H}(t)$ that depends explicitly on the time. If $\mathcal{H}(t)$ is slowly varying with t, we expect the system to be well described by state functions $u_n(t)$, which satisfy

$$\mathcal{H}(t)u_n(t) = E_n(t)u(t),$$

where other variables are present but not indicated explicitly. Expand the exact wave function $\psi(t)$ in terms of the $u_n(t)$ as

$$\psi(t) = \sum_n a_n(t)u_n(t) \exp\left[-\frac{i}{\hbar}\int_0^t E_n(t')\,dt'\right]$$

and derive an expression for a_k in terms of $\omega_{kn}(t')$, where

$$\omega_{kn}(t') = \frac{1}{\hbar}[E_k(t') - E_n(t')]$$

and the dot implies first partial derivative with respect to t.

(b) By taking the first time derivative of $\mathcal{H}(t)u_n(t) = E_n(t)u_n(t)$, show that

$$\langle u_k | \dot{u}_n \rangle = -(\hbar\omega_{kn})^{-1} \left\langle u_k \left| \frac{\partial \mathcal{H}}{\partial t} \right| u_n \right\rangle, \qquad k \neq n.$$

Show also that the phase of the functions u_n may be chosen so that

$$\langle u_n | \dot{u}_n \rangle = 0.$$

Use these results in your answer to (a) to show that

$$\dot{a}_k = \sum_n{}' \frac{a_n}{\hbar\omega_{kn}} \left\langle u_k \left| \frac{\partial \mathcal{H}}{\partial t} \right| u_n \right\rangle \exp\left[i \int_0^t \omega_{kn}(t')\, dt' \right],$$

where the prime on the summation reminds us that $n \neq k$. This result is exact.

(c) Now consider the case in which $\mathcal{H}$ is slowly varying with time so that $\partial\mathcal{H}/\partial t$ is essentially constant. It is also reasonable to assume that in the expression for $\dot{a}_k$ all the other quantities on the right-hand side are constant. Assume that, at $t = 0$, $a_n = \delta_{nm}$ and show that

$$a_k \simeq (i\hbar\omega_{km}^2)^{-1} \left\langle u_k \left| \frac{\partial \mathcal{H}}{\partial t} \right| u_m \right\rangle (e^{i\omega_{km}t} - 1).$$

This is a form of the adiabatic approximation. Discuss the implications of this result. Does it require that $\mathcal{H}$ be small? Is it restricted to only small times $t > 0$? Why or why not? Make up a physical example where this approach might make sense (not the cat in a box of mice).

5.5 Einstein Coefficients (***)

In Sec. 5.4 it was shown that the transition rate $R_{n\ell m \to n'\ell'm'}$ is proportional to the average energy density evaluated at the frequency $\omega_{n'n}$. Let us define the constant of proportionality as $B_{n'n}$; that is, $R_{n\ell m \to n'\ell'm'} = B_{n'n}\rho(\omega_{n'n})$. $B_{n'n}$ is called the *coefficient of absorption* (for $E_{n'} > E_n$). The constant $B_{nn'}$ is called the *coefficient of stimulated emission.*

(a) Derive an expression for $B_{n'n}$ in terms of $\langle \psi_{n'\ell'm'} | \mathbf{r} | \psi_{n\ell m} \rangle$.

(b) Show that $B_{n'n} = B_{nn'}$.

(c) When the system (atom + field) is in thermal equilibrium, the number of photons absorbed per second from the field is equal to the number of photons emitted per second into it. To achieve equilibrium an additional coefficient, $A_{nn'}$, called the *coefficient of spontaneous emission,* must be introduced so that

$$R_{n' \to n} = A_{nn'} + B_{n'n}\rho(\omega_{n'n}).$$

Using the density of electromagnetic radiation in equilibrium with a blackbody

$$\rho(\omega_{n'n}) = \frac{8\pi h \nu_{n'n}^3}{c^3} \frac{1}{e^{-\hbar\omega_{n'n}/k_B T} - 1}$$

(where k_B is the Boltzmann constant), show that $A_{nn'} \propto \omega_{nn'}^3$. [HINT: The number of atoms in the initial state N_n is related to the number of atoms in the final state $N_{n'}$ by

$$N_{n'} = N_n e^{\hbar\omega_{n'n}/k_BT}].$$

The coefficients $A_{nn'}$ and $B_{nn'}$ are called *Einstein A and B coefficients*. (See H. G. Kuhn, *Atomic Spectra*. New York: Academic, 1969, pp. 66 ff.)

6
Spin

Had I been present at the creation, I would have given some useful hints for the better ordering of the universe.

Alphonso X, the Learned King of Spain, 1252–1284

The average physicist can perhaps be forgiven a certain amount of paranoia regarding the physical universe, for it often seems that no sooner does he display his latest Theory than the universe is found to feature an additional Effect that renders said theory incomplete if he is lucky (e.g., Erwin Schroedinger) or incorrect if he is not (e.g., Niels Bohr). This element of mystery is one of the things that makes physics exciting.

The Schroedinger wave theory of the hydrogenic atom, the principal features of which have been discussed in the last several chapters, seems at this point to be complete, at least for the case of an isolated atom. However, several pieces of experimental data have been ignored so far. We shall now look briefly at these troublesome observations and see small but significant flaws begin to appear in Schroedinger's beautiful theory. Fortunately, there are two ways to handle them—one involving a rather bizarre ad hoc hypothesis that must be appended to Schroedinger theory, the other requiring a complete reconstruction of the wave theory of matter, using Einstein's theory of relativity. We shall choose the first approach, the simpler of the two. To help motivate the introduction of the new postulates, we shall briefly survey the historical development of spin.

6.1 A HISTORICAL DIGRESSION

In the early 1900s, following a period of intense activity in the development of quantum mechanics, various problems arose.[1] There were two main sources of difficulty, the first of which was apparent in existing spectral data. Taking a close look at the lowest line of the Balmer series of hydrogen (6562.8 Å), physicists discovered that it actually consisted of two distinct lines separated by a mere 0.3 Å. Further study revealed that some lines in the spectra of many atoms actually had several components. These lines were said to have a *multiplet structure*; a line with two components was called a *doublet*, one with three components a *triplet*, and so on.

The second difficulty was discovered when atoms were exposed to external magnetic fields. The Schroedinger theory predicted that the energy levels of a hydrogen atom in an external magnetic field, say $\mathbf{B} = B_z\hat{\mathbf{z}}$, would be split according to the value of the magnetic quantum number m appropriate to the level. In particular, the energy of a particular level in such a field is

$$E_{nm} = -\frac{1}{n^2}\left(\frac{e^2}{2a_0}\right) + mg_\ell \beta B_z, \tag{6.1}$$

where m is one of the integral values between $-\ell$ and $+\ell$, β is the Bohr magneton—$\beta = e\hbar/(2mc)$—and $g_\ell = 1$. Similar predictions were made for energy levels of other atoms in the magnetic field. Always a particular multiplet structure was predicted, since, for any n, the magnetic quantum number could assume a variety of values. The effect of magnetic fields on spectral lines is called the *Zeeman effect.*

Exercise 6.1 Derive Eq. (6.1) by showing that the eigenfunctions of the system Hamiltonian $\mathcal{H} = \mathcal{H}^{(0)} - \mathbf{M}_\ell \cdot \mathbf{B}$ are simply the eigenfunctions of the isolated one-electron atom Hamiltonian $\mathcal{H}^{(0)}$—that is, the functions $\psi_{n\ell m}(\mathbf{r})$ of Chapter 3. Obtain the corresponding eigenvalues E_{nm}.

However, although the alkali atoms seemed well behaved in this respect, certain other atoms displayed multiplets not predicted by Schroedinger theory. (This result was referred to as the *anomalous Zeeman effect.* We shall return to it in Sec. 7.5.)

[1]See the historical references at the end of Chapter 1.

The Stern-Gerlach Experiment

Next we must examine the results of some experiments that were carried out by Stern and Gerlach and reported in the period 1921–1924.[2] They sent a beam of silver atoms through an inhomogeneous (nonuniform) magnetic field.[3] For purposes of discussion, let us consider a field that is directed along $\hat{\mathbf{z}}$ and nonuniform only with respect to variations in z, so that

$$\frac{\partial B_z}{\partial x} = 0, \qquad \frac{\partial B_z}{\partial y} = 0, \quad \text{but} \quad \frac{\partial B_z}{\partial z} \neq 0. \tag{6.2}$$

It was known that if the atoms in the beam possessed a magnetic moment with a nonzero component in the z direction, then a force would be exerted on the beam, causing it to deflect. The average force exerted on the atoms turns out to be[4]

$$\bar{\mathbf{F}} = \overline{(M_\ell)_z}\left(\frac{\partial B_z}{\partial z}\right)\hat{\mathbf{z}}, \tag{6.3}$$

where we have assumed that the inhomogeneous magnetic field is symmetric with respect to the yz plane. This result, by the way, is valid either classically or quantum mechanically.

If the atom were a classical magnetic dipole, $\overline{(M_\ell)_z}$ could take on all values in the continuous range [see Eq. (3.100)]

$$-\frac{g_\ell\beta}{\hbar}|\mathbf{L}| \leq \overline{(M_\ell)_z} \leq \frac{g_\ell\beta}{\hbar}|\mathbf{L}|, \tag{6.4}$$

and we would expect a continuous band to appear. Of course, the atom must actually be treated quantum mechanically. Thus the average $\overline{(M_\ell)_z}$ is the expectation value

$$\overline{(M_\ell)_z} = \langle\psi_{n\ell m}|(M_\ell)_z|\psi_{n\ell m}\rangle = -\frac{g_\ell\beta}{\hbar}\langle L_z\rangle. \tag{6.5}$$

Recall from Sec. 3.10 that since L_z is sharp, $(M_\ell)_z$ is sharp. Therefore $\overline{(M_\ell)_z}$ can take on $2\ell + 1$ possible distinct values corresponding to the allowed values of m; that is, $\overline{(M_\ell)_z} = -mg_\ell\beta$. The important point is that since

[2]The original references for these important papers are O. Stern and W. Gerlach, *Z. Physik* **7**, 249(1921), *Z. Physik* **8**, 110, *Z. Physik* **9**, 349(1922), and *Ann. Physik* **74**, 673 (1924).

[3]Stern and Gerlach actually used a beam of ground-state silver (Ag) atoms, but for reasons to be explained later (Chapter 9), these atoms may be regarded (roughly) as "heavy" one-electron atoms. In 1927 Phipps and Taylor obtained similar results by doing the Stern-Gerlach experiment on ground-state hydrogen atoms.

[4]For an explanation and derivation of Eq. (6.3), see Robert M. Eisberg, *Fundamentals of Modern Physics* (New York: Wiley, 1961), Chap. 11.

$2\ell + 1$ is odd for any integer ℓ, Schroedinger theory predicts an odd number of components—for example, for the ground state of the "one-electron atom" $\ell = 0$, so we expect only one component.

To the surprise of the experimentalists, the beam of atoms split, not into a continuous band, as predicted by classical mechanics, not into a set of odd $(2\ell + 1)$ components, as predicted by Schroedinger theory, but into two distinct components corresponding to magnetic moments $\pm\beta$. The same result was observed (by Phipps and Taylor) even for s states of hydrogen atoms, for which $\ell = m = 0$ and no splitting is expected.

In one sense, these results "verified" quantum mechanics; a discrete series of components was seen, rather than a continuous band. But the number of components seemed wrong. To appreciate the nature of these observations, we must consider their implication for wave mechanics. We are forced to come to one of two possible conclusions:

1. Either $\langle L_z \rangle$ is not really zero but is $\frac{1}{2}\hbar$ for the ground state of hydrogen, or
2. There exists some other kind of angular momentum whose existence was completely overlooked in the derivation of the Schroedinger theory of the one-electron atom.

Conclusion (1) implies that our theory of angular momentum is in error. On the other hand, conclusion (2) suggests that the electron possesses a degree of freedom which was overlooked in the Schroedinger theory.

6.2 THE POSTULATE OF ELECTRON SPIN

Refusing to give up the theory of orbital angular momentum, we shall consider conclusion (2). Actually, it was Goudsmit and Uhlenbeck in 1925 who took the daring and original step of postulating that the electron possesses some sort of *intrinsic angular momentum* similar to orbital angular momentum in certain of its properties. But this angular momentum differs from the familiar orbital angular momentum in that it is an internal degree of freedom of the electron and thus is completely independent of the electron's environment. It can be cautiously regarded as an internal motion vaguely reminiscent of the rotation of a body about an axis through its center. For this reason, it was given the highly suggestive name *spin angular momentum*. Of course, we know that the electron is not actually a classical spinning particle whizzing around the nucleus. Instead it must be described by quantum mechanics.

Let us now restate the postulate of Goudsmit and Uhlenbeck more precisely. The electron is assumed to possess a fourth degree of freedom—spin—which gives rise to a spin magnetic moment. To make our theory conform with experiment, we assume that the components of this new magnetic moment along the direction of **B** can assume only two discrete values, $\pm\beta$. Moreover, *this new spin angular momentum is assumed to behave formally in essentially the same fashion as does orbital angular momentum.* This statement means that we can tentatively postulate some properties of spin angular momentum (and can derive others) by analogy with orbital angular momentum.

Of course, there was no initial guarantee that this apparently ad hoc stratagem would work. Physicists gradually gained confidence in it by performing countless numbers of experiments, the results of which it successfully explained. Then, in 1928, Dirac rederived quantum mechanics, including the requirements of Einstein's special theory of relativity. In his book, *Principles of Quantum Mechanics*,[5] he showed that the existence and all the formalism of electron spin evolve directly out of this relativistic theory and need not be postulated at all. However, Dirac's theory is extremely complicated and, in most cases, rather difficult to apply. In this chapter we shall simply accept the postulate of electron spin and leave more formal derivations to other books. Let us see if we can develop a mathematical formulation of this theory that might be applied to problems.

6.3 THE THEORY OF ELECTRON SPIN

We begin by introducing operators **S** and $\mathbf{M}_s$ for the spin angular momentum and spin magnetic moment, respectively. Reasoning by formal analogy with **L** and $\mathbf{M}_\ell$, we propose that **S** and $\mathbf{M}_s$ are related by [see Eq. (3.100)]

$$\mathbf{M}_s = -\frac{g_s\beta}{\hbar}\mathbf{S}, \tag{6.6}$$

where the value of g_s must be chosen to make the theory agree with experiment. Cautiously continuing by analogy, we now write eigenvalue equations for S^2 and S_z [see Eqs. (3.79) and (3.80)]:

$$S^2\chi_{sm_s} = s(s+1)\hbar^2\chi_{sm_s}, \tag{6.7}$$

$$S_z\chi_{sm_s} = m_s\hbar\chi_{sm_s}, \tag{6.8}$$

where the χ_{sm_s} are the eigenfunctions of S^2 and S_z with eigenvalues $s(s+1)\hbar^2$ and $m_s\hbar$, respectively. The function χ_{sm_s} describes a "spin state" of the electron, in which the magnitude of the spin is $\sqrt{s(s+1)}\hbar$ and the projection of the spin angular momentum along the $\hat{z}$ axis is $m_s\hbar$.

[5] P. A. M. Dirac, *Principles of Quantum Mechanics*, 4th ed. (revised). (New York: Clarendon Press, 1958). An extraordinary book, this volume is well worth repeated perusal by the serious student of quantum theory.

So far the theory is strictly formal; we do not know what sort of entities the eigenfunctions χ are or what form the operators S^2 and S_z take. We do not even know what coordinates χ depends on. One point is certain: χ *cannot* depend on x, y, or z, since the spin of the electron is assumed to be independent of its spatial distribution. There seems to be no way to picture these functions in ordinary space.

If the theory of spin angular momentum is to formally resemble that of orbital angular momentum, we would expect the new spin quantum numbers s and m_s labeling χ to take on only discrete values. Moreover, for a given value of s, m_s should take on the $(2s + 1)$ integral values $-s, -s + 1, \ldots, s - 1, s$ [see Eqs. (2.61) and (2.62)]. However, the experimental observations leading us to spin in the first place indicate that the magnitude of the spin magnetic moment can assume only two values, $\pm\beta$. Consequently, we must further postulate that the only allowed value for s is

$$s = \tfrac{1}{2}, \tag{6.9}$$

so that the allowed values of m_s are

$$m_s = -\tfrac{1}{2}, \tfrac{1}{2}. \tag{6.10}$$

From these postulates and Eq. (6.6), we see that the only allowed eigenvalues of the operator corresponding to the z component of the spin magnetic moment are $\pm g_s(\beta/2)$. However, the observed values for the magnetic moment were $\pm\beta$; so the value of the so-called *spin g factor* must be[6]

$$g_s = 2. \tag{6.11}$$

The essence of the theory of electron spin is contained in Eqs. (6.6) to (6.10). But what of the mathematical form of the operators **S** and S^2 and the strange eigenfunctions χ_{sm_s}? Regardless of their form, these entities are defined by the eigenvalue equations (6.7) and (6.8), and there are, in fact, several possible ways to specify them. We choose to adopt a particularly convenient one due to Wolfgang Pauli, in which we write the operators **S** and S^2 as 2×2 matrices and the eigenfunctions χ_{sm_s} as two-component column vectors.[7]

[6]Accurate measurements reveal that g_s is slightly different from 2, its observed value being 2.00232. This tiny discrepancy is explained by the theory of quantum electrodynamics and will not be pursued here.

[7]If you are wondering how an eigenfunction can be represented by a column vector, you are to be commended! These suggestions should not seem too unreasonable, however. Since m_s can only take on two values, $\pm\frac{1}{2}$, we might expect the eigenfunction χ_{sm_s} in $S_z\chi_{sm_s} = m_s\hbar\chi_{sm_s}$ to have two components. Then in an eigenvalue equation for a two-component eigenvector, the operator must be a 2×2 matrix.

6.4 THE PAULI SPIN MATRICES AND SPIN OPERATORS[8]

We shall now introduce explicit forms for the operators S^2 and S_z and then show that **S** so defined does indeed behave like the orbital angular momentum (that is, **S** satisfies commutation relations analogous to those satisfied by **L**).

Thus let us write the operator $\mathbf{S} = \hat{\mathbf{x}}S_x + \hat{\mathbf{y}}S_y + \hat{\mathbf{z}}S_z$ as

$$\mathbf{S} = \tfrac{1}{2}\hbar(\hat{\mathbf{x}}\sigma_x + \hat{\mathbf{y}}\sigma_y + \hat{\mathbf{z}}\sigma_z). \tag{6.12}$$

where $\hat{\mathbf{x}}$, $\hat{\mathbf{y}}$, and $\hat{\mathbf{z}}$ are unit vectors and σ_x, σ_y, σ_z are the *Pauli spin matrices*, defined to be

$$\sigma_x \equiv \begin{pmatrix} 0 & 1 \\ 1 & 0 \end{pmatrix}, \qquad \sigma_y \equiv \begin{pmatrix} 0 & -i \\ i & 0 \end{pmatrix}, \qquad \sigma_z \equiv \begin{pmatrix} 1 & 0 \\ 0 & -1 \end{pmatrix}. \tag{6.13}$$

Then the operator for the square of the spin angular momentum is

$$S^2 = S_x^2 + S_y^2 + S_z^2 = \tfrac{3}{4}\hbar^2 1, \tag{6.14}$$

where 1 is the unit matrix,

$$1 \equiv \begin{pmatrix} 1 & 0 \\ 0 & 1 \end{pmatrix}. \tag{6.15}$$

The operator for the z component of **S** is simply

$$S_z = \frac{\hbar}{2}\sigma_z. \tag{6.16}$$

Notice that S_x, S_y, S_z, and S^2 are each Hermitian.

With these definitions, the eigenvalue equations (6.7) and (6.8) become *matrix equations*

$$S^2\chi_{sm_s} = s(s+1)\hbar^2\chi_{sm_s}, \tag{6.17}$$

$$S_z\chi_{sm_s} = m_s\hbar\chi_{sm_s}, \tag{6.18}$$

where χ_{sm_s} is a two-component column vector and $s = \frac{1}{2}$, $m_s = \pm\frac{1}{2}$. The two eigenvectors χ_{sm_s} that satisfy these equations are

$$\chi_{1/2,1/2} = \begin{pmatrix} 1 \\ 0 \end{pmatrix} \equiv \alpha, \tag{6.19}$$

$$\chi_{1/2,-1/2} \equiv \begin{pmatrix} 0 \\ 1 \end{pmatrix} \equiv \beta. \tag{6.20}$$

[8]In this section we shall presuppose a rudimentary knowledge of matrices and their properties. See, for example, Henry Margenau and George M. Murphy, *The Mathematics of Physics and Chemistry*, 2nd ed. (London: D. Van Nostrand Co. Ltd., 1956), Chap. 10.

In fact, substituting α and β into Eqs. (6.17) and (6.18), we have

$$S^2\alpha = \tfrac{3}{4}\hbar^2\alpha \tag{6.21a}$$

$$S^2\beta = \tfrac{3}{4}\hbar^2\beta, \tag{6.21b}$$

and

$$S_z\alpha = +\tfrac{1}{2}\hbar\alpha \tag{6.22a}$$

$$S_z\beta = -\tfrac{1}{2}\hbar\beta. \tag{6.22b}$$

Exercise 6.2 Verify Eqs. (6.21) and (6.22) by matrix multiplication.

Since α corresponds to $m_s\hbar = +\hbar/2$, it is sometimes referred to as the "spin-up eigenfunction." Similarly, β is called the "spin-down eigenfunction".[9] Thus the introduction of the Pauli spin matrices has enabled us to give a precise operational meaning to the spin operators S^2 and S_z (and, of course, S_x and S_y) and to the spin eigenfunctions χ_{sm_s}.

Commutation Relations

Recall that the orbital angular momentum operator satisfied the commutation relations [see Eqs. (2.63) to (2.65)]

$$[L_x, L_y] = i\hbar L_z, \qquad [L_y, L_z] = i\hbar L_x, \qquad [L_z, L_x] = i\hbar L_y, \tag{6.23}$$

$$[L_x, L^2] = 0, \qquad [L_y, L^2] = 0, \qquad [L_z, L^2] = 0. \tag{6.24}$$

Since S is also an angular momentum, it might be expected to satisfy commutation relations of the same form.

Some simple matrix multiplications reveal that the Pauli spin matrices σ_x, σ_y, and σ_z satisfy the relations

$$[\sigma_x, \sigma_y] = 2i\sigma_z, \qquad [\sigma_y, \sigma_z] = 2i\sigma_x, \qquad [\sigma_z, \sigma_x] = 2i\sigma_y \tag{6.25a}$$

$$[\sigma_x, \sigma_y]_+ = 0, \qquad [\sigma_y, \sigma_z]_+ = 0, \qquad [\sigma_z, \sigma_x]_+ = 0, \tag{6.25b}$$

where the anticommutator $[A, B]_+ \equiv AB + BA$ was introduced in Eq. (1.33).

Exercise 6.3 Verify Eqs. (6.25).

From Eqs. (6.25) and the definition of S in terms of σ_x, σ_y, and σ_z, Eq. (6.12), it follows immediately that S satisfies the commutation relations

$$[S_x, S_y] = i\hbar S_z, \qquad [S_y, S_z] = i\hbar S_x, \qquad [S_z, S_x] = i\hbar S_y \tag{6.26}$$

$$[S_x, S^2] = 0, \qquad [S_y, S^2] = 0, \qquad [S_z, S^2] = 0. \tag{6.27}$$

[9] Do not confuse the eigenfunction β with the Bohr magneton β.

These results are completely analogous to the commutation relations for the orbital magnetic momentum. [Alternatively, we could have chosen to postulate that Eqs. (6.26) and (6.27) be satisfied for spin operators and derived the Pauli spin matrices as a consequence.]

The analogy between **S** and **L** can only be carried so far. L^2 and **L** are differential operators that operate on functions of the spatial coordinates θ and φ. On the other hand, S^2 and **S** are (in the Pauli theory) represented by 2×2 matrices that operate on column vectors in some sort of mathematical "spin space." It seems that the commutation relations satisfied by **S** and **L** are, in some sense, characteristic of angular momentum operators in general. In fact, these relations can be used as the basis for a very general theory of angular momentum.[10]

Matrix Elements

As we have seen in Chapters 2 to 5, occasionally we are required to calculate matrix elements of L^2 and the components of **L**. For example, in calculating the expectation value of L^2, we evaluate

$$\langle \psi_{n\ell m} | L^2 | \psi_{n\ell m} \rangle \equiv \int \psi^*_{n\ell m}(\mathbf{r}) L^2 \psi_{n\ell m}(\mathbf{r})\, d\mathbf{r}. \tag{6.28}$$

Similarly, we shall sometimes need matrix elements of spin angular momentum operators.

These matrix elements are defined by simple matrix multiplication—for example,

$$\langle \chi | S^2 | \chi' \rangle \equiv \chi^\dagger S^2 \chi', \tag{6.29}$$

where the row vector $\chi^\dagger$ is the *adjoint* of χ, the complex conjugate of the transpose of χ:

$$\chi^\dagger \equiv (\chi^T)^*. \tag{6.30}$$

For the eigenfunctions α and β, we have from Eqs. (6.21)

$$\langle \alpha | S^2 | \alpha \rangle = \tfrac{3}{4}\hbar^2 \quad \text{and} \quad \langle \beta | S^2 | \beta \rangle = \tfrac{3}{4}\hbar^2, \tag{6.31a}$$

$$\langle \alpha | S^2 | \beta \rangle = 0 \quad \text{and} \quad \langle \beta | S^2 | \alpha \rangle = 0. \tag{6.31b}$$

We sometimes write matrix elements in a form that explicitly reminds us of the value of the quantum numbers s and m_s,

$$\langle \alpha | S^2 | \beta \rangle = \langle \underset{\substack{\uparrow \\ s}}{\tfrac{1}{2}}, \underset{\substack{\uparrow \\ m_s}}{+\tfrac{1}{2}} | S^2 | \tfrac{1}{2}, -\tfrac{1}{2} \rangle. \tag{6.32}$$

[10] A clear presentation of this often-difficult theory can be found in M. E. Rose, *Elementary Theory of Angular Momentum* (New York: Wiley, 1957).

Using this notation, we can write the matrix elements of the spin operators:

$$\langle \tfrac{1}{2}, m_s | S^2 | \tfrac{1}{2}, m_s' \rangle = \tfrac{3}{4}\hbar^2 \, \delta_{m_s, m_s'}, \tag{6.33}$$

$$\langle \tfrac{1}{2}, m_s | S_z | \tfrac{1}{2}, m_s' \rangle = m_s \hbar \, \delta_{m_s, m_s'}, \tag{6.34}$$

where $\delta_{m_s, m_s'}$ is the Kronicker delta of Eq. (1.27).

Exercise 6.4 Derive Eqs. (6.33) and (6.34).

Raising and Lowering Operators

It is useful to introduce operators S_+ and S_-, defined by analogy with L_+ and L_- as [see Eqs. (2.66)]

$$S_\pm \equiv S_x \pm iS_y. \tag{6.35}$$

(Note that $S_+^\dagger = S_-$ and $S_-^\dagger = S_+$.) Operating on an eigenfunction χ_{sm_s} with $S_\pm$ yields an eigenfunction of S^2 and S_z with eigenvalues $s(s+1)\hbar^2$ and $(m_s \pm 1)\hbar$. Thus the effect of $S_\pm$ is to "raise" or "lower" the value of m_s:

$$S_\pm \chi_{1/2, m_s} = \hbar \chi_{1/2, m_s \pm 1}. \tag{6.36}$$

By definition, we say that

$$S_+ \chi_{1/2, 1/2} = 0 \quad \text{and} \quad S_- \chi_{1/2, -1/2} = 0. \tag{6.37}$$

The matrix elements of $S_\pm$ are simply

$$\langle \tfrac{1}{2}, m_s' | S_\pm | \tfrac{1}{2}, m_s \rangle = \hbar \sqrt{(\tfrac{1}{2} \mp m_s)(\tfrac{3}{2} \pm m_s)} \, \delta_{m_s', m_s \pm 1}. \tag{6.38}$$

Exercise 6.5 (a) Use the definition of S_x and S_y, in terms of the Pauli spin matrices, to verify Eqs. (6.36) and (6.38).
(b) Show that $S_+^\dagger = S_-$ and $S_-^\dagger = S_+$.

It is not easy to say much about the physical significance of the results of this section. The commutation relations of Eqs. (6.26) specify that it is not possible to find a state of the electron in which two components of **S** are simultaneously sharp. However, Eq. (6.27) tells us that there are states in which the square of the magnitude of **S** and a single component are simultaneously well defined. For example, α is simultaneously an eigenvector of the operators S^2 and S_z, just as $Y_{\ell m}(\theta, \varphi)$ is simultaneously an eigenfunction of the operators L^2 and L_z.

6.5 SOME CONCLUDING REMARKS

In this chapter we have briefly examined the unusual idea of electron spin. In the experimental observations reported in Sec. 6.1, we found facts that began to cast doubt on the Schroedinger theory, and it was necessary to adopt the hypothesis of Goudsmit and Uhlenbeck, set forth in Sec. 6.2. We postulated a new observable—electron spin—represented by the operator **S**, which could assume two independent orientations corresponding to $m_s = +\frac{1}{2}$ and $m_s = -\frac{1}{2}$.

It is important to notice that since the eigenfunctions χ_{sm_s} do not depend on spatial variables, differential operators do not affect them. These functions satisfy the eigenvalue equations (6.7) and (6.8). We should also keep in mind that spin is an intrinsic angular momentum, a fundamental quantum mechanical property of the electron independent of its environment (like mass and charge), and that its properties are formally analogous to those of orbital angular momentum.

The peculiar nature of spin angular momentum is reflected in the fact that, unlike orbital angular momentum, it has no classical counterpart. Recall that the magnitude of the orbital angular momentum is given by $|\mathbf{L}| = \sqrt{\ell(\ell+1)}\hbar$. In the classical limit $\hbar \longrightarrow 0$, we only need take ℓ "big enough" ($\ell \longrightarrow \infty$) to be left with a nonzero angular momentum. However, the magnitude of the spin angular momentum is $|\mathbf{S}| = \sqrt{s(s+1)}\hbar$, and since the value of s is restricted to $+\frac{1}{2}$, if $\hbar \longrightarrow 0$ it follows that $|\mathbf{S}| \longrightarrow 0$. *Spin disappears in the classical limit.*

The electron is by no means unique in possessing spin angular momentum. Indeed, it is one of a host of other particles, generically termed *fermions*, which have half-odd-integral values of s. Other fermions are protons and neutrons. In addition, there is another class of particles, the *bosons*, which have integral spin. An example of a boson is the deuteron (see Prob. 3.2). It turns out that there are dramatic differences in the properties of fermions and bosons, and we shall explore them in a later chapter (Chapter 8).

We are now prepared for the final stage in our study of the hydrogenic atom; we must reexamine the theory of Chapter 3 in the light of the new development of spin.

7

Spin in the Hydrogenic Atom

Upon a slight conjecture I have ventured on a dangerous journey, and I already behold the foothills of new lands. Those who have the courage to continue the search will set foot upon them . . .

Immanual Kant, 1775

In Chapter 3 the time-independent Schroedinger equation for an isolated hydrogenic atom was solved. In particular, we first separated the center-of-mass and relative motion and then solved for the eigenfunctions and eigenvalues of the equation of relative motion, which was found to be

$$\mathcal{H}^{(0)}\psi^{(0)}_{n\ell m_\ell}(\mathbf{r}) = E^{(0)}_n\psi^{(0)}_{n\ell m_\ell}(\mathbf{r}), \tag{7.1}$$

where a superscript (0) is appended to $\mathcal{H}$, $\psi_{n\ell m_\ell}(\mathbf{r})$, and E_n as a reminder that these quantities neglect spin.[1] The Hamiltonian $\mathcal{H}^{(0)}$ is

$$\mathcal{H}^{(0)} = -\frac{\hbar^2}{2\mu}\frac{1}{r^2}\frac{\partial}{\partial r}\left(r^2\frac{\partial}{\partial r}\right) + \frac{1}{2\mu r^2}L^2 - \frac{Ze^2}{r}. \tag{7.2}$$

[1]We have also used m_ℓ for the magnetic quantum number so that it will not be confused with the spin quantum number m_s.

The stationary-state eigenfunctions $\psi^{(0)}_{nlm_l}(\mathbf{r})$ describe the spatial distribution of the electron; that is, $|\psi^{(0)}_{nlm_l}(\mathbf{r})|^2\,d\mathbf{r}$ is the probability that the electron will be found in volume element $d\mathbf{r}$ at position $\mathbf{r}$.

In this analysis, we viewed the hydrogenic atom as consisting of two *structureless* particles, a proton and an electron, whose interaction is given by the familiar coulomb potential energy.

But we learned in Chapter 6 that the electron is not structureless; it possesses a fourth degree of freedom, its intrinsic or spin angular momentum. Thus it is not enough to specify the spatial distribution of the electron; we must also specify its spin state. We do this via the spin eigenfunction χ_{sm_s}. This function satisfies the eigenvalue equations

$$S^2\chi_{sm_s} = s(s+1)\hbar^2\chi_{sm_s} \tag{7.3a}$$

$$S_z\chi_{sm_s} = m_s\hbar\chi_{sm_s}. \tag{7.3b}$$

The two spin states of the electron are "spin up" $\chi_{sm_s} = \chi_{1/2,1/2} = \alpha$ and "spin down" $\chi_{sm_s} = \chi_{1/2,-1/2} = \beta$.
The objective of this chapter is to reformulate the theory of the hydrogenic atom to take account of the spin of the electron.[2]

7.1 NEW HYDROGENIC EIGENFUNCTIONS

What might the eigenfunctions describing stationary states of the hydrogen atom be if we do not ignore the spin of the electron? Since spin operators and eigenfunctions do not depend on the spatial coordinates, we might take as our first guess simple product functions of the form

$$\psi^{(0)}_{nlm_lm_s}(\mathbf{r}) \equiv \psi^{(0)}_{nlm_l}(\mathbf{r})\chi_{sm_s}, \tag{7.4}$$

where χ_{sm_s} is either α or β, depending on whether the spin of the electron is "up" or "down." At first, this product form seems reasonable; $\psi^{(0)}_{nlm_lm_s}(\mathbf{r})$ is indeed an eigenfunction of $\mathcal{H}^{(0)}$ with eigenvalue $E^{(0)}_n$. It is also an eigenfunction of L^2, L_z, S^2, and S_z.

Unfortunately, the choice of Eq. (7.4) is not correct. The reason is that the Hamiltonian for the hydrogenic atom which takes into account spin is not $\mathcal{H}^{(0)}$. As we shall see in Sec. 7.2, the spin and spatial attributes of the electron interact, giving rise to an additional potential energy term $\mathcal{H}_{so}$ that must be included in $\mathcal{H}$; that is,

$$\mathcal{H} = \mathcal{H}^{(0)}1 + \mathcal{H}_{so}, \tag{7.5}$$

[2]The nucleus can also have spin, but we shall ignore the effects of nuclear spin in this book.

where $\mathcal{H}$ is now a matrix operator, and 1 is the 2×2 unit matrix of Eq. (6.15). The new term $\mathcal{H}_{so}$ is proportional to $\mathbf{S} \cdot \mathbf{L} = S_x L_x + S_y L_y + S_z L_z$. Hence $[\mathcal{H}_{so}, S_z] \neq 0$ and $[\mathcal{H}_{so}, L_z] \neq 0$ so that $\psi^{(0)}_{n\ell m_\ell m_s}$ is not an eigenfunction of $\mathcal{H}$.

Well, then, what are the eigenfunctions of $\mathcal{H}$? Let us call them $\psi_E(\mathbf{r})$ for now; they satisfy the Schroedinger equation

$$\mathcal{H}\psi_E(\mathbf{r}) = E\psi_E(\mathbf{r}). \tag{7.6}$$

Since $\mathcal{H}$ is a 2×2 matrix operator, $\psi_E(\mathbf{r})$ is a two-component column vector. Therefore it can be written as[3]

$$\psi_E(\mathbf{r}) = \begin{pmatrix} \varphi_{1/2}(\mathbf{r}) \\ \varphi_{-1/2}(\mathbf{r}) \end{pmatrix}, \tag{7.7}$$

or

$$\psi_E(\mathbf{r}) = \varphi_{1/2}(\mathbf{r})\alpha + \varphi_{-1/2}(\mathbf{r})\beta, \tag{7.8}$$

where we have used the definitions of α and β [Eqs. (6.19) and (6.20)] in the last step; $\varphi_{1/2}(\mathbf{r})$ and $\varphi_{-1/2}(\mathbf{r})$ are as-yet-unknown functions of $\mathbf{r}$. Although we have not yet determined the precise form of these functions their interpretation is clear; $|\varphi_{1/2}(\mathbf{r})|^2\, d\mathbf{r}$ is the probability that the electron will be found in a volume element $d\mathbf{r}$ centered on $\mathbf{r}$ with spin "up." Similarly, $|\varphi_{-1/2}(\mathbf{r})|^2\, d\mathbf{r}$ is the probability that the electron will be found in $d\mathbf{r}$ at $\mathbf{r}$ with spin "down."

Since $[\mathcal{H}, S_z] \neq 0$, $\psi_E(\mathbf{r})$ is not an eigenfunction of S_z. Thus we know that in general neither $\varphi_{1/2}(\mathbf{r})$ nor $\varphi_{-1/2}(\mathbf{r})$ can be zero for all $\mathbf{r}$. Moreover, we cannot label the new eigenfunction $\psi_E(\mathbf{r})$ by the quantum number m_s, since $\psi_E(\mathbf{r})$ is the sum of two terms, each of which corresponds to a different m_s. [Since both terms correspond to $s = \frac{1}{2}$, we could label $\psi_E(\mathbf{r})$ by the quantum number s.]

This knowledge is rather limited. Clearly, we must solve the time-independent Schroedinger equation (7.6) for $\psi_E(\mathbf{r})$ and E. Unfortunately, this equation is very complicated, and approximation methods are required.

Let us turn now to a derivation of the new potential energy term $\mathcal{H}_{so}$. We shall see that $\mathcal{H}_{so}$ is small, an observation that will lead us to use time-independent perturbation theory to solve for approximate wave functions and energies of the hydrogenic atom.

7.2 DERIVATION OF THE SPIN-ORBIT INTERACTION TERM

The term $\mathcal{H}_{so}$ arises because the electron has associated with it a spin magnetic moment $\mathbf{M}_s = -(g_s\beta/\hbar)\mathbf{S}$ [see Eq. (6.6)]. This magnetic moment interacts with a magnetic field $\mathbf{B}$ at the electron; in the electron's rest frame, this field

[3]Notice that Eq. (7.8) is just an expansion of $\psi_E(\mathbf{r})$, a vector in a two-dimensional space, in terms of α and β, two linearly independent vectors that span the space. The coefficients are $\varphi_{1/2}(\mathbf{r})$ and $\varphi_{-1/2}(\mathbf{r})$, and both are functions of the spatial coordinates.

is due to the motion of the positively charged nucleus around the electron. Thus we have $\mathcal{H}_{so} = -\mathbf{M}_s \cdot \mathbf{B}$. We shall now amplify these introductory remarks and derive an explicit expression for the operator $\mathcal{H}_{so}$. In particular, we shall consider the problem "classically" and obtain an expression for the potential energy of interaction V_{so}; replacing observables with their operators in V_{so}, we shall then write an expression for $\mathcal{H}_{so}$.

Classical Analysis

Suppose that we view the electron as moving in a circular orbit around the nucleus with velocity $\mathbf{v}$.[4] Let us fix the coordinate frame on the electron rather than on the nucleus, remembering to transform back to a frame fixed on the nucleus at the end. The electron sees the nucleus of charge $+Ze$ at position $-\mathbf{r}$ moving in a circular orbit with velocity $-\mathbf{v}$. The electron experiences an electric field due to the nucleus,

$$\mathbf{E}(\mathbf{r}) = \frac{Ze}{r^2}\hat{\mathbf{r}} = \frac{Ze}{r^3}\mathbf{r}. \tag{7.9}$$

To the electron, this electric field appears to be moving along with the nucleus. Now, a charged particle in a time-varying electric field experiences a magnetic field whose magnitude is given by the classical expression[5]

$$\mathbf{B} = -\frac{1}{c}\mathbf{v} \times \mathbf{E}. \tag{7.10}$$

The momentum of the electron in the rest frame of the nucleus is $\mathbf{p} = m_e\mathbf{v}$. Hence we can write the magnetic field as

$$\mathbf{B} = \frac{1}{m_e c}\mathbf{E} \times \mathbf{p}. \tag{7.11}$$

Substituting Eq. (7.9) into (7.11), we obtain

$$\mathbf{B} = \frac{Ze}{m_e c r^3}\mathbf{r} \times \mathbf{p}, \tag{7.12}$$

or

$$\mathbf{B} = \frac{Ze}{m_e c r^3}\mathbf{L}. \tag{7.13}$$

The magnetic field interacts with the spin magnetic moment of the electron, giving rise to a potential energy V_{so}; that is,

$$V_{so} \equiv -\mathbf{M}_s \cdot \mathbf{B}. \tag{7.14}$$

[4]This assumption is not absolutely necessary, but it does simplify the ensuing analysis and leads to the correct answer.

[5]See Edward M. Purcell, *Electricity and Magnetism* (New York: McGraw-Hill, 1965), Chap. 5.

Since $\mathbf{M}_s = -(g_s\beta/\hbar)\mathbf{S}$, we can write this result as

$$V_{so} = \frac{g_s\beta}{\hbar}\mathbf{S}\cdot\mathbf{B}, \tag{7.15}$$

or

$$V_{so} \simeq \frac{Ze^2}{m_e^2c^2r^3}\mathbf{S}\cdot\mathbf{L}, \tag{7.16}$$

where we have used $g_s = 2$, $\beta = e\hbar/2\mu c$, and have set $\mu \simeq m_e$.

Remember that this result was derived in a coordinate frame fixed to the electron; we must transform it to a frame fixed to the nucleus. As a consequence of this transformation, we find that the preceding result for V_{so} is too large by a factor of 2. Thus Eq. (7.16) must be multiplied by $\frac{1}{2}$. This mysterious factor of $\frac{1}{2}$ arises because $\mathbf{M}_s$ undergoes a precession *even in the absence of a magnetic field.*[6] Accepting this purely kinematic relativistic effect, we see that the final answer for V_{so} is

$$V_{so} = \frac{Ze^2}{2m_e^2c^2}\left(\frac{1}{r^3}\right)\mathbf{S}\cdot\mathbf{L}. \tag{7.17}$$

Quantum Mechanical Operator

It is now a simple matter to substitute into Eq. (7.17) the operators for spin and orbital angular momentum to obtain

$$\boxed{\mathcal{H}_{so} = \frac{Ze^2}{2m_e^2c^2}\left(\frac{1}{r^3}\right)\mathbf{S}\cdot\mathbf{L}.} \quad \text{Spin-orbit interaction Hamiltonian} \tag{7.18}$$

Notice that $\mathcal{H}_{so}$ is proportional to $1/r^3$ and to $\mathbf{S}\cdot\mathbf{L}$. This is the term that must be added to $\mathcal{H}^{(0)}$ in order to obtain the full Hamiltonian for the one-electron atom.

If we were to ignore $\mathcal{H}_{so}$ in a treatment of hydrogen ($Z = 1$), the energies of the system would simply be the familiar hydrogenic energies $E_n^{(0)} = -(13.596)/n^2$ eV; for example, $E_1^{(0)} = -13.6$ eV, $E_2^{(0)} = -3.4$ eV. How big is $\mathcal{H}_{so}$ compared to these energies? From Appendix 3 we know the expectation value of $1/r^3$; let us make the rough estimate $\langle r^{-3}\rangle_n \simeq 1/(n^3a_0^3)$, where a_0 is the first Bohr radius, $a_0 = \hbar/\mu e^2 = 0.529$ Å. The expectation value of $\mathbf{S}\cdot\mathbf{L}$ is of the order of $\hbar^2$. Hence the term $\mathcal{H}_{so}$ is roughly of the order of 10^{-4} eV.

[6]Called Thomas precession, this relativistic effect was first presented by L. H. Thomas in *Nature* **117**, 514(1926). Its derivation is somewhat tedious but not difficult. Two different discussions may be found in Robert M. Eisberg, *Fundamentals of Modern Physics* (New York: Wiley, 1961), Chap. 11, and Robert B. Leighton, *Principles of Modern Physics* (New York: McGraw-Hill, 1959), Chap. 5. Eisberg derives Thomas precession by a straightforward relativistic velocity transformation. Leighton uses an argument based on time dilation.

Exercise 7.1 (a) Using the values of the fundamental constants given in Appendix 1, verify that $\langle \mathcal{H}_{so} \rangle \simeq 10^{-4}$ eV.

(b) Obtain a numerical estimate of the magnitude of the field **B** due to the nuclear motion around the electron (in gauss).

Clearly, $\mathcal{H}_{so}$ is small and *can be treated as a perturbation*. Thus we propose to obtain the new wave functions $\psi_j(r)$ and energies E_j to first order from the familiar formulas [see Eqs. (4.21) and (4.23)]

$$E_j = E_j^{(0)} + \langle \psi_j^{(0)} | \mathcal{H}_{so} | \psi_j^{(0)} \rangle, \tag{7.19}$$

$$\psi_j(\mathbf{r}) = \psi_j^{(0)}(\mathbf{r}) + \sum_{k \neq j} \frac{\langle \psi_k^{(0)} | \mathcal{H}_{so} | \psi_j^{(0)} \rangle}{E_j^{(0)} - E_k^{(0)}} \psi_k^{(0)}(\mathbf{r}). \tag{7.20}$$

The zeroth-order energies $E_j^{(0)}$ are merely the hydrogenic energies obtained in Chapter 3, and we could use the functions $\psi^{(0)}_{n\ell m_\ell m_s}(\mathbf{r}) = \psi^{(0)}_{n\ell m_\ell}(\mathbf{r})\chi_{sm_s}$ as zeroth-order wave functions $\psi_j^{(0)}$ in these equations. However, doing so leads to an unpleasant algebraic problem; because of the degeneracy of the functions $\psi^{(0)}_{n\ell m_\ell m_s}(\mathbf{r})$ it would seem that we must use the techniques of Sec. 4.6. Since there are $2n^2$ degenerate functions corresponding to each $E_n^{(0)}$, we must cope with enormous determinants and algebraic equations for all but the smallest values of n. We could do so, of course, thereby obtaining (along with the perturbed energies) new degenerate zeroth-order wave functions, each of which would be a linear combination of the old functions $\psi^{(0)}_{n\ell m_\ell m_s}(\mathbf{r})$; see Prob. 7.3.

Instead let us see if the physical properties of the hydrogenic atom can be used to determine the form of new zeroth-order eigenfunctions that simplify the application of perturbation theory.

7.3 COUPLING OF SPIN AND ORBITAL ANGULAR MOMENTA

In Sec. 7.1 we saw that, for a hydrogen atom, the projections of **L** and **S** along the $\hat{z}$ axis, L_z and S_z, are no longer well defined when we take the spin of the electron into account. Hence we cannot use the quantum numbers m_ℓ and m_s to label eigenstates of the full Hamiltonian $\mathcal{H}$.

The Total Angular Momentum

These results are a consequence of the fact that the spin-orbit interaction is proportional to $\mathbf{S} \cdot \mathbf{L}$. From a more physical point of view, we can think of the effective magnetic field due to the orbital motion of the electron as exerting a torque on the spin angular momentum. This torque couples **L** and **S** in such a way that both vectors precess about a resultant vector **J**, defined as

$$\mathbf{J} \equiv \mathbf{L} + \mathbf{S}. \tag{7.21}$$

The total angular momentum, **J**, is the vector sum of the orbital and spin angular momenta. [Equation (7.21) is analogous to classical angular momentum coupling, in which the total angular momentum for a system is the vector sum of the angular momenta of all parts of the system.]

Equation (7.21) defines a new quantum mechanical operator **J**, the *total angular momentum*. In classical physics the total angular momentum of a system is a constant of the motion provided that the system is free of external forces. In quantum physics **J** is a constant of the motion if each of its components commutes with the Hamiltonian. In the absence of external torques, the Hamiltonian is $\mathcal{H} = \mathcal{H}^{(0)}1 + \mathcal{H}_{so}$, and **J** is a constant of the motion, since

$$[J_x, \mathcal{H}] = 0, \qquad [J_y, \mathcal{H}] = 0, \qquad [J_z, \mathcal{H}] = 0. \tag{7.22}$$

Exercise 7.2 Verify Eqs. (7.22).

In terms of a vector model, we can visualize **L** and **S** precessing with the same angular frequency about the vector **J** as shown in Fig. 7.1. We must add our usual cautionary note to this picture. This is merely a useful device, *not* a physical picture of the "motion" of these vectors.

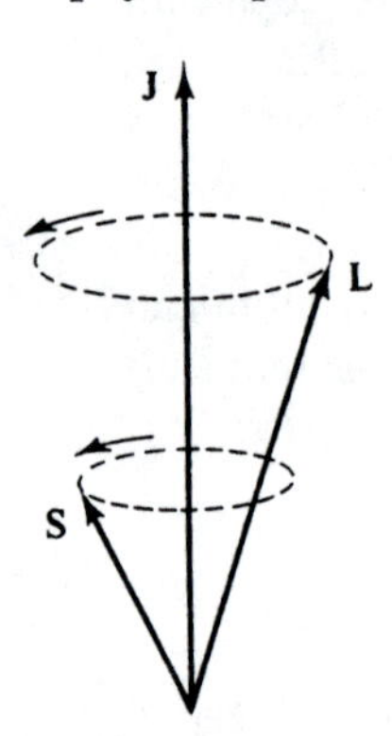

Figure 7.1 Precession of **L** and **S** about the total angular momentum vector **J**.

Properties of J

Since **J** is the sum of **L** and **S**, it is a matrix operator. To see what **J** really looks like, let us use Eqs. (2.35) and (6.12) for the operators **L** and **S** in order to write out its components:

$$J_x = i\hbar\left(\sin\varphi\frac{\partial}{\partial\theta} + \operatorname{ctn}\theta\cos\varphi\frac{\partial}{\partial\varphi}\right)\begin{pmatrix}1 & 0\\0 & 1\end{pmatrix} + \frac{\hbar}{2}\begin{pmatrix}0 & 1\\1 & 0\end{pmatrix}, \tag{7.23a}$$

$$J_y = -i\hbar\left(\cos\varphi\frac{\partial}{\partial\theta} - \operatorname{ctn}\theta\sin\varphi\frac{\partial}{\partial\varphi}\right)\begin{pmatrix}1 & 0\\0 & 1\end{pmatrix} + \frac{\hbar}{2}\begin{pmatrix}0 & -i\\i & 0\end{pmatrix}, \tag{7.23b}$$

$$J_z = -i\hbar\frac{\partial}{\partial\varphi}\begin{pmatrix}1 & 0\\0 & 1\end{pmatrix} + \frac{\hbar}{2}\begin{pmatrix}1 & 0\\0 & -1\end{pmatrix}, \tag{7.23c}$$

where $\mathbf{J} = \hat{\mathbf{x}}J_x + \hat{\mathbf{y}}J_y + \hat{\mathbf{z}}J_z$. Each of the components J_x, J_y, and J_z contains 2×2 matrices and differential operators.

As always, the essence of an angular momentum operator lies in its commutation relations. From the familiar commutation relations for **L** and **S**, we find[7]

$$[J_x, J_y] = i\hbar J_z, \qquad [J_y, J_z] = i\hbar J_x, \qquad [J_z, J_x] = i\hbar J_y. \tag{7.24}$$

Therefore eigenfunctions of $\mathcal{H}$ for which more than one component of **J** is sharp do not exist. A more accurate vector model than that of Fig. 7.1 would have **L** and **S** precessing about **J**, which itself precesses around one axis, say the $\hat{z}$ axis. Such a picture appears in Fig. 7.2.

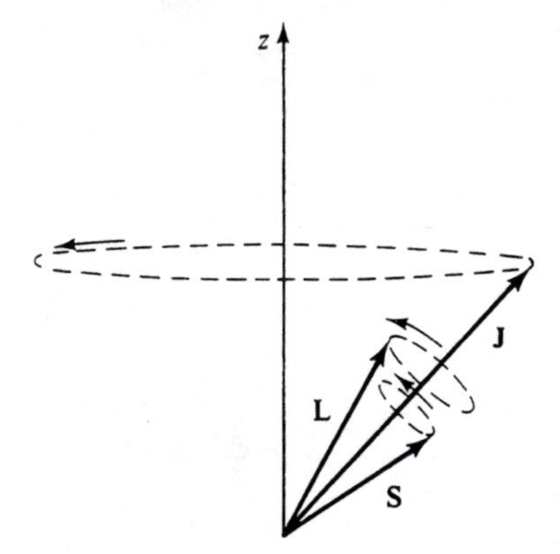

Figure 7.2 Vector model of the coupling of **L** and **S** and precession of **L**, **S**, and **J**.

The operator for the square of the total angular momentum, J^2, is

$$J^2 = J_x^2 + J_y^2 + J_z^2. \tag{7.25}$$

Exercise 7.3 Use Eqs. (7.23) to write out J^2 explicitly.

It is not hard to verify that J^2 commutes with $\mathcal{H}$ and with each component of **J**; that is,

$$[\mathcal{H}, J^2] = 0, \tag{7.26a}$$

$$[J_x, J^2] = 0, \qquad [J_y, J^2] = 0, \qquad [J_z, J^2] = 0. \tag{7.26b}$$

Finally, we note that **J** commutes with L^2 and S^2:

$$[J_x, L^2] = 0, \qquad [J_y, L^2] = 0, \qquad [J_z, L^2] = 0, \tag{7.27a}$$

[7]Commutation relations of this form seem to appear every time we discuss an angular momentum. In more formal theories it is precisely these relationships that define a quantum-mechanical angular momentum operator. See M. E. Rose, *Elementary Theory of Angular Momentum* (New York: Wiley, 1957).

$$[J_x, S^2] = 0, \qquad [J_y, S^2] = 0, \qquad [J_z, S^2] = 0. \tag{7.27b}$$

Therefore J^2 commutes with L^2 and S^2:

$$[J^2, L^2] = [J^2, S^2] = 0. \tag{7.28}$$

Exercise 7.4 Derive Eqs. (7.26), (7.27), and from them Eq. (7.28).

These commutation relations, together with Eq. (7.22), indicate that all the operators $\mathcal{H}$, L^2, S^2, J^2, and J_z commute with one another. Therefore we can find simultaneous eigenfunctions of all these operators. (We could have chosen J_x or J_y instead of J_z; the choice of J_z is the most convenient one.) Let us call these eigenfunctions $\psi_{n\ell jm_j}(\mathbf{r})$. (Remember that $s = \frac{1}{2}$.) They are two-component vector functions that satisfy the eigenvalue equations

$$\mathcal{H}\psi_{n\ell jm_j}(\mathbf{r}) = E_{n\ell jm_j}\psi_{n\ell jm_j}(\mathbf{r}), \tag{7.29}$$

$$L^2\psi_{n\ell jm_j}(\mathbf{r}) = \ell(\ell + 1)\hbar^2\psi_{n\ell jm_j}(\mathbf{r}), \tag{7.30}$$

$$S^2\psi_{n\ell jm_j}(\mathbf{r}) = \tfrac{3}{4}\hbar^2\psi_{n\ell jm_j}(\mathbf{r}), \tag{7.31}$$

$$J^2\psi_{n\ell jm_j}(\mathbf{r}) = j(j + 1)\hbar^2\psi_{n\ell jm_j}(\mathbf{r}), \tag{7.32}$$

$$J_z\psi_{n\ell jm_j}(\mathbf{r}) = m_j\hbar\psi_{n\ell jm_j}(\mathbf{r}). \tag{7.33}$$

We have introduced two new quantum numbers, j and m_j. They are defined by the eigenvalue equations (7.32) and (7.33). Since $\mathbf{J} = \mathbf{L} + \mathbf{S}$, the quantum numbers j and m_j can take on the values[8]

$$j = \ell + \tfrac{1}{2}, \ell - \tfrac{1}{2} \tag{7.34a}$$

$$m_j = -j, -j + 1, \ldots, j - 1, j \tag{7.34b}$$

for a given ℓ.

Back To Perturbation Theory

Let us look at Eq. (7.29). It is *precisely* the time-independent Schroedinger equation for the hydrogenic atom (including spin), Eq. (7.6). This is the equation that we wish to solve, using perturbation theory, for the eigenfunctions $\psi_{n\ell jm_j}(\mathbf{r})$ and eigenvalues $E_{n\ell jm_j}$. Are we any closer to being able to do so?

Yes, indeed. The key point is that J^2 and J_z commute with $\mathcal{H}^{(0)}$ as well as with $\mathcal{H}$; that is,

$$[J^2, \mathcal{H}^{(0)}] = 0 \quad \text{and} \quad [J_z, \mathcal{H}^{(0)}] = 0. \tag{7.35}$$

[8]This follows from the rules of vector coupling, since $s = \frac{1}{2}$. See the reference in footnote 10 of this chapter.

This result follows from the definitions of $\mathcal{H}^{(0)}$ and $\mathbf{J}$ [Eqs. (7.2) and (7.21)]. Since L^2 commutes with each component of $\mathbf{S}$ and $\mathbf{L}$, we have $[J_z, \mathcal{H}^{(0)}] = 0$, and since J^2 can be written

$$J^2 = \mathbf{J}\cdot\mathbf{J} = (\mathbf{S}+\mathbf{L})\cdot(\mathbf{S}+\mathbf{L}), \tag{7.36}$$

or

$$J^2 = S^2 + L^2 + 2(\mathbf{S}\cdot\mathbf{L}), \tag{7.37}$$

we have $[J^2, \mathcal{H}^{(0)}] = 0$. Hence there exist simultaneous eigenfunctions of the operators $\mathcal{H}^{(0)}, L^2, S^2, J^2$, and J_z. They are eigenfunctions of $\mathcal{H}^{(0)}$ (zeroth-order eigenfunctions), so let us call them $\psi^{(0)}_{n\ell jm_j}(\mathbf{r})$. They satisfy the zeroth-order time-independent Schroedinger equation

$$\mathcal{H}^{(0)}\psi^{(0)}_{n\ell jm_j}(\mathbf{r}) = E^{(0)}_n\psi^{(0)}_{n\ell jm_j}(\mathbf{r}) \tag{7.38}$$

and the eigenvalue equations (7.30) to (7.33). They do *not* satisfy Eq. (7.29).

Suppose that we use these zeroth-order functions in the perturbation theory equations (7.19) and (7.20), watching carefully for degeneracy problems. The first-order energy of the state $(n\ell jm_j)$ would be

$$E_{n\ell jm_j} \simeq E^{(0)}_n + \langle\psi^{(0)}_{n\ell jm_j}|\mathcal{H}_{so}|\psi^{(0)}_{n\ell jm_j}\rangle, \tag{7.39}$$

and the first-order perturbed wave function of this state would be

$$\psi_{n\ell jm_j}(\mathbf{r}) \simeq \psi^{(0)}_{n\ell jm_j}(\mathbf{r}) + \sum_{(n'\ell' j'm_j')\neq(n\ell jm_j)} \frac{\langle\psi^{(0)}_{n'\ell' j'm_j'}|\mathcal{H}_{so}|\psi^{(0)}_{n\ell jm_j}\rangle}{E^{(0)}_n - E^{(0)}_{n'}}\psi^{(0)}_{n'\ell' j'm_j'}(\mathbf{r}). \tag{7.40}$$

where we notice that the sum over $(n'\ell' j'm_j')$ does not include $(n\ell jm_j)$. But the perturbation Hamiltonian $\mathcal{H}_{so}$ can be written

$$\mathcal{H}_{so} = \frac{Ze^2}{4m_e^2c^2}\left(\frac{1}{r^3}\right)[J^2 - S^2 - L^2], \tag{7.41}$$

where we have used Eq. (7.37). Since the set of unperturbed functions $\{\psi^{(0)}_{n\ell jm_j}\}$ is orthonormal and each function in the set is an eigenfunction of J^2, J_z, L^2, and S^2, the matrix elements in Eq. (7.40) become

$$\langle\psi^{(0)}_{n'\ell' j'm_j'}|\mathcal{H}_{so}|\psi^{(0)}_{n\ell jm_j}\rangle = \langle\psi^{(0)}_{n'\ell jm_j}|\mathcal{H}_{so}|\psi^{(0)}_{n\ell jm_j}\rangle\,\delta_{j'j}\,\delta_{\ell'\ell}\,\delta_{m_j'm_j}. \tag{7.42}$$

Then the first-order perturbed wave function is simply

$$\psi_{n\ell jm_j}(\mathbf{r}) \simeq \psi^{(0)}_{n\ell jm_j}(\mathbf{r}) + \sum_{n'\neq n} \frac{\langle\psi^{(0)}_{n'\ell jm_j}|\mathcal{H}_{so}|\psi^{(0)}_{n\ell jm_j}\rangle}{E^{(0)}_n - E^{(0)}_{n'}}\psi^{(0)}_{n'\ell jm_j}(\mathbf{r}). \tag{7.43}$$

Since the summation in this expression contains only matrix elements between states $(n\ell jm_j)$ and $(n'\ell jm_j)$, which have different zeroth-order energies $E^{(0)}_{n'} \neq$

$E_n^{(0)}$, we encounter no degeneracy problems and can use Eq. (7.43) directly to calculate $\psi_{n\ell jm_j}(\mathbf{r})$ as desired. [Similarly, we could calculate the energy correct to second order by using Eq. (4.33).]

Let us summarize briefly. We have introduced spin into the theory of the hydrogenic atom, finding that the spin interacts with the spatial attributes of the electron, specifically with an effective magnetic field due to its orbital angular momentum. This fact leads to a new term $\mathcal{H}_{so}$, which must be added to $\mathcal{H}^{(0)}$ to give the full Hamiltonian of the isolated atom. It also leads to a coupling of **L** and **S** to form the total angular momentum **J**. As a result of this interaction of the two angular momenta, S_z and L_z are not sharp for eigenfunctions of $\mathcal{H}$.

We found justification for using perturbation theory in the fact that $\mathcal{H}_{so}$ is small. In order to avoid the degeneracies that characterize the zeroth-order functions $\psi^{(0)}_{n\ell m_\ell m_s}(\mathbf{r})$, we chose to use the alternative zeroth-order functions $\psi^{(0)}_{n\ell jm_j}(\mathbf{r})$.

Two tasks remain: to find explicit expressions for the functions $\psi^{(0)}_{n\ell jm_j}(\mathbf{r})$ and to use these functions to evaluate approximate energies and wave functions of the hydrogenic atom.

The New Zeroth-Order Functions

Let us tackle the first task (the second is the subject of Sec. 7.4). The functions $\psi^{(0)}_{n\ell jm_j}(\mathbf{r})$ are two-component vectors; each component is a function of **r**. We have available another set of two-component vectors, each component of which is a function of **r**—namely, $\{\psi^{(0)}_{n\ell m_\ell m_s}(\mathbf{r})\}$. Moreover, this set is complete. Therefore we can expand each function $\psi^{(0)}_{n\ell jm_j}(\mathbf{r})$ in the form[9]

$$\psi^{(0)}_{n\ell jm_j}(\mathbf{r}) = \sum_{m_\ell m_s} a_{m_\ell m_s} \psi^{(0)}_{n\ell m_\ell m_s}(\mathbf{r}), \tag{7.44}$$

where the expansion coefficients $a_{m_\ell m_s}$ are real numbers called *Clebsch-Gordan coefficients*. They are often written so as to display all the angular momentum quantum numbers:

$$a_{m_\ell m_s} \equiv \langle \ell \tfrac{1}{2} m_\ell m_s | \ell \tfrac{1}{2} j m_j \rangle. \tag{7.45a}$$

These coefficients can be looked up in tables like those in Appendix 4 or derived explicitly.[10] In general, the Clebsch-Gordan coefficient satisfies

[9] We do not sum over n or ℓ in this expansion, since both $\psi^{(0)}_{n\ell jm_j}(\mathbf{r})$ and $\psi^{(0)}_{n\ell m_\ell m_s}(\mathbf{r})$ are eigenfunctions of $\mathcal{H}^{(0)}$ and L^2. We only include functions $\psi^{(0)}_{n\ell m_\ell m_s}(\mathbf{r})$ in the summation that have the same principal and orbital angular momentum quantum numbers as those of the function $\psi^{(0)}_{n\ell jm_j}(\mathbf{r})$ that we are expanding.

[10] See, for example, Albert Messiah, *Quantum Mechanics*, Vol. 2 (Amsterdam: North-Holland Publishing Co., 1966), pp. 560–563.

$$\langle \ell\, s\, m_\ell\, m_s | \ell\, s\, j\, m_j \rangle = 0 \quad \text{unless} \quad \begin{cases} \Delta(\ell, s, j) \\ \quad \text{and} \\ m_j = m_\ell + m_s, \end{cases} \tag{7.45b}$$

where $\Delta(\ell, s, j)$ is shorthand notation for the restriction $|\ell - s| \leq j \leq \ell + s$. Equation (7.45b) holds for any allowed indices ℓ, s, m_ℓ, m_s, j, and m_j. Equation (7.44), then, is the desired expression for $\psi^{(0)}_{n\ell jm_j}(\mathbf{r})$. (See Prob. 7.2 for an example of the calculation of Clebsch-Gordan coefficients for the $n = 2$ states of hydrogen.) Next, we shall substitute the expansion (7.44) into Eqs. (7.39) and (7.43) in order to calculate first-order approximations to $\psi^{(0)}_{n\ell jm_j}(\mathbf{r})$ and $E_{n\ell jm_j}$.

7.4 SPIN-ORBIT INTERACTION ENERGY CORRECTION

The unperturbed hydrogenic energies are [see Eq. (3.62)]

$$E^{(0)}_n = -\frac{Z^2}{n^2}\left(\frac{e^2}{2a_0}\right). \tag{7.46}$$

To first order, the perturbed hydrogenic energies are given by Eq. (7.39); that is,

$$E_{n\ell jm_j} = E^{(0)}_n + \langle \psi^{(0)}_{n\ell jm_j} | \mathcal{H}_{so} | \psi^{(0)}_{n\ell jm_j} \rangle, \tag{7.47}$$

where we are using the eigenfunctions of Eq. (7.44) as the zeroth-order wave functions. For any particular state $(n\ell jm_j)$, we can look up the appropriate Clebsch-Gordan coefficients and form

$$\psi^{(0)}_{n\ell jm_j}(\mathbf{r}) = \sum_{m_\ell=-\ell}^{\ell} \sum_{m_s=-1/2}^{1/2} \langle \ell\, \tfrac{1}{2}\, m_\ell\, m_s | \ell\, \tfrac{1}{2}\, j\, m_j \rangle \psi^{(0)}_{n\ell m_\ell m_s}(\mathbf{r}). \tag{7.48}$$

The matrix element in (7.47) is easily evaluated if we use the form [see Eq. (7.41)]

$$\mathcal{H}_{so} = \frac{Ze^2}{4m_e^2c^2}\left(\frac{1}{r^3}\right)[J^2 - L^2 - S^2]. \tag{7.49}$$

After a little algebra, we obtain (for $\ell \neq 0$)

$$E_{n\ell j} = E^{(0)}_n + \frac{Z^2 |E^{(0)}_n| \alpha^2}{2n} \frac{[j(j+1) - \ell(\ell+1) - \frac{3}{4}]}{\ell(\ell + \frac{1}{2})(\ell + 1)}, \tag{7.50}$$

where we have introduced the *fine-structure constant* α, defined by

$$\alpha \equiv \frac{e^2}{\hbar c} \simeq \frac{1}{137}. \tag{7.51}$$

Exercise 7.5 Derive Eq. (7.50).

Notice that the first-order perturbed energy depends on the quantum numbers n, ℓ, j, and $s\ (=\frac{1}{2})$ but not on m_j; this fact is reasonable, since $\mathcal{H}_{so}$ in Eq. (7.49) does not explicitly contain J_z. Thus the effect of the spin-orbit interaction is to lift some of the degeneracy of the nth level of the "spinless" hydrogenic atom. The level that in Chapter 3 was n^2-fold degenerate with energy $E_n^{(0)}$ and wave functions $\psi_{n\ell jm_j}^{(0)}(\mathbf{r})$ or $\psi_{n\ell m_\ell m_s}^{(0)}(\mathbf{r})$ is revealed to consist of a number of separate levels when spin is taken into account. Each of these new levels has energy $E_{n\ell j}$ given by (7.50) and is $(2j + 1)$-fold degenerate.

We could go on to calculate the wave functions $\psi_{n\ell jm_j}(\mathbf{r})$ to first order by simply substituting the expansion (7.48) for $\psi_{n\ell jm_j}^{(0)}(\mathbf{r})$ into (7.43) and evaluating the largest terms in the summation. Similarly, we could calculate higher-order corrections to the energy, using the standard equations of time-independent perturbation theory.

With one exception, this completes our revision of the theory of the one-electron atom. It turns out that there is one additional correction to the energy, the so-called *relativistic correction*. A derivation of this correction term requires the use of Dirac theory, which is beyond the scope of this text.[11] The point to be retained is that this correction term is of roughly the same order as the spin-orbit correction, $\sim 10^{-4}$ eV, and is simply added to $E_{n\ell j}$ to obtain highly accurate values for the energies of the one-electron atom that agree very well with spectroscopic observations.[12]

7.5 THE ZEEMAN EFFECT

In Chapters 4 and 5 we explored the behavior of a hydrogenic atom (without spin) in the presence of an external field. Now that we have seen how to correct the Schroedinger theory to include electron spin, let us consider an example in which the hydrogenic atom is exposed to an external field. In particular, we shall discuss the effect on the atom of a static, homogeneous, external magnetic field; this is called the *anomalous Zeeman effect*.

In general, the term Zeeman effect refers to the splitting of energy levels that occurs when an atom is placed in an external magnetic field. If the field $\mathbf{B}$ is very strong, then the potential energy term due to the spin-orbit interaction $\mathcal{H}_{so}$ is small compared to the term due to the interaction of the atom

[11] See, however, Robert M. Eisberg, *Fundamentals of Modern Physics* (New York: Wiley, 1961), pp. 353–355, for a derivation of the correction term for $\ell > 0$.

[12] Still another effect, the so-called *hyperfine interaction*, takes into account the interaction of the spin of the electron with the spin of the *nucleus*, which we have ignored. The correction term is of the order of 10^{-7} eV and thus is too small to be observed in most spectral observations. See, for example, H. G. Kuhn, *Atomic Spectra* (New York: Academic 1969), Chap. 6.

with the field. However, if the field is weak, then the two interaction terms can be of comparable magnitude and both must be included as perturbations in the full Hamiltonian.

If the field is homogeneous and directed in the z direction, we can write

$$\mathbf{B} = B_z\hat{\mathbf{z}}. \tag{7.52}$$

The potential energy due to the interaction of the magnetic moment of the electron with the magnetic field is $-\mathbf{M}\cdot\mathbf{B}$. Since the electron possesses both an orbital magnetic moment $\mathbf{M}_\ell = -(g_\ell\beta/\hbar)\mathbf{L}$ and spin magnetic moment $\mathbf{M}_s = -(g_s\beta/\hbar)\mathbf{S}$, the total one-electron magnetic moment is

$$\mathbf{M} = \mathbf{M}_\ell + \mathbf{M}_s, \tag{7.53}$$

or

$$\mathbf{M} = -\frac{\beta}{\hbar}(\mathbf{L} + 2\mathbf{S}), \tag{7.54}$$

where we have used $g_\ell = 1$ and $g_s = 2$. [Recall that β is the Bohr magneton, $\beta = e\hbar/2\mu c$, of Eq. (3.101).] Since $\mathbf{B}$ is directed parallel to $\hat{\mathbf{z}}$, the potential energy term due to the interaction of the atom's magnetic moment with the field is $-M_zB_z$.

If we choose the zeroth-order Hamiltonian to be that of the isolated hydrogenic atom *without spin*, Eq. (7.2), then the perturbation Hamiltonian must contain both the term due to the spin-orbit interaction and the term $-M_zB_z$; that is,

$$\mathcal{H}^{(1)} = \mathcal{H}_{so} - M_zB_z. \tag{7.55}$$

The full Hamiltonian for the one-electron atom in the field $\mathbf{B}$ is then

$$\mathcal{H} = \mathcal{H}^{(0)} + \mathcal{H}^{(1)}. \tag{7.56}$$

We shall treat the problem via time-independent perturbation theory, using the functions $\psi^{(0)}_{n\ell jm_j}(\mathbf{r})$ as zeroth-order eigenfunctions. The energy of state $(n\ell jm_j)$, correct to first order, which we shall call $E_{n\ell jm_j}$, is

$$E_{n\ell jm_j} \simeq E_{n\ell j} - B_z\langle\psi^{(0)}_{n\ell jm_j}|M_z|\psi^{(0)}_{n\ell jm_j}\rangle, \tag{7.57}$$

where $E_{n\ell j}$ is given by Eq. (7.50).

All that remains is the evaluation of the matrix element of M_z. The z component of the total one-electron magnetic moment is

$$\begin{aligned} M_z &= -\frac{\beta}{\hbar}(L_z + 2S_z) \\ &= -\frac{\beta}{\hbar}(J_z + S_z), \end{aligned} \tag{7.58}$$

where we have used the fact that $J_z = L_z + S_z$. Therefore the energy $E_{n\ell jm_j}$ can be written

$$E_{n\ell jm_j} \simeq E_{n\ell j} + \frac{\beta B_z}{\hbar}\langle\psi^{(0)}_{n\ell jm_j}|J_z + S_z|\psi^{(0)}_{n\ell jm_j}\rangle. \tag{7.59}$$

The result, correct to first order, is (see Prob. 7.4)

$$E_{n\ell jm_j} \simeq E_{n\ell j} + g\beta B_z m_j, \tag{7.60}$$

where g is the so-called *Landé g factor*,

$$g = 1 + \frac{j(j+1) + s(s+1) - \ell(\ell+1)}{2j(j+1)}. \tag{7.61}$$

Thus the application of an external magnetic field in the z-direction lifts the one remaining degeneracy in the energy of the hydrogenic atom. Each level now has an energy $E_{n\ell jm_j}$ and a corresponding wave function $\psi_{n\ell jm_j}(\mathbf{r})$, which can be obtained by perturbation theory. The splitting of the m_j degeneracy induced by the magnetic field is of magnitude $g\beta B_z m_j$; notice that this splitting increases with increasing B_z.

So long as the m_j splitting ("Zeeman splitting") is small compared to the j splitting ("spin-orbit splitting"), the procedure outlined here is valid and Eq. (7.60) is a good approximation. However, at very large B_z, the term $-M_z B_z$ becomes more important than $\mathcal{H}_{so}$ and Eq. (7.60) is incorrect. [The effect of a strong magnetic field on the states of a hydrogen atom is called *the Paschen-Back effect* (see Prob. 7.4f).]

When several perturbations are acting simultaneously, it is often best to treat them individually, in decreasing order of importance. Thus, in our example, for small fields B_z, we first allowed for the spin-orbit coupling by choosing $\psi^{(0)}_{n\ell jm_j}(\mathbf{r})$ as the basis functions. If B_z is very large, a better choice would be $\psi^{(0)}_{n\ell m_\ell m_s}(\mathbf{r})$. In cases where the two perturbations are of approximately equal importance, either choice is appropriate.

7.6 THE SAGA OF GOOD AND BAD QUANTUM NUMBERS

A great profusion of quantum numbers has appeared in our study of the hydrogenic atom. Before bringing that study to a close and moving on to multielectron atoms, let us take a last look at the situation vis-à-vis these quantum numbers.

We saw in Sec. 7.1 that, in the presence of the spin-orbit interaction, L_z and S_z are not sharp, and the corresponding quantum numbers m_ℓ and m_s cannot be used to label the eigenfunctions of the Hamiltonian $\mathcal{H}$; we say that m_ℓ and m_s are no longer good quantum numbers.

A quantum number is simply a parameter that labels eigenvalues of a quantum mechanical operator. If this operator is a constant of the motion[13] then the wave function of a particular stationary state of the system may be chosen in such a way that it is simultaneously an eigenfunction of the operator. In such a case, the quantum number is called a *good quantum number*, since it can be used to characterize the state. Good quantum numbers are usually chosen as labels for the eigenfunctions of the Hamiltonian of the system.

For example, consider again the one-electron atom, first ignoring the spin-orbit interaction. We have the following constants of the motion and corresponding candidates for good quantum numbers:

Constants of the motion	$\mathcal{H}^{(0)}$	L^2	L_z	S^2	S_z	J^2	J_z
"Good" quantum numbers	n	ℓ	m_ℓ	s	m_s	j	m_j

Since we are ignoring $\mathcal{H}_{so}$, both $\psi^{(0)}_{n\ell jm_j}(\mathbf{r})$ and $\psi^{(0)}_{n\ell m_\ell m_s}(\mathbf{r})$ are perfectly good Hamiltonian eigenfunctions. Note that L_x, L_y, S_x, S_y, J_x, and J_y are constants of the motion, and the eigenvalues of these operators could also be used as good quantum numbers. However, L_z and L_x do not commute and so do not share simultaneous eigenfunctions. Each of the operators L_x and L_z shares a different eigenfunction with the Hamiltonian. We prefer to work with eigenfunctions of $\mathcal{H}$ and the z-components of the angular momentum operators.

When the spin-orbit interaction is included, L_z and S_z are no longer constants of the motion. Consequently, m_ℓ and m_s are no longer good quantum numbers and cannot be used to label eigenfunctions of $\mathcal{H}$.

A moment's thought will reveal a perplexing anomaly. Due to the presence of the r^{-3} term in $\mathcal{H}_{so}$, $\mathcal{H}$ does not commute with $\mathcal{H}^{(0)}$. Therefore when the spin-orbit interaction *is* included, $\mathcal{H}^{(0)}$ itself is not a constant of the motion, and the principal quantum number n is *not* a good quantum number! How, then, do we justify using it to label the eigenfunctions $\psi_{n\ell jm_j}(\mathbf{r})$?

The answer is that *n is approximately a good quantum number*. To see what this statement means, consider the first-order wave functions that would be obtained from a perturbation theory calculation based on Eq. (7.43). It is clear from this equation that n is "spoiled" as a good quantum number, since eigenfunctions of different values of n are mixed by the perturbation. However, the terms in the infinite summation that are proportional to $\langle\psi^{(0)}_{n'\ell jm_j}|\mathcal{H}_{so}|\psi^{(0)}_{n\ell jm_j}\rangle$ are certainly no larger than $\langle\psi^{(0)}_{n\ell jm_j}|\mathcal{H}_{so}|\psi^{(0)}_{n\ell jm_j}\rangle$, which is

[13]Recall that a constant of the motion of a system is a Hermitian operator that commutes with the Hamiltonian of the system.

simply the first-order correction to the energy [see Eq. (7.47)]. From Eq. (7.50) we see that

$$\langle \psi^{(0)}_{n\ell jm_j} | \mathcal{H}_{so} | \psi^{(0)}_{n\ell jm_j} \rangle \simeq \frac{Z^2\alpha^2}{2n} | E^{(0)}_n |, \tag{7.62}$$

where

$$\alpha^2 \simeq 10^{-4}. \tag{7.63}$$

Since the energy denominators in the summation terms in Eq. (7.43) are of the order of $E^{(0)}_n$, the mixing of states with $n' \neq n$ is very small, $\psi^{(0)}_{n\ell jm_j}(\mathbf{r})$ is a good approximation to the Hamiltonian eigenfunction $\psi_{n\ell jm_j}(\mathbf{r})$, and n is approximately a good quantum number.

PROBLEMS

7.1 Classical Model for Precession of L and S about J (**)

The purpose of this problem is to show that the electron spin and orbital motion, when considered as classical angular momenta, appear to precess about their resultant **J**. (In quantum mechanics this coupling also takes place, but it is described in a different way.) We shall ignore Thomas precession.

(a) Consider the model described in the text, in which the magnetic field due to relative nuclear motion exerts a torque on the spin magnetic moment. Use Newton's Second Law to derive an expression for $d\mathbf{S}/dt$ in terms of **S**, **J**, constants, and the orbital radius r.

(b) Take **J** as defining the $\hat{z}$ axis and assume that $d\mathbf{J}/dt = 0$. Find and solve differential equations for S_x, S_y, and S_z.

(c) Discuss your results, showing that **S** precesses with an angular frequency that is independent of time. Obtain a numerical estimate for this frequency for the $2p$ state of hydrogen, assuming reasonable values for all unspecified parameters.

(d) Carry out the same analysis for the orbital angular momentum **L**.

7.2 Construction of Eigenfunctions of J^2 and J_z for the $n = 2$ States of Hydrogen—Use of Raising and Lowering Operators (***)

Consider a hydrogenic (one-electron) atom with principal quantum number $n = 2$.

(a) If we neglect spin-orbit coupling, the states of this atom can be specified equally well either by the set of quantum numbers $(n\ \ell\ m_\ell\ m_s)$ or by the set $(n\ \ell\ j\ m_j)$. Write down the set of quantum numbers $(2\ \ell\ m_\ell\ m_s)$ for each possible $n = 2$ eigenstate of this atom.

(b) Write down all possible eigenstates of this atom, using the alternative description $(2\ \ell\ j\ m_j)$.

(c) Now you have specified the eight possible states of the atom in two different ways. Therefore each state in one representation must be a linear combination of the states in the other representation. (Why?) Expressing this fact in an equation, we write, for each eigenfunction $\psi^{(0)}_{2\ell jm_j}$,

$$\psi^{(0)}_{2\ell jm_j} = \sum_{m_\ell}\sum_{m_s} a_{m_\ell m_s}\psi^{(0)}_{2\ell m_\ell m_s},$$

where $a_{m_\ell m_s}$ are the expansion coefficients. Use the raising and lowering operators $J_\pm$, $L_\pm$, and $S_\pm$, along with the orthonormality property of the wave functions, to determine the expansion coefficients in each of these expansions. [HINT: Define raising and lowering operators $J_\pm = J_x \pm iJ_y$. These operators obey the same rules as do $L_\pm$ and $S_\pm$.]

7.3 Another Look at the $n = 2$ States of Hydrogen (****)

In this problem we want to calculate the first-order correction to the $n = 2$ unperturbed energy of the hydrogen atom due to the spin-orbit interaction. However, instead of the formula derived in the text, we shall use the techniques of Sec. 4.6.

(a) Using the zeroth-order eigenfunctions $\psi^{(0)}_{n\ell m_\ell m_s}$ (where $n = 2$), set up the problem and calculate the matrix elements needed for the secular equation; show how the secular determinant can be broken into a block structure. (The largest block you will have to deal with is 2×2.) You must think carefully about how to order the states in such a way that these blocks appear. [HINT: You will find it convenient to prove and use the identity

$$\mathbf{S}\cdot\mathbf{L} = \tfrac{1}{2}(S_-L_+ + L_-S_+ + 2L_zS_z).]$$

(b) Find all eight roots of the secular determinant. (Some are very simple to write down.) Verify that your results agree with those given in the text.

(c) Find all eight new zeroth-order eigenfunctions corresponding to the energies obtained in part (b). Show that these functions are precisely the same as the eigenfunctions $\psi^{(0)}_{n\ell jm_j}(\mathbf{r})$ for $n = 2$ and thus are eigenfunctions of L^2, S^2, J^2 and J_z.

(d) Write the expression for the second-order correction to the energies of part (b) due to the $3p$ states. Leave your answer in terms of the radial integrals. Indicate explicitly which of the $3p$ states are effective in the perturbation.

7.4 The Zeeman Effect (****)

Consider a hydrogen atom in a uniform external magnetic field $\mathbf{B} = B_z\hat{\mathbf{z}}$, directed along the $\hat{z}$ axis. Our goal is to find the first-order corrections to the energies of the $2p$ ($n = 2$, $\ell = 1$) states of atomic hydrogen due to the combined effects of the spin-orbit interaction and the interaction of the magnetic moment of the atom and the magnetic field $\mathbf{B}$. We shall use the techniques of Sec. 4.6 to handle the degeneracy of the unperturbed wave functions.

(a) Take for the zeroth-order Hamiltonian $\mathcal{H}^{(0)}$ as in Eq. (7.2) and for the zeroth-order eigenfunctions $\psi^{(0)}_{n\ell jm_j}(\mathbf{r})$, $n = 2$, $\ell = 1$. How many zeroth-order states are there for $n = 2$, $\ell = 1$, and which ones are degenerate?

(b) The combined perturbations may be written

$$\mathcal{H}^{(1)} = \mathcal{H}_{so} - \mathbf{M}\cdot\mathbf{B} = \mathcal{H}_{so} - M_zB_z,$$

where $\mathbf{M}$ and $\mathcal{H}_{so}$ are given by Eqs. (7.54) and (7.18) (take $Z = 1$). Indicate which of the operators L^2, S^2, J^2, and J_z commute with $\mathcal{H}^{(1)}$ and hence with $\mathcal{H} = \mathcal{H}^{(0)} + \mathcal{H}^{(1)}$. What does this tell you about the nature of the eigenfunctions of $\mathcal{H}$, as compared with the zeroth-order eigenfunctions $\psi^{(0)}_{n\ell jm_j}(\mathbf{r})$?

(c) Write expressions for all the matrix elements of $\mathcal{H} = \mathcal{H}^{(0)} + \mathcal{H}^{(1)}$ in terms of the zeroth-order energy eigenvalues, the "spin-orbit" energy, and the matrix elements

$$\langle \psi^{(0)}_{n\ell jm_j} | S_z | \psi^{(0)}_{n\ell j'm_j'} \rangle.$$

Using the results of part (b), determine how to order the states so that the secular determinant breaks up into a simple block structure. (Be sure to label carefully each row and column of the determinant with the appropriate values of j and m_j.)

(d) Calculate the required matrix elements of S_z by writing the function $\psi^{(0)}_{n\ell jm_j}(\mathbf{r})$ in terms of the $\psi^{(0)}_{n\ell m_\ell m_s}(\mathbf{r})$. (Use the table of Clebsch-Gordan coefficients provided in Appendix 4.)

(e) Solve for all the roots, thereby obtaining the energy eigenvalues of $\mathcal{H}$ correct to first order. Based on your results, draw an energy level diagram that shows how the energy levels change as the perturbations $\mathcal{H}_{so}$ and $-M_zB_z$ are included. [Draw three sets of energy levels: one for $\mathcal{H}^{(0)}$, one for $\mathcal{H}^{(0)} + \mathcal{H}_{so}$, and, finally, one for $\mathcal{H} = \mathcal{H}^{(0)} + \mathcal{H}_{so} - M_zB_z$.]

(f) Consider the "strong field" case where the magnetic field is very large, so that the spin-orbit coupling is relatively unimportant. Use the energy expressions derived in part (e) to obtain simple limiting forms for the energy eigenvalues in the strong field case. Show how you could obtain this result without the use of degenerate perturbation theory. (This is called the Paschen-Back effect.)

8

Introduction to the Quantum Mechanics of the Multielectron Atom

"The difficulty is to detach the framework of fact—of absolute undeniable fact—from the embellishments of theorists and reporters. Then, having established ourselves upon this sound basis, it is our duty to see what inferences may be drawn and what are the special points upon which the whole mystery turns."
Arthur Conan Doyle, Sherlock Holmes to Dr. Watson in "Silver Blaize," The Complete Sherlock Holmes *(Garden City, N.Y.: Doubleday & Company).*

The full theory of the one-electron atom, which was finally developed in Chapter 7, is the basis of the quantum theory of multielectron atoms. In the next three chapters we shall explore various multielectron atoms, both increasing our knowledge of atomic physics and preparing for the study of molecules, which begins in Chapter 11. As might be imagined, all multielectron atoms must be treated approximately; the Schroedinger equation for such systems simply cannot be solved exactly.

In this chapter we shall first present the elements of the Schroedinger theory for a multielectron atom and briefly discuss some features of the problem not encountered in one-electron atoms. Next, we begin the study of multielectron atoms as such by considering two useful models in one dimension; these models enable us to understand many properties of true three-dimensional atoms in a relatively simple context.

8.1 THE MANY-PARTICLE PROBLEM IN THREE DIMENSIONS

It is useful to begin by considering the general problem of a system with many (more than two) particles (e.g., electrons and nuclei) so that we can become familiar with the type of operators and wave functions that will appear throughout the remainder of this book. (Most of the results of this section are merely generalizations of equations in Chapter 1.) Describing a state of a system of, say, N particles is no trivial matter, for the wave function Ψ must represent the entire system. Therefore Ψ will, in general, be a function of $3N$ spatial coordinates and the time. It must also indicate the spins of the particles. The system itself is described by the Hamiltonian $\mathcal{H}$, which is the sum of the operators for the total kinetic energy and the total potential energy.

The Schroedinger Equation

For a system of N particles with coordinates $\mathbf{r}_j$, masses m_j $(j = 1, \ldots, N)$, and total potential energy $V(\mathbf{r}_1, \ldots, \mathbf{r}_N, t)$, the Hamiltonian is

$$\mathcal{H} = T + V \tag{8.1}$$

or

$$\mathcal{H} = \sum_{j=1}^{N} \left(-\frac{\hbar^2}{2m_j}\nabla_j^2\right) + V(\mathbf{r}_1, \ldots, \mathbf{r}_N, t), \tag{8.2}$$

where ∇_j^2 is the Laplacian for the jth particle; that is,

$$\nabla_j^2 = \frac{\partial^2}{\partial x_j^2} + \frac{\partial^2}{\partial y_j^2} + \frac{\partial^2}{\partial z_j^2}. \tag{8.3}$$

Every physically realizable state of the system can be represented by a wave function $\Psi(\mathbf{r}_1, \ldots, \mathbf{r}_N, t)$ that satisfies the time-dependent Schroedinger equation

$$\mathcal{H}\Psi(\mathbf{r}_1, \ldots, \mathbf{r}_N, t) = i\hbar\frac{\partial}{\partial t}\Psi(\mathbf{r}_1, \ldots, \mathbf{r}_N, t). \tag{8.4}$$

The probability density is defined as

$$\rho(\mathbf{r}_1, \ldots, \mathbf{r}_N, t) \equiv \Psi^*(\mathbf{r}_1, \ldots, \mathbf{r}_N, t)\Psi(\mathbf{r}_1, \ldots, \mathbf{r}_N, t), \tag{8.5}$$

so that $\rho(\mathbf{r}_1, \ldots, \mathbf{r}_N, t)\, d\mathbf{r}_1\, d\mathbf{r}_2 \cdots d\mathbf{r}_N$ is the probability of finding particle 1 at $\mathbf{r}_1$ in volume element $d\mathbf{r}_1$ and particle 2 at $\mathbf{r}_2$ in $d\mathbf{r}_2$ and . . . and particle N at $\mathbf{r}_N$ in volume element $d\mathbf{r}_N$, all at time t. The wave function Ψ can be normalized so that

$$\int \rho(\mathbf{r}_1, \ldots, \mathbf{r}_N, t)\, d\mathbf{r}_1\, d\mathbf{r}_2 \cdots d\mathbf{r}_N = 1. \tag{8.6}$$

Stationary States

A stationary state of the system is a special state, one in which the energy is sharp. Stationary states exist only if the total potential energy is independent of time. The wave function of a stationary state of the system can be written

$$\Psi(\mathbf{r}_1, \ldots, \mathbf{r}_N, t) = \psi_E(\mathbf{r}_1, \ldots, \mathbf{r}_N)e^{-i(E/\hbar)t}, \tag{8.7}$$

where ψ_E is an eigenfunction of the Hamiltonian $\mathcal{H}$ and therefore satisfies the time-independent Schroedinger eigenvalue equation

$$\mathcal{H}\psi_E(\mathbf{r}_1, \ldots, \mathbf{r}_N) = E\psi_E(\mathbf{r}_1, \ldots, \mathbf{r}_N). \tag{8.8}$$

The eigenvalue E is the energy of the system in this state.

Several additional properties characterize a stationary state. In particular, the probability density ρ, the probability flux density $\mathbf{j}$, and the expectation values of all time-independent operators are independent of time. For a time-independent operator, G, the expectation value is

$$\langle G \rangle = \underbrace{\int \cdots \int}_{(N)} \Psi^*(\mathbf{r}_1, \ldots, \mathbf{r}_N, t) G \Psi(\mathbf{r}_1, \ldots, \mathbf{r}_N, t)\, d\mathbf{r}_1\, d\mathbf{r}_2 \cdots d\mathbf{r}_N, \tag{8.9}$$

where we integrate over all spatial coordinates of the N particles.

The Multielectron Atom

Our main focus at present will be the N-electron atom.[1] The nucleus of charge Ze is taken as the origin of coordinates, the location of each electron is specified by coordinate $\mathbf{r}_j$, and the distance between each pair of electrons, r_{ij}, is defined as

$$r_{ij} \equiv |\mathbf{r}_i - \mathbf{r}_j|. \tag{8.10}$$

The Hamiltonian of the N-electron atom, neglecting magnetic interactions (such as the spin-orbit interaction), is

$$\mathcal{H} = -\frac{\hbar^2}{2m_e}\sum_{j=1}^{N} \nabla_j^2 + V(\mathbf{r}_1, \ldots, \mathbf{r}_N), \tag{8.11}$$

where m_e is the mass of an electron and $V(\mathbf{r}_1, \ldots, \mathbf{r}_N)$ is the total potential energy,

$$V(\mathbf{r}_1, \ldots, \mathbf{r}_N) = \sum_{j=1}^{N}\left(-\frac{Ze^2}{r_j}\right) + \sum_{j=1}^{N}\sum_{i>j}^{N}\left(\frac{e^2}{r_{ij}}\right). \tag{8.12}$$

[1]The N-electron atom is actually an $(N+1)$-particle system if we take the nucleus into account. However, the nucleus is much heavier than the electrons and can be approximately considered as infinitely massive (see Sec. 3.2).

$V(\mathbf{r}_1, \ldots, \mathbf{r}_N)$ is simply the sum of the (negative) contributions arising from the attractive coulomb interaction of each electron with the nucleus and the (positive) contribution due to the repulsive coulomb interaction between each pair of electrons.[2] Notice that V is independent of time.

8.2 IDENTICAL PARTICLES AND EXCHANGE SYMMETRY

A one-electron atom consists of two clearly distinguishable particles—an electron of mass m_e and charge $-e$ and a nucleus of mass m_n and charge $+Ze$. On the other hand, an N-electron atom consists of a nucleus and N electrons. All the electrons have the same mass and charge; they are *identical particles*, and therefore one electron cannot be distinguished from another.

In quantum mechanics the word "indistinguishable" has a precise meaning.[3] Suppose that we interchange the coordinates (and spins) of two particles. If it is not possible to determine by a physical measurement that a change in the system has been made, then the particles are said to be *indistinguishable*. Thus all measurable quantities must be unchanged by the interchange of indistinguishable particles (e.g., the probability density and the expectation value of a Hermitian operator).

If the coordinates of two electrons, say $\mathbf{r}_1$ and $\mathbf{r}_2$, are interchanged, the Hamiltonian $\mathcal{H}$ of Eq. (8.11) remains unchanged. We say "the Hamiltonian of a system is invariant with respect to the interchange of any two electrons." This property of $\mathcal{H}$ is called *exchange symmetry*.

We can express the exchange symmetry of the system mathematically by introducing the *particle interchange operators* P_{ij}. For particular i and j, P_{ij} simply interchanges the coordinates (and spins) of particles i and j so that, for example,

$$P_{12}\Psi(\mathbf{r}_1, \mathbf{r}_2, \mathbf{r}_3, \ldots, \mathbf{r}_N, t) = \Psi(\mathbf{r}_2, \mathbf{r}_1, \mathbf{r}_3, \ldots, \mathbf{r}_N, t). \tag{8.13}$$

Thus, in general, we say that the Hamiltonian of a system is invariant with respect to the interchange of any two identical particles; that is, $\mathcal{H}$ commutes with all the operators $\{P_{ij}\}$:

$$\boxed{[\mathcal{H}, P_{ij}] = 0.} \quad \text{Exchange symmetry} \tag{8.14}$$

[2]You should convince yourself that the summation in the second term of Eq. (8.12) is correct. The point is that we must include each interaction only once. This term may also be written as

$$\frac{1}{2}\sum_{j=1}^{N}\sum_{\substack{i=1\\(i\neq j)}}^{N}\left(\frac{e^2}{r_{ij}}\right) = \frac{e^2}{r_{12}} + \frac{e^2}{r_{13}} + \frac{e^2}{r_{23}} + \cdots.$$

[3]Robert H. Dicke and James P. Wittke, *Introduction to Quantum Mechanics* (Reading, Mass: Addison-Wesley, 1960), Chap. 17.

Example 8.1
Two Particles in Identical One-Dimensional Potentials

Consider a simple one-dimensional system consisting of two particles of mass m in identical harmonic oscillator potentials. If the particles interact with one another via a potential energy $V'(x_1 - x_2) = \frac{1}{2}k(x_1 - x_2)^2$, the Hamiltonian is (assume spinless particles)

$$\mathcal{H} = -\frac{\hbar^2}{2m}\frac{\partial^2}{\partial x_1^2} - \frac{\hbar^2}{2m}\frac{\partial^2}{\partial x_2^2} + \frac{1}{2}m\omega^2 x_1^2 + \frac{1}{2}m\omega^2 x_2^2 + \frac{1}{2}k(x_1 - x_2)^2. \tag{8.15}$$

Since operation on $\mathcal{H}$ by P_{12} simply reverses the subscripts 1 and 2, we can write the operator equation

$$P_{12}\mathcal{H} = \mathcal{H}P_{12} \tag{8.16}$$

or

$$[P_{12}, \mathcal{H}] = 0. \tag{8.17}$$

The operator P_{ij} is an example of a symmetry operation. For any system, the *symmetry operations* are defined by operators that leave the Hamiltonian of the system unchanged—that is, that commute with $\mathcal{H}$. In general, a symmetry operation operates on all coordinates of all particles in the system. All the operators P_{12}, P_{23}, P_{3N}, etc., are symmetry operators of the N-electron atom. Another frequently encountered symmetry operation is inversion $\imath$, which simply inverts the coordinates of all particles; that is,

$$\imath\Psi(\mathbf{r}_1, \mathbf{r}_2, \ldots, \mathbf{r}_N, t) = \Psi(-\mathbf{r}_1, -\mathbf{r}_2, \ldots, -\mathbf{r}_N, t). \tag{8.18}$$

We shall return to the subject of symmetry operations in the study of molecules (Chapter 15).

Exchange Degeneracy

Let us see what consequences the exchange symmetry of the Hamiltonian has for the eigenfunctions $\psi_E(\mathbf{r}_1, \mathbf{r}_2, \ldots, \mathbf{r}_N)$. These functions satisfy the time-independent Schroedinger equation

$$\mathcal{H}\psi_E = E\psi_E. \tag{8.19}$$

Operating on this equation with P_{ij}, we obtain

$$P_{ij}\mathcal{H}\psi_E = P_{ij}E\psi_E, \tag{8.20}$$

which, since $[\mathcal{H}, P_{ij}] = 0$, can be written

$$\mathcal{H}P_{ij}\psi_E = EP_{ij}\psi_E. \tag{8.21}$$

Now, $P_{ij}\psi_E$ is a new function, call it χ_E. Equation (8.21) tells us that χ_E is an eigenfunction of $\mathcal{H}$ with eigenvalue E. Therefore χ_E and ψ_E are degenerate functions. Given any eigenfunction ψ_E, we can operate on it with all the particle interchange operators and thereby generate a whole set of degenerate eigenfunctions of $\mathcal{H}$; this property of the eigenfunctions of $\mathcal{H}$ is called *exchange degeneracy*.

Among the many possible ψ_E which satisfy Eq. (8.21), there exist functions which have the property[4]

$$P_{ij}\psi_E = \gamma\psi_E, \tag{8.22}$$

where γ is a constant. Operating twice with P_{ij} leaves the operand unchanged, so we have

$$P_{ij}(P_{ij}\psi_E) = \gamma^2\psi_E = \psi_E, \tag{8.23}$$

which implies that

$$\gamma = \pm 1. \tag{8.24}$$

Experimental evidence verifies that Eq. (8.22) holds for all systems found in nature (if both spin and spatial coordinates are exchanged by P_{ij}) and the choice of $\gamma = +1$ or -1 depends in a very special way on the nature of the system. Therefore we can classify the eigenfunctions of a system as either symmetric or antisymmetric.[5] A *symmetric* wave function is one that remains unchanged under the interchange of two identical particles,

$$P_{ij}\psi_E = \psi_E \qquad \text{(symmetric)}. \tag{8.25}$$

An *antisymmetric* wave function is one that changes sign under the interchange of two identical particles,

$$P_{ij}\psi_E = -\psi_E \qquad \text{(antisymmetric)}. \tag{8.26}$$

[4]See A. Messiah, *Quantum Mechanics*, Vol. II (Amsterdam: North-Holland Pub. Co., and New York: John Wiley & Sons, Inc., 1965), p. 582.

[5]This method of classification actually applies to the total wavefunction $\Psi(\mathbf{r}_1, \mathbf{r}_2, \ldots, \mathbf{r}_N, t)$ as well as to Hamiltonian eigenfunctions. Since $P_{ij}\rho(\mathbf{r}_1, \mathbf{r}_2, \ldots, \mathbf{r}_N, t) = \rho(\mathbf{r}_1, \mathbf{r}_2, \ldots, \mathbf{r}_N, t)$, the wave function Ψ for any physically realizable state of the system must satisfy $P_{ij}\Psi = \gamma\Psi$. The results (8.25) and (8.26) follow immediately for stationary states.

It is further observed that the behavior of the wavefunction under pairwise particle interchange depends on the spin of the particles which are being interchanged. All particles are divided into two classes:

1. *bosons*: particles with integral spin ($s_i = 0, 1, 2, \ldots$),
2. *fermions*: particles with half-odd-integral spin ($s_i = \frac{1}{2}, \frac{3}{2}, \frac{5}{2}, \ldots$).

The wave function is symmetric under interchange of particles i and j if (and only if) i and j are identical bosons. The wave function is antisymmetric if (and only if) the particles being interchanged are identical fermions.

A large number of elementary particles belong to each class. For example, electrons, protons, and neutrons are known to be fermions with spin $= \frac{1}{2}$. The photon is a boson with spin $s = 1$, and the π-meson is a boson with spin $s = 0$.

> **Exercise 8.1** Prove that the exchange symmetry of a wavefunction Ψ is time independent; for example, if the wave function describing a state of a system at some particular time is symmetric, then it will necessarily be symmetric at all other times.

Notice that Eqs. (8.25) and (8.26) describe the properties of ψ_E only under the interchange of two bosons or two fermions. The eigenfunctions of a system containing both bosons and fermions satisfy these equations but do not exhibit any particular symmetry under interchange of a boson and a fermion. (This fact comes as no surprise, however, since these particles have different spins and thus are not identical.)

Our attention will focus primarily on systems of fermions (e.g., the multielectron atom), in which case the wave functions will be antisymmetric. We shall see that Eq. (8.26) provides the basis for the famous Pauli exclusion principle, which states that no two identical fermions in a system can occupy the same quantum state, where a quantum state is specified by the quantum numbers, spatial and spin, of the fermion. We shall discuss this principle in detail in Sec. 8.4.

8.3 TWO IDENTICAL PARTICLES IN ONE DIMENSION

The remarks of the previous sections concerning properties of several particles in three dimensions have been of a general nature. We shall now begin our study of the operational aspects of multielectron problems. Many of the difficulties, conceptual and notational, that must be dealt with in analyzing real systems appear in a simpler context in one-dimensional models. We shall therefore study briefly the quantum mechanical properties of two interacting identical particles in one dimension.

For such a system of particles (bosons or fermions) of mass m, the Hamiltonian is

$$\mathcal{H} = -\frac{\hbar^2}{2m}\frac{\partial^2}{\partial x_1^2} + V(x_1) - \frac{\hbar^2}{2m}\frac{\partial^2}{\partial x_2^2} + V(x_2) + V'(x_1 - x_2), \quad (8.27)$$

where $V'(x_1 - x_2)$ describes the interaction of the two particles. We shall assume that $V'(x_1 - x_2) > 0$ for all values of x_1 and x_2, that $V'(x_1 - x_2)$ attains its maximum value at $x_1 - x_2 = 0$ and decreases as $|x_1 - x_2|$ increases, and that $V'(x_1 - x_2) = V'(x_2 - x_1)$.

We seek stationary-state eigenfunctions of $\mathcal{H}$, solutions of

$$\mathcal{H}\psi_E(x_1, x_2) = E\psi_E(x_1, x_2). \quad (8.28)$$

The presence of the additional term $V'(x_1 - x_2)$ makes exact solution of (8.28) impossible. However, if $V'(x_1 - x_2)$ is small, then we can apply perturbation theory to the problem. Thus we separate $\mathcal{H}$ as $\mathcal{H} = \mathcal{H}^{(0)} + \mathcal{H}^{(1)}$, where

$$\mathcal{H}^{(0)} = -\frac{\hbar^2}{2m}\frac{\partial^2}{\partial x_1^2} + V(x_1) - \frac{\hbar^2}{2m}\frac{\partial^2}{\partial x_2^2} + V(x_2) \quad (8.29)$$

is the zeroth-order Hamiltonian and

$$\mathcal{H}^{(1)} = V'(x_1 - x_2) \quad (8.30)$$

is the perturbation Hamiltonian.[6]

The time-independent Schroedinger equation for $\mathcal{H}^{(0)}$ is

$$\mathcal{H}^{(0)}\psi^{(0)}_{n_1n_2}(x_1, x_2) = E^{(0)}_{n_1n_2}\psi^{(0)}_{n_1n_2}(x_1, x_2), \quad (8.31)$$

where the subscripts n_1 and n_2 represent quantum numbers to be specified, and where spin coordinates are implicit. Suppose that we can solve the one-particle Schroedinger equations (we assume the one-particle states are nondegenerate),

$$\left[-\frac{\hbar^2}{2m}\frac{d^2}{dx_1^2} + V(x_1)\right]\psi^{(0)}_{n_1}(x_1) = E^{(0)}_{n_1}\psi^{(0)}_{n_1}(x_1), \quad (8.32a)$$

$$\left[-\frac{\hbar^2}{2m}\frac{d^2}{dx_2^2} + V(x_2)\right]\psi^{(0)}_{n_2}(x_2) = E^{(0)}_{n_2}\psi^{(0)}_{n_2}(x_2). \quad (8.32b)$$

Then one function that satisfies Eq. (8.31) is $\psi^{(0)}_{n_1n_2}(x_1, x_2) = \psi^{(0)}_{n_1}(x_1)\psi^{(0)}_{n_2}(x_2)$;

[6]In Example 8.1 we considered two particles in identical oscillator potentials, where

$$V(x_1) = \tfrac{1}{2}m\omega^2 x_1^2, \qquad V(x_2) = \tfrac{1}{2}m\omega^2 x_2^2, \qquad V'(x_1 - x_2) = \tfrac{1}{2}k(x_1 - x_2)^2.$$

this is the product of two single-particle functions, or "orbitals." The corresponding energy is $E_{n_1n_2}^{(0)} = E_{n_1}^{(0)} + E_{n_2}^{(0)}$. (If the particles possess spin, assume here that the two orbitals have identical spin quantum numbers.)

Unfortunately, the function $\psi_{n_1}^{(0)}(x_1)\psi_{n_2}^{(0)}(x_2)$ is not physically admissible, for the eigenfunctions of $\mathcal{H}^{(0)}$ must be either symmetric or antisymmetric under pairwise particle interchange. How, then, do we form eigenfunctions of $\mathcal{H}^{(0)}$ with this property?

Exchange degeneracy supplies the answer. Consider the product function $\psi_{n_1}^{(0)}(x_2)\psi_{n_2}^{(0)}(x_1)$. This is also an eigenfunction of $\mathcal{H}^{(0)}$; it is obtained by operating on $\psi_{n_1}^{(0)}(x_1)\psi_{n_2}^{(0)}(x_2)$ with P_{12}. Since $\psi_{n_1}^{(0)}(x_1)\psi_{n_2}^{(0)}(x_2)$ and $\psi_{n_1}^{(0)}(x_2)\psi_{n_2}^{(0)}(x_1)$ are degenerate, we can combine them to form two other degenerate functions

$$ {}^S\psi_{n_1n_2}^{(0)} \equiv N_S[\psi_{n_1}^{(0)}(x_1)\psi_{n_2}^{(0)}(x_2) + \psi_{n_2}^{(0)}(x_1)\psi_{n_1}^{(0)}(x_2)], \tag{8.33a} $$

$$ {}^A\psi_{n_1n_2}^{(0)} \equiv N_A[\psi_{n_1}^{(0)}(x_1)\psi_{n_2}^{(0)}(x_2) - \psi_{n_2}^{(0)}(x_1)\psi_{n_1}^{(0)}(x_2)], \tag{8.33b} $$

where N_S and N_A are normalization constants. Both ${}^S\psi_{n_1n_2}^{(0)}$ and ${}^A\psi_{n_1n_2}^{(0)}$ correspond to $E_{n_1n_2}^{(0)} = E_{n_1}^{(0)} + E_{n_2}^{(0)}$. The first function is symmetric under particle interchange; the second is antisymmetric. Thus we use ${}^S\psi_{n_1n_2}^{(0)}$ as zeroth-order wave functions in perturbation theory calculations if the two particles are identical bosons.

[Notice that if the two identical fermions have the same quantum numbers, so that $n_1 = n_2$, then ${}^A\psi_{n_1n_2}^{(0)} = 0$; that is, the wave function describing the system is identically zero. (Recall that we have assumed the spin quantum numbers were identical in this simple example.) Therefore no two identical fermions can occupy the same quantum state. This fact reflects the Pauli exclusion principle, alluded to previously.]

Spin-0 Bosons

Suppose that the particles are bosons with spins $s_1 = s_2 = 0$. Since the spins of both particles are zero, we can treat them as structureless particles. Since $P_{12}\psi_E = +\psi_E$ for any eigenfunction of the system, the properly symmetrized zeroth-order functions are

$$ {}^S\psi_{n_1n_2}^{(0)} = \frac{1}{\sqrt{2}}[\psi_{n_1}^{(0)}(x_1)\psi_{n_2}^{(0)}(x_2) + \psi_{n_2}^{(0)}(x_1)\psi_{n_1}^{(0)}(x_2)], \tag{8.34} $$

Symmetric two-boson eigenfunction

where the factor of $1/\sqrt{2}$ normalizes the function.[7] Notice that, in accordance with the indistinguishability of the two particles, we cannot tell from

[7]If $n_1 = n_2$, we must multiply this function by an additional factor of $1/\sqrt{2}$ to enforce normalization.

${}^S\psi^{(0)}_{n_1n_2}$ which particle, the one at x_1 or the one at x_2, has quantum number n_1 and which has quantum number n_2.

We are now ready to proceed with the perturbation theory calculations. The ground state of the system is the state with the lowest possible energy; that is, with quantum numbers $n_1 = n_2 = 1$. For this state, the zeroth-order wave function is

$$ {}^S\psi^{(0)}_{11} = \psi^{(0)}_1(x_1)\psi^{(0)}_1(x_2). \tag{8.35}$$

To first-order, the perturbed energy is

$$E_{11} = 2E^{(0)}_1 + \langle {}^S\psi^{(0)}_{11} | V'(x_1 - x_2) | {}^S\psi^{(0)}_{11}\rangle. \tag{8.36}$$

The matrix element in (8.36) is

$$\langle {}^S\psi^{(0)}_{11} | V'(x_1 - x_2) | {}^S\psi^{(0)}_{11}\rangle$$
$$= \int_{-\infty}^{\infty}\int_{-\infty}^{\infty} |\psi^{(0)}_1(x_1)|^2\, V'(x_1 - x_2)\, |\psi^{(0)}_1(x_2)|^2\, dx_1\, dx_2.$$

To express the perturbed energy in a more convenient form, let us define the *direct integral* $J_{nn'}$ as

$$J_{nn'} \equiv \int_{-\infty}^{\infty}\int_{-\infty}^{\infty} |\psi^{(0)}_n(x_1)|^2\, V'(x_1 - x_2)\, |\psi^{(0)}_{n'}(x_2)|^2\, dx_1\, dx_2. \tag{8.37}$$

Then the first-order energy of the ground state is just

$$E_{11} = 2E^{(0)}_1 + J_{11}. \tag{8.38}$$

Let us turn now to the first excited state, which we shall choose to be the state with $n_1 = 1$, $n_2 = 2$. The zeroth-order wave function is

$${}^S\psi^{(0)}_{12} = \frac{1}{\sqrt{2}}[\psi^{(0)}_1(x_1)\psi^{(0)}_2(x_2) + \psi^{(0)}_2(x_1)\psi^{(0)}_1(x_2)], \tag{8.39}$$

so the first-order energy is

$$E_{12} = E^{(0)}_1 + E^{(0)}_2 + \langle {}^S\psi^{(0)}_{12} | V'(x_1 - x_2) | {}^S\psi^{(0)}_{12}\rangle. \tag{8.40}$$

Writing out the matrix element, we have

$$\langle {}^S\psi^{(0)}_{12} | V'(x_1 - x_2) | {}^S\psi^{(0)}_{12}\rangle$$
$$= \int_{-\infty}^{\infty}\int_{-\infty}^{\infty} |\psi^{(0)}_1(x_1)|^2\, V'(x_1 - x_2)\, |\psi^{(0)}_2(x_2)|^2\, dx_1\, dx_2$$
$$+ \int_{-\infty}^{\infty}\int_{-\infty}^{\infty} \psi^{(0)}_1(x_1)^*\psi^{(0)}_2(x_1)V'(x_1 - x_2)\psi^{(0)}_2(x_2)^*\psi^{(0)}_1(x_2)\, dx_1\, dx_2. \tag{8.41}$$

The first term in (8.41) is simply J_{12} according to the definition of $J_{nn'}$. If we further define the *exchange integral* $K_{nn'}$ as

$$K_{nn'} \equiv \int_{-\infty}^{\infty}\int_{-\infty}^{\infty} \psi_n^{(0)}(x_1)^*\psi_{n'}^{(0)}(x_1)V'(x_1 - x_2)\psi_{n'}^{(0)}(x_2)^*\psi_n^{(0)}(x_2)\, dx_1\, dx_2, \tag{8.42}$$

the perturbed energy of the first excited state can be written as

$$E_{12} = E_1^{(0)} + E_2^{(0)} + J_{12} + K_{12}. \tag{8.43}$$

Let us look a little closer at the direct and exchange integrals. $J_{nn'}$ has the same form as the energy of two classical charge distributions with densities $|\psi_n^{(0)}(x_1)|^2$ and $|\psi_{n'}^{(0)}(x_2)|^2$ interacting via potential energy $V'(x_1 - x_2)$. Since $V'(x_1 - x_2) > 0$ by assumption, it follows that $J_{nn'} > 0$. It is more difficult to interpret $K_{nn'}$ classically. It could be thought of as the energy of interaction of an "exchange charge" density $\psi_n^{(0)}(x)^*\psi_{n'}^{(0)}(x)$ with itself, the interaction potential energy again being $V'(x_1 - x_2)$. This suggests that $K_{nn'} > 0$ for $V'(x_1 - x_2) > 0$. However, we should be wary of carrying this picture too far. In fact, the exchange integral has no clear classical counterpart.[8]

All that remains is the evaluation of the integrals J_{11}, J_{12}, and K_{12} (see Probs. 8.1 and 8.2). The effect of these two correction terms on the energy is illustrated schematically in Fig. 8.1. We could also use the perturbation

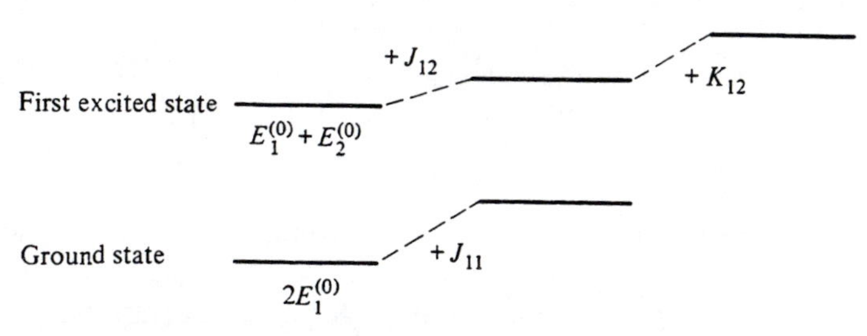

Figure 8.1 Schematic representation of the shift in energy due to the direct and exchange integrals for two bosons in a one-dimensional potential (assuming $K_{12} > 0$). See Eq. (8.43).

theory results of Chapter 4 to calculate first-order wave functions and second-order corrections to the energies of various states, but we prefer to move on to the two-fermion problem.

Spin-$\frac{1}{2}$ Fermions

Consider now a system consisting of two particles of spin $\frac{1}{2}$ in one dimension. Of course, the particles can no longer be treated as structureless, since their

[8] See John C. Slater, *Quantum Theory of Atomic Structure*, Vol. 1 (New York: McGraw-Hill, 1960), p. 282 for a discussion of the exchange integral and App. 19 for a proof that it must be positive.

spin must explicitly be taken into account. The subscripts n_1 and n_2 will represent only spatial quantum numbers; the spin quantum numbers will be indicated explicitly. The zeroth-order wave function ${}^A\psi^{(0)}_{n_1n_2}$ must be antisymmetric, and consists of sums of products of single-particle orbitals and spin eigenvectors α or β. For example one such product would be $\psi^{(0)}_{n_1}(x_1)\psi^{(0)}_{n_2}(x_2)\cdot\alpha(1)\beta(2)$. We form the antisymmetrized sum from each of the four independent products with the coordinates of particle one associated with quantum number n_1 by operating on the products with the operator $1/\sqrt{2}\,(1 - P_{12})$. Doing so, we obtain four functions,[9]

$${}^A\psi^{(0)}_{n_1n_2;++} \equiv \frac{1}{\sqrt{2}}[\psi^{(0)}_{n_1}(x_1)\alpha(1)\psi^{(0)}_{n_2}(x_2)\alpha(2) - \psi^{(0)}_{n_2}(x_1)\alpha(1)\psi^{(0)}_{n_1}(x_2)\alpha(2)], \tag{8.44a}$$

$${}^A\psi^{(0)}_{n_1n_2;+-} \equiv \frac{1}{\sqrt{2}}[\psi^{(0)}_{n_1}(x_1)\alpha(1)\psi^{(0)}_{n_2}(x_2)\beta(2) - \psi^{(0)}_{n_2}(x_1)\beta(1)\psi^{(0)}_{n_1}(x_2)\alpha(2)], \tag{8.44b}$$

$${}^A\psi^{(0)}_{n_1n_2;-+} \equiv \frac{1}{\sqrt{2}}[\psi^{(0)}_{n_1}(x_1)\beta(1)\psi^{(0)}_{n_2}(x_2)\alpha(2) - \psi^{(0)}_{n_2}(x_1)\alpha(1)\psi^{(0)}_{n_1}(x_2)\beta(2)], \tag{8.44c}$$

$${}^A\psi^{(0)}_{n_1n_2;--} \equiv \frac{1}{\sqrt{2}}[\psi^{(0)}_{n_1}(x_1)\beta(1)\psi^{(0)}_{n_2}(x_2)\beta(2) - \psi^{(0)}_{n_2}(x_1)\beta(1)\psi^{(0)}_{n_1}(x_2)\beta(2)]. \tag{8.44d}$$

In these equations we use the subscripts $++$, $--$, $+-$, $-+$ to distinguish the various antisymmetric wave functions. Thus the function in (8.44d) has both spins "down" ($m_{s_1} = -\frac{1}{2}$, $m_{s_2} = -\frac{1}{2}$) so we write ${}^A\psi^{(0)}_{n_1n_2;--}$, and so on.

Notice that each of these functions is antisymmetric under P_{12} and is an eigenfunction of $\mathcal{H}^{(0)}$. Since $\mathcal{H}^{(0)}$ contains no spin-dependent terms, these four functions are degenerate, with energy $E^{(0)}_{n_1} + E^{(0)}_{n_2}$. They are perfectly valid zeroth-order wave functions.

In the special case that $n_1 = n_2$, the functions ${}^A\psi^{(0)}_{n_1n_2;++}$ and ${}^A\psi^{(0)}_{n_1n_2;--}$ are identically zero. Thus two fermions can have the same spatial quantum numbers only if their spin quantum numbers m_s are different (i.e., we have verified the Pauli exclusion principle for this case). [Notice also that ${}^A\psi^{(0)}_{n_1n_2;+-} = -\,{}^A\psi^{(0)}_{n_1n_2;-+}$ if $n_1 = n_2$.]

We could proceed now with perturbation calculations as in the spin-0 boson example. However, it turns out to be convenient to form new degenerate zeroth-order eigenfunctions from the functions of Eqs. (8.44).

[9] In the literature one sometimes sees α_1 written for $\alpha(1)$, etc. Attempting to keep the proliferation of subscripts to a tolerable minimum, we have chosen not to use this convention.

Spin Coupling

In Chapter 7 we coupled the orbital and spin angular momenta of the one-electron atom to form a total angular momentum vector **J**. This coupling led to the introduction of the zeroth-order functions $\psi^{(0)}_{n\ell jm_j}(\mathbf{r})$.

Our one-dimensional fermions have no orbital angular momentum defined, but the two spins $\mathbf{S}_1$ and $\mathbf{S}_2$ can be coupled to give a total spin **S**,

$$\mathbf{S} \equiv \mathbf{S}_1 + \mathbf{S}_2. \tag{8.45}$$

Continuing by analogy with the coupling of **L** and **S**, we write the eigenvalue equations

$$S^2\chi_{SM_S} = S(S+1)\hbar^2\chi_{SM_S}, \tag{8.46}$$

$$S_z\chi_{SM_S} = M_S\hbar\chi_{SM_S}. \tag{8.47}$$

The functions χ_{SM_S} are simultaneous eigenfunctions of the operators S^2 and S_z. The quantum numbers S and M_S for the total spin and its projection along the $\hat{z}$ axis can take on values consistent with Eq. (8.45) and $M_S = -S, \ldots, S$; that is,

$$S = 0 \qquad M_S = 0 \tag{8.48}$$

$$S = 1 \qquad M_S = -1, 0, +1. \tag{8.49}$$

Suppose that we seek new zeroth-order eigenfunctions [of $\mathcal{H}^{(0)}$] that are also eigenfunctions of S^2 and S_z. We shall denote them by $\psi^{(0)}_{n_1n_2;SM_S}$, dropping the superscript A. Since we know the degenerate zeroth-order functions $\psi^{(0)}_{n_1n_2;m_{s_1}m_{s_2}}$ of Eqs. (8.44), we shall use them to expand $\psi^{(0)}_{n_1n_2;SM_S}$:

$$\psi^{(0)}_{n_1n_2;SM_S} = \sum_{m_{s_1}m_{s_2}} a_{m_{s_1}m_{s_2}}\psi^{(0)}_{n_1n_2;m_{s_1}m_{s_2}}. \tag{8.50}$$

In this expansion, $a_{m_{s_1}m_{s_2}}$ is a Clebsch-Gordan coefficient [see Eqs. (7.45)], which is written

$$a_{m_{s_1}m_{s_2}} = \langle s_1\, s_2\, m_{s_1} m_{s_2} | s_1\, s_2\, S\, M_S\rangle. \tag{8.51}$$

These coefficients are zero unless

$$M_S = m_{s_1} + m_{s_2} \tag{8.52a}$$

and

$$|s_1 - s_2| \leq S \leq s_1 + s_2; \tag{8.52b}$$

the nonzero coefficients may be found in the tables of Appendix 4 (see also Prob. 8.3).

In order to see how to construct the new functions $\psi^{(0)}_{n_1n_2;SM_s}$, consider $\psi^{(0)}_{n_1n_2;11}$—that is, the case $S = 1$, $M_S = 1$. From Eq. (8.52a) we see that of the four coefficients in (8.51), only $\langle \frac{1}{2}\,\frac{1}{2}\,\frac{1}{2}\,\frac{1}{2}|\frac{1}{2}\,\frac{1}{2}\,1\,1\rangle$ is nonzero. Thus we obtain

$$\psi^{(0)}_{n_1n_2;11} = \psi^{(0)}_{n_1n_2;++}. \tag{8.53}$$

Similarly, we find that

$$\psi^{(0)}_{n_1n_2;1,-1} = \psi^{(0)}_{n_1n_2;--}. \tag{8.54}$$

As another example, consider $S = 1$, $M_S = 0$. The new function is

$$\psi^{(0)}_{n_1n_2;10} = \langle \tfrac{1}{2}\tfrac{1}{2}\tfrac{1}{2} \; -\tfrac{1}{2}|\tfrac{1}{2}\tfrac{1}{2}\,1\,0\rangle\,\psi^{(0)}_{n_1n_2;+-} + \langle \tfrac{1}{2}\tfrac{1}{2} \; -\tfrac{1}{2}\tfrac{1}{2}|\tfrac{1}{2}\tfrac{1}{2}\,1\,0\rangle\,\psi^{(0)}_{n_1n_2;-+}. \tag{8.55}$$

Using Appendix 4 to evaluate the coefficients, we obtain

$$\psi^{(0)}_{n_1n_2;10} = \frac{1}{\sqrt{2}}[\psi^{(0)}_{n_1n_2;+-} + \psi^{(0)}_{n_1n_2;-+}]. \tag{8.56}$$

Similarly, we can show that[10]

$$\psi^{(0)}_{n_1n_2;00} = \frac{1}{\sqrt{2}}[\psi^{(0)}_{n_1n_2;+-} - \psi^{(0)}_{n_1n_2;-+}]. \tag{8.57}$$

The functions in Eqs. (8.53), (8.54), (8.56), and (8.57) are the new degenerate zeroth-order eigenfunctions. Notice that these new functions are also normalized and orthogonal to one another.

Before returning to the fermion perturbation calculations, let us rewrite our new zeroth-order eigenfunctions in a more compact form. Substituting the definitions of (8.44) into the preceding results, we obtain

$$\psi^{(0)}_{n_1n_2;00} = \frac{1}{\sqrt{2}}[\psi^{(0)}_{n_1}(x_1)\psi^{(0)}_{n_2}(x_2) + \psi^{(0)}_{n_2}(x_1)\psi^{(0)}_{n_1}(x_2)]\chi_{00}, \tag{8.58a}$$

$$\psi^{(0)}_{n_1n_2;1M_S} = \frac{1}{\sqrt{2}}[\psi^{(0)}_{n_1}(x_1)\psi^{(0)}_{n_2}(x_2) - \psi^{(0)}_{n_2}(x_1)\psi^{(0)}_{n_1}(x_2)]\chi_{1M_S},$$
$$M_S = +1, 0, -1, \tag{8.58b}$$

where we have introduced the new spin functions χ_{SM_S}, defined by

$$\chi_{00} \equiv \frac{1}{\sqrt{2}}[\alpha(1)\beta(2) - \beta(1)\alpha(2)], \tag{8.59a}$$

$$\chi_{11} \equiv \alpha(1)\alpha(2), \tag{8.59b}$$

$$\chi_{10} \equiv \frac{1}{\sqrt{2}}[\alpha(1)\beta(2) + \beta(1)\alpha(2)], \tag{8.59c}$$

$$\chi_{1,-1} \equiv \beta(1)\beta(2). \tag{8.59d}$$

[10]Alternatively, we could obtain Eq. (8.57) from (8.56) by requiring that $\psi^{(0)}_{n_1n_2;10}$ and $\psi^{(0)}_{n_1n_2;00}$ be orthogonal.

Exercise 8.2 (a) Derive Eqs. (8.58) in the manner suggested.

(b) Verify that the functions χ_{SM_S} are eigenfunctions of S^2 and S_z with eigenvalues $S(S+1)\hbar^2$ and $M_S\hbar$, respectively.

Finally, we should note that either set of functions $\{\psi^{(0)}_{n_1n_2;m_{s_1}m_{s_2}}\}$ or $\{\psi^{(0)}_{n_1n_2;SM_S}\}$ is a valid set of eigenfunctions of $\mathcal{H}^{(0)}$.

Fermion Perturbation Calculations

Equations (8.58) are new degenerate eigenfunctions of $\mathcal{H}^{(0)}$, S^2, S_1^2, S_2^2, and S_z, which we intend to use in a perturbation theory calculation. When the interaction potential energy $V'(x_1 - x_2)$ is taken into account, we shall see that the three states with $S = 1$ shift in energy away from $E_{n_1}^{(0)} + E_{n_2}^{(0)}$ but remain degenerate; the state with $S = 0$ also shifts away from the zeroth-order energy but in the opposite direction and thus is no longer degenerate with the $S = 1$ states. The three $S = 1$ states are called *triplet states.* Notice that triplet functions all have the same antisymmetric spatial part and different symmetric spin parts. The $S = 0$ state is called the *singlet state.* The singlet function has an antisymmetric spin part and a symmetric spatial part.[11]

The ground state of the two-fermion system corresponds to $n_1 = n_2 = 1$; its unperturbed energy is $2E_1^{(0)}$. We see from Eqs. (8.58b) that, for $n_1 = n_2$, the triplet functions are identically equal to zero; that is, $\psi^{(0)}_{11;1M_S} = 0$. Therefore the unperturbed ground-state eigenfunction is the singlet function $\psi^{(0)}_{11;00}$. The spin function χ_{00} is $1/\sqrt{2}\,[\alpha(1)\beta(2) - \beta(1)\alpha(2)]$, so for this state the values of m_s for the two particles are different. We say that the spins are "paired" or "antiparallel" in the singlet state.

The perturbed energy of the ground state, to first order, is

$$E_{11} = 2E_1^{(0)} + \langle \psi^{(0)}_{11;00} | V'(x_1 - x_2) | \psi^{(0)}_{11;00} \rangle, \tag{8.60}$$

or

$$E_{11} = 2E_1^{(0)} + J_{11}, \tag{8.61}$$

where $J_{nn'}$ is the direct integral of Eq. (8.37). Notice that the ground-state energy is the same for a system of two interacting fermions as for a system of two interacting bosons (provided that the system Hamiltonians are the same). We also see that the zeroth-order ground state is represented by the same spatial function in the fermion and boson problems.

[11] It is important to understand the effect of spin on these wave functions. We could have considered spin from the first and formed the four spin functions that appear in Eqs. (8.59)—the three symmetric triplet ($S = 1$) functions and the antisymmetric singlet ($S = 0$) function. Then, to satisfy the requirement that the wave function be antisymmetric with respect to particle interchange, we would have been forced to multiply these spin functions by antisymmetric and symmetric spatial functions for triplet and singlet states, respectively.

Turning to the *first excited state* of the two-fermion system, $n_1 = 1$, $n_2 = 2$, we find that the perturbed energy

$$E_{12} = E_1^{(0)} + E_2^{(0)} + \langle \psi_{12;SM_S}^{(0)} | V'(x_1 - x_2) | \psi_{12;SM_S}^{(0)} \rangle \qquad (8.62)$$

is different for the singlet and triplet states. None of the antisymmetric eigenfunctions $\psi_{12;SM_S}^{(0)}$ is identically zero in this case, so we must use them all. However, the three triplet functions have the same spatial dependence, and thus we need to evaluate only one triplet matrix element and the singlet matrix element. Substitution of the appropriate functions into Eq. (8.62) yields

$$E_{12} = E_1^{(0)} + E_2^{(0)} + J_{12} \pm K_{12}, \qquad (8.63)$$

where the upper sign corresponds to the singlet state ($S = 0$, $M_S = 0$) and the lower sign to the triplet states ($S = 1$, $M_S = -1, 0, +1$). The exchange integral $K_{nn'}$ is defined in Eq. (8.42).

Exercise 8.3 Derive Eq. (8.63).

This result differs from the first-order boson energy of Eq. (8.43). In particular, the contribution due to the exchange integral partially lifts the degeneracy. Schematically, the situation is as shown in Fig. 8.2, where we

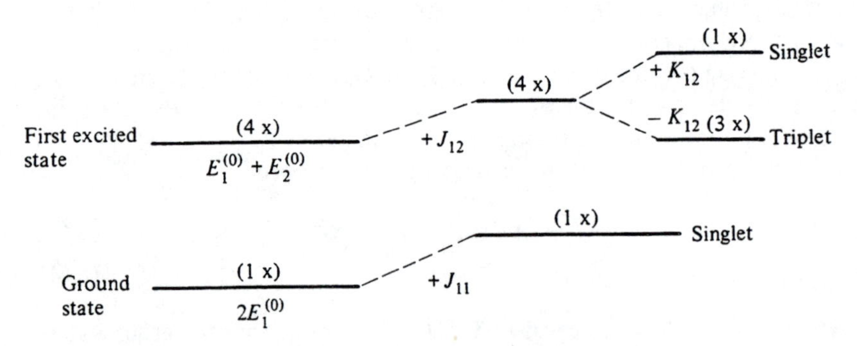

Figure 8.2 Schematic representation of perturbation corrections to the energies of the ground and first excited states of a two-fermion system (assuming $K_{12} > 0$). See Eq. (8.63).

note that the degeneracy with respect to M_S remains even after the first-order corrections are taken into account.

Spin Pairing

Figure 8.2 shows that the energies of the two-fermion system depend on the value of the total spin of the system, even though the Hamiltonian $\mathcal{H}$ does not explicitly include spin. Therefore this effect is not due to any type of magnetic interaction, such as the spin-orbit interaction of Chapter 7.[12] We can understand the effect of the total spin on the energies by examining the triplet and singlet functions, Eqs. (8.58), for the special case of the two particles very close to one another—that is, for $x_1 \simeq x_2$.

In particular, since the triplet spin functions χ_{1M_S} are symmetric, the triplet spatial functions are necessarily antisymmetric. Thus this spatial function is small for $x_1 \simeq x_2$, which, for most interactions, is where $V'(x_1 - x_2)$ is expected to be largest. Two identical fermions with spins unpaired (parallel), either both "up" or both "down," have a very small probability of being found close together. Figuratively speaking, we say that they "repel" one another. (In fact, the triplet function vanishes when $x_1 = x_2$.) Notice that this "repulsion" is a spin-dependent effect and is not due to the coulomb repulsion between two particles of like charge. On the other hand, the singlet spin function χ_{00} is antisymmetric, so the singlet spatial function is symmetric. This spatial function does not become small for $x_1 \simeq x_2$. Two identical fermions with paired (antiparallel) spins are not prohibited from being close together by the form of the wave function. We say that they "attract" one another.

The first-order perturbed energy E_{12} is larger in the singlet than in the triplet case, since the repulsive interaction between the two identical fermions is greater in the former than in the latter case. Even though the energy-correction terms depend only on the spatial part of the zeroth-order wave functions, the symmetry of the spatial part is determined by the value of the total spin quantum number S. It is in this way that S affects the energy.

Summarizing our conclusion, we say that *fermions having parallel* (*unpaired*) *spins tend to repel one another, whereas those with antiparallel* (*paired*) *spins tend to attract one another.*[13]

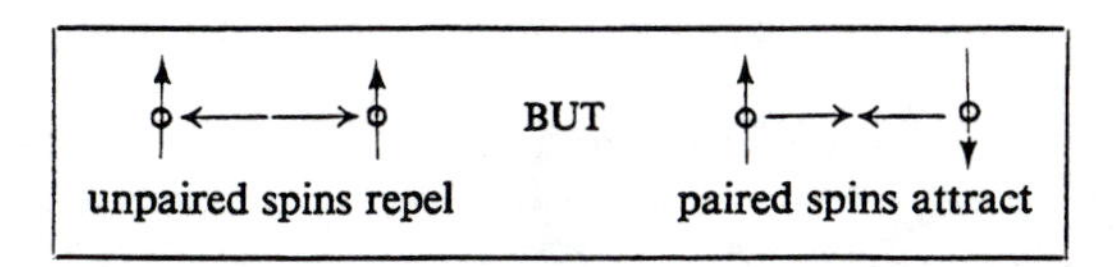

This terminology refers to precisely the exchange symmetry effects discussed in this section. The effect is called *spin pairing* and is a purely quantum mechanical effect completely unrelated to the coulomb interaction between the two particles.

[12] In real multielectron atoms there will, of course, be a magnetic interaction between two magnetic dipoles (e.g., two spinning electrons). However, the energy due to this "spin-spin interaction" is very small (orders of magnitude smaller than $K_{nn'}$ in most cases).

[13] We are using the word "attract" here to mean that the two electrons are not excluded by the Pauli principle from occupying the same point in space. Thus paired spins make a larger energy contribution than do unpaired spins.

8.4 A ONE-DIMENSIONAL N-ELECTRON "ATOM"

Many of the results of Sec. 8.3 will be used in Chapter 10, where the helium atom is studied in some detail. But first it is useful to see how these new ideas introduced generalize to N-electron atoms. Thus we shall consider here a one-dimensional *model* of an atom. The model consists of N electrons, any two of which, say electrons i and j, interact with one another via the potential energy $V'(x_i - x_j)$.[14] We shall treat these electron-electron interactions collectively as a perturbation.

The full Hamiltonian is

$$\mathcal{H} = \mathcal{H}^{(0)} + \mathcal{H}^{(1)}, \tag{8.64}$$

where the zeroth-order Hamiltonian is

$$\mathcal{H}^{(0)} = \sum_{i=1}^{N} \left[-\frac{\hbar^2}{2m_e} \frac{\partial^2}{\partial x_i^2} + V(x_i) \right] \tag{8.65}$$

and the perturbation Hamiltonian is

$$\mathcal{H}^{(1)} = \sum_{i=1}^{N} \sum_{j>i}^{N} V'(|x_i - x_j|) \tag{8.66}$$

In Eq. (8.65), $V(x_i)$ is the single-particle potential energy of the ith particle.

The stationary-state eigenfunctions of this N-electron "atom" satisfy the time-independent Schroedinger equation

$$\mathcal{H}\psi_E = E\psi_E, \tag{8.67}$$

where ψ_E depends on the coordinates and spins of the N electrons. Since the N electrons are indistinguishable, $\mathcal{H}$ commutes with all the particle interchange operators P_{ij}; that is,

$$[\mathcal{H}, P_{ij}] = 0, \qquad i, j = 1, 2, \ldots, N; \quad i \neq j. \tag{8.68}$$

(We should again emphasize that P_{ij} interchanges spatial coordinates and spin.) Since the electrons are fermions, ψ_E must be antisymmetric under pairwise particle interchange:

$$P_{ij}\psi_E = -\psi_E, \qquad i, j = 1, 2 \ldots, N; \quad i \neq j. \tag{8.69}$$

[14]Of course, this reduces to the case considered in See. 8.3 for $N = 2$. You should refer back frequently to that section to see parallels where they occur.

Independent Particle Model

As a zeroth-order approximation, let us ignore the electron-electron interaction terms $\mathcal{H}^{(1)}$. We call this approximation the *independent particle model* of the "atom" because, in effect, each electron behaves independently of all the other electrons. The zeroth-order Schroedinger equation is

$$\mathcal{H}^{(0)}\psi_{\gamma}^{(0)} = E_{\gamma}^{(0)}\psi_{\gamma}^{(0)}, \tag{8.70}$$

where the subscript γ represents *all* quantum numbers of the N electrons. Thus for the case $N = 2$, $\gamma = (n_1, n_2)$ [see Eq. (8.31)].

Suppose that we know solutions to the N single-electron equations

$$\left[-\frac{\hbar^2}{2m_e}\frac{d^2}{dx_1^2} + V(x_1)\right]\psi_{n_1}^{(0)}(x_1) = E_{n_1}^{(0)}\psi_{n_1}^{(0)}(x_1), \tag{8.71a}$$

$$\left[-\frac{\hbar^2}{2m_e}\frac{d^2}{dx_2^2} + V(x_2)\right]\psi_{n_2}^{(0)}(x_2) = E_{n_2}^{(0)}\psi_{n_2}^{(0)}(x_2). \tag{8.71b}$$

$$\vdots \qquad\qquad\qquad \vdots$$

There is one such equation for each electron. The subscripts $n_1, n_2, \ldots$ represent the spatial quantum numbers of the electrons. Remember that each electron also has a particular value of m_s that must also be specified. Each single-particle function, or *orbital*, is the product of a spatial function and a spin function $\chi_{1/2,m_s}$, either α or β, depending on the value of m_s.

Momentarily neglecting the requirement of indistinguishability, we can write, for zeroth-order eigenfunctions of $\mathcal{H}^{(0)}$, product functions composed of these orbitals; for example,

$$\psi_{\gamma}^{(0)}(x_1, x_2, \ldots, x_N) = \psi_{n_1}^{(0)}(x_1)\chi_{1/2,m_{s_1}}(1)\psi_{n_2}^{(0)}(x_2)\chi_{1/2,m_{s_2}}(2)\cdots \psi_{n_N}^{(0)}(x_N)\chi_{1/2,m_{s_N}}(N). \tag{8.72}$$

The energy of this function is just the sum of the zeroth-order energies $E_{n_1}^{(0)} + E_{n_2}^{(0)} + \cdots + E_{n_N}^{(0)}$. Such functions violate the fundamental property that the N electrons are indistinguishable. Hence Eq. (8.72) is not a valid representation of physical states of the system. We can obtain another eigenfunction of $\mathcal{H}^{(0)}$ by operating on $\psi_{\gamma}^{(0)}$ with one of the operators P_{ij} and thereby interchanging the coordinates and spins of particles i and j. A new function of the form $\psi_{\gamma}^{(0)} - P_{ij}\psi_{\gamma}^{(0)}$ will be degenerate with $\psi_{\gamma}^{(0)}$ (because of exchange degeneracy).[15] However, it will be antisymmetric only under interchange of particles i and j. We want a function that is antisymmetric under

[15] $P_{ij}\psi_{\gamma}^{(0)}$ is orthogonal to $\psi_{\gamma}^{(0)}$; why?

interchange of any two electrons. We can construct just such a function by forming a linear combination of all the degenerate functions that can be obtained by repeated pairwise particle interchanges:

$$\psi_\gamma^{(0)} \equiv \frac{1}{\sqrt{N!}} \sum_P (-1)^P P \psi_\gamma^{(0)}(x_1, x_2, \ldots, x_N), \tag{8.73}$$

where the sum is taken over all possible permutations of the N electrons. There are $N!$ such permutations and thus $N!$ terms in $\psi_\gamma^{(0)}$. In Eq. (8.73) the symbol $(-1)^P$ is defined to be $+1$ or -1, depending on whether the permutation is even or odd, respectively.[16] The multiplicative factor $1/\sqrt{N!}$ normalizes the wave function. The function $\psi_\gamma^{(0)}$ satisfies the requirements of indistinguishability because all possible labelings of the N electrons by quantum numbers are represented in the sum.

Exercise 8.4 (a) Use Eq. (8.73) to generate the antisymmetric wave function for $N = 2$. Your result should agree with Eq. (8.33b).
(b) Use Eq. (8.73) to generate $\psi_\gamma^{(0)}$ for $N = 3$.

The wave functions $\psi_\gamma^{(0)}$ in Eq. (8.73) can be used as zeroth-order wave functions in a perturbation theory calculation. They can be conveniently represented by a Slater determinant. For an N-electron "atom", the *Slater determinant* is

$$\psi_\gamma^{(0)} = \frac{1}{\sqrt{N!}} \begin{vmatrix} \varphi_{n_1 m_{s_1}}^{(0)}(x_1) & \varphi_{n_1 m_{s_1}}^{(0)}(x_2) & \cdots & \varphi_{n_1 m_{s_1}}^{(0)}(x_N) \\ \varphi_{n_2 m_{s_2}}^{(0)}(x_1) & \varphi_{n_2 m_{s_2}}^{(0)}(x_2) & \cdots & \varphi_{n_2 m_{s_2}}^{(0)}(x_N) \\ \vdots & \vdots & & \\ \varphi_{n_N m_{s_N}}^{(0)}(x_1) & \varphi_{n_N m_{s_N}}^{(0)}(x_2) & \cdots & \varphi_{n_N m_{s_N}}^{(0)}(x_N) \end{vmatrix}, \tag{8.74a}$$

where the functions $\varphi_{n_1 m_{s_1}}^{(0)}(x_1) \equiv \psi_{n_1}^{(0)}(x_1)\chi_{1/2, m_{s_1}}(1)$, and so on have been temporarily introduced. To avoid writing all the elements of the determinant, we use the notation

$$\psi_\gamma^{(0)} \equiv |\varphi_{n_1 m_{s_1}}^{(0)}(x_1) \quad \varphi_{n_2 m_{s_2}}^{(0)}(x_2) \quad \cdots \quad \varphi_{n_N m_{s_N}}^{(0)}(x_N)|, \tag{8.74b}$$

where, by convention, only the diagonal terms are explicitly written. The parallel bars are a reminder that a determinant is to be formed and normalized.

[16] P is a permutation operator; it permutes the coordinates and spins of the N electrons. An *even permutation* is defined as one that is equivalent to an even number of two-particle interchanges; an *odd permutation* is one that is equivalent to an odd number of two-particle interchanges.

The function $\psi_\gamma^{(0)}$ is a Hamiltonian eigenfunction that represents a physically realizable stationary state of the N-electron "atom" in the independent particle model. Because of the presence in the wave function of the various permuted product functions, it is no longer possible to say which electron has a particular set of quantum numbers. This fact is consistent with our notion of the electrons as indistinguishable. Although we specify the electrons in individual terms of $\psi_\gamma^{(0)}$ (e.g., "the electron at x_1"), our theory must not include any way to distinguish one electron from another physically. Thus it does not make sense to discuss the quantum numbers of, say, "the electron at x_1."

The Pauli Exclusion Principle

The next topic is the choice of spin quantum numbers m_s—that is, the assignment of a spin ("up" or "down") to each electron. Clearly, there are an enormous number of alternative ways to assign the spins! For example, all particles could have spin "up," or all but one could have spin "up" and that one with spin "down" could be *any* of the N electrons, and so forth. Each assignment leads to a different wave function $\psi_\gamma^{(0)}$, and all these wave functions are degenerate, since $\mathcal{H}^{(0)}$ does not explicitly contain spin.

This formidable array of wave functions is considerably diminished by application of the Pauli exclusion principle, which was first enunciated in 1925. It states that *no two identical fermions in a physical system can have the same set of quantum numbers.*

We can easily demonstrate the connection between the Pauli exclusion principle and the antisymmetrization requirement. Suppose, in violation of the principle, that two of the N electrons do have the same quantum numbers—for example, $n_1 = n_2$ and $m_{s_1} = m_{s_2} = \frac{1}{2}$. Then the wave function of the "atom" is

$$\psi_\gamma^{(0)} = |\psi_{n_1}^{(0)}(x_1)\alpha(1) \quad \psi_{n_1}^{(0)}(x_2)\alpha(2) \quad \cdots \quad \psi_{n_N}^{(0)}(x_N)\chi_{1/2, m_{s_N}}|. \tag{8.75}$$

Thus at least two of the rows of the Slater determinant are identically equal. But this fact implies that the determinant is zero and hence that $\psi_\gamma^{(0)} = 0$. Therefore if $n_1 = n_2$, the two spin quantum numbers m_{s_1} and m_{s_2} must be unequal lest the wave function vanish. The Pauli exclusion principle—the symmetry restriction on the wave functions—has "forced" electrons in identical spatial orbitals to be paired with their spins antiparallel.

Configurations

It is convenient to use a shorthand notation for orbitals. If the ith electron is in the state with spatial quantum number $n_i = 1$, we denote the orbital $1s(i)$, where the letter s is NOT the spin quantum number. We use it strictly to establish a useful notation for the three-dimensional atom.[17] If $n_i = 2$, we denote the orbital $2s(i)$ and so on.

A particular assignment of the N electrons to N orbitals is called a *configuration* of the atom. Thus one configuration that is allowed by the Pauli exclusion principle is $1s(1)\,\overline{1s(2)} \cdots ms(N)$, where m represents the spatial quantum number of the Nth electron and the overbar $1s(2)$ indicates that the electron in the orbital $1s(2)$ has spin "down." Remember that this configuration corresponds to a fully antisymmetrized wave function—that is, a normalized Slater determinant.

Ground State

The concepts of the Pauli exclusion principle and spin pairing provide the key to understanding the structure of "atoms." Consider the zeroth-order ground-state wave functions for the N-electron "atom." The ground state corresponds to the lowest allowed assignment of quantum numbers. Clearly, we cannot have all $n_1 = n_2 = n_3 = \cdots = n_N = 1$ or the wave function will vanish. The best we can do is to pair the spins of the N electrons so that two electrons have spatial quantum numbers $n_1 = n_2 = 1$ and spins $m_s = +\frac{1}{2}$ and $m_s = -\frac{1}{2}$, respectively, and so on. If we think of each spatial quantum number n_i as defining a "shell" with energy $2E_{n_i}^{(0)}$, then we can picture the "atom" as built up by placing electrons, two at a time, in shells, beginning with the shell of lowest energy and continuing until all the electrons are used up. This process is symbolically represented for the case $N = 6$ by the energy level diagram in Fig. 8.3. A shell with two paired electrons in it is said to be "filled." We shall see in Chapter 9 that this seemingly simple-minded picture generalizes into the shell model of the atom, one of the basic structural principles of atomic physics.

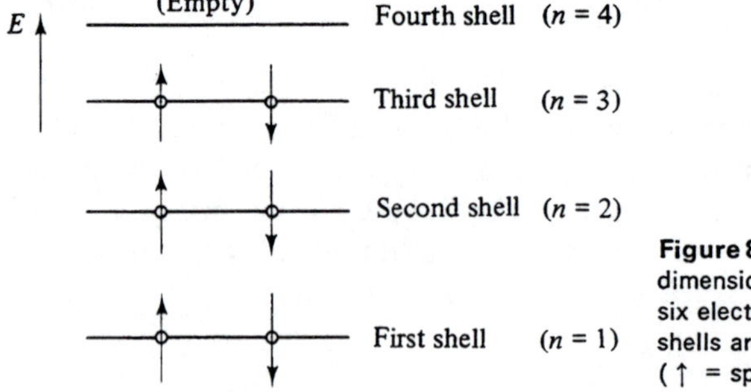

Figure 8.3 Filling of shells in a one-dimensional model of an atom with six electrons ($N = 6$). The first three shells are filled; the fourth is empty. (↑ = spin up; ↓ = spin down).

[17]The letter s can be taken to mean that $\ell = 0$; in one dimension, an electron has no angular momentum.

Spin Coupling

We saw in Sec. 8.3 that the total spin S plays an important role in determining the degeneracies and energies of the states of a two-fermion system. This point is also true, in general, for N-electron "atoms." The total spin is defined by [see Eq. (8.45)]

$$\boxed{\mathbf{S} \equiv \sum_{i=1}^{N} \mathbf{S}_i.} \tag{8.76}$$

The quantum number corresponding to S_z is M_S, which must satisfy [see Eq. (8.52a)]

$$M_S = \sum_{i=1}^{N} m_{s_i}. \tag{8.77}$$

This tells us, for example, that the only allowed value of M_S for the ground state of an "atom" with an even number of electrons is $M_S = 0$. Thus S can only take on the value $S = 0$, and the ground state is a nondegenerate singlet state in which all the spins are paired.

Excited States

We can generate an excited state of an N-electron "atom" by simply shifting one (or more) of the electrons into an unoccupied level. Let us consider the case $N = 6$ and excite one electron from the third to the fourth shell, as in Fig. 8.4. Since the excited electron and the remaining third-shell electron can each have either spin "up" or "down", four degenerate wave functions result. This fourfold degeneracy obviously arises because of the freedom we

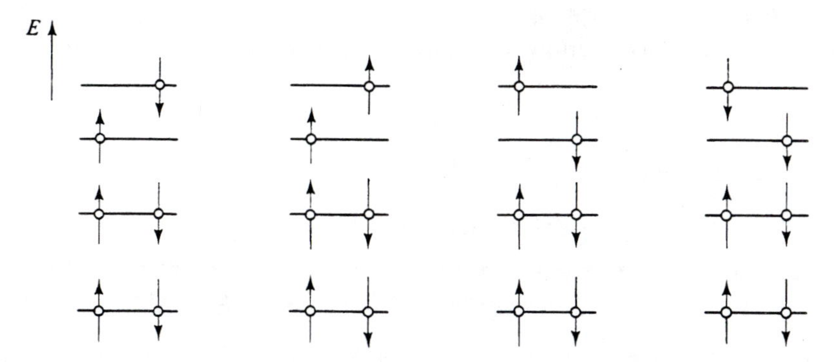

Figure 8.4 The first excited state of the six-electron one-dimensional "atom". The schematic energy level diagram shows four degenerate possible assignments of spin. (↑ = spin up; ↓ = spin down).

have in assigning spins for this state; it will correspond to the number of possible values for the quantum numbers S and M_S. From Eq. (8.77) it follows that the allowed values of S and M_S are

$$\begin{array}{llll} S=0 & M_S=0 & \text{singlet} & (1\times) \\ S=1 & M_S=-1,0,1 & \text{triplet} & \underline{(3\times)} \\ & & & 4\times = \text{degeneracy.} \end{array}$$

Thus the first excited state is fourfold degenerate.

Exercise 8.5 Show that the ground state of an N-electron "atom" with N odd is a doublet corresponding to one electron outside a "closed shell." What are the allowed values of S and M_S for this state?

Perturbation Calculations

Yet to be discussed are the first-order perturbation calculations for this system. The approach is identical to that of Sec. 8.3; we take the perturbation Hamiltonian to be

$$\mathcal{H}^{(1)} \equiv \sum_{i=1}^{N} \sum_{j>i}^{N} V'(x_i - x_j), \tag{8.78}$$

provided that this term is small. However, little is to be learned by doing the calculation; the equations are similar to those of the previous section. For example, the first-order correction to the ground-state energy is simply

$$\left\langle \psi_1^{(0)} \middle| \sum_{i=1}^{N} \sum_{j>i}^{N} V'(x_i - x_j) \middle| \psi_1^{(0)} \right\rangle. \tag{8.79}$$

Rather than become involved in the details of such calculations, we shall move on to the generalization of the ideas of this section to real three-dimensional atoms.

PROBLEMS

8.1 Coupled Harmonic Oscillator: Exact Solution (**)

Consider a system of two particles of masses m_1 and m_2, each of which is in a harmonic oscillator potential. The particles interact with each other via a harmonic interaction. The total Hamiltonian is

$$\mathcal{H} = \frac{p_1^2}{2m_1} + \frac{p_2^2}{2m_2} + \frac{1}{2}m_1\omega^2 x_1^2 + \frac{1}{2}m_2\omega^2 x_2^2 + \frac{1}{2}k(x_1 - x_2)^2.$$

Consider the case $k > 0$ (attractive).

(a) By transforming to center-of-mass and relative coordinates, show that an exact solution may be obtained and give expressions for the Hamiltonian eigenfunctions and energy eigenvalues. Describe the energy spectrum (draw an energy level diagram) for the two cases $k \ll \mu\omega^2$ and $k \gg \mu\omega^2$, where μ = reduced mass.

(b) Now take the two particles to be indistinguishable spin-0 bosons of mass m. Give expressions for the stationary-state wave functions and energies. Discuss the question of exchange degeneracy. For the case $k \ll \mu\omega^2$, compare the energy spectrum with that of part (a). (Do not actually ignore k.)

(c) Repeat part (b) for the case of two spin-$\frac{1}{2}$ fermions. Characterize the states as "singlets" or "triplets." Give an intuitive argument for the relative positions of the singlet and triplet energy levels for $k \ll \mu\omega^2$. Compare the energy spectrum for $k \ll \mu\omega^2$ with that for bosons and distinguishable particles.

8.2 Two Interacting Particles in a Box: Perturbation Theory (***)

Consider two indistinguishable particles of mass m in an infinite potential well of width L and interacting with each other via a simple potential of the form $V_{12}(x_1, x_2) = v\,\delta(x_1 - x_2)$ (assume that $v > 0$). The total Hamiltonian is

$$\mathcal{H} = -\frac{\hbar^2}{2m}\frac{\partial^2}{\partial x_1^2} - \frac{\hbar^2}{2m}\frac{\partial^2}{\partial x_2^2} + V(x_1) + V(x_2) + V_{12}(x_1, x_2),$$

where

$$V(x_i) = \begin{cases} 0, & |x_i| < \dfrac{L}{2}, \\ \infty, & |x_i| > \dfrac{L}{2}, \end{cases} \qquad i = 1, 2.$$

(a) Identifying V_{12} as the perturbation Hamiltonian, write expressions for the zeroth-order energies and wave functions for the ground configuration and a general excited configuration of a system of two spin-0 bosons. Draw a zeroth-order energy level diagram, labeling levels with appropriate quantum numbers, and discuss exchange degeneracy. Describe qualitatively how you expect the perturbation V_{12} to change the spectrum of energy levels.

(b) Repeat part (a) for the system of two spin-$\frac{1}{2}$ fermions. Label the energy levels as "singlets" or "triplets."

(c) Using zeroth-order wave functions of the correct symmetry, calculate the first-order correction to the ground-state energy for the systems of parts (a) and (b).

(d) Calculate the first-order corrections to the first excited state energy for the systems of parts (a) and (b).

8.3 Application of Raising and Lowering Operators: Coupled Spin Functions (**)

In this problem we shall use the raising and lowering operators $S_\pm$ for the total spin $\mathbf{S} = \mathbf{S}_1 + \mathbf{S}_2$ in order to obtain the set of functions $\{{}^A\psi^{(0)}_{n_1n_2;SM_S}\}$ from the set $\{{}^A\psi^{(0)}_{n_1n_2;m_{s_1}m_{s_2}}\}$.

(a) Verify that the functions $\{{}^A\psi^{(0)}_{n_1n_2;m_{s_1}m_{s_2}}\}$ are eigenfunctions of S_z. What are the eigenvalues?

(b) Verify that ${}^A\psi^{(0)}_{n_1n_2;++}$ and ${}^A\psi^{(0)}_{n_1n_2;--}$ are eigenfunctions of S^2 but that ${}^A\psi^{(0)}_{n_1n_2;+-}$ and ${}^A\psi^{(0)}_{n_1n_2;-+}$ are not. What are the eigenvalues in the former case?

(c) Use the results of parts (a) and (b) to form the wave functions ${}^A\psi^{(0)}_{n_1n_2;11}$ and ${}^A\psi^{(0)}_{n_1n_2;1,-1}$.

(d) Operate on ${}^A\psi^{(0)}_{n_1n_2;11}$ with S_- to obtain an expression for ${}^A\psi^{(0)}_{n_1n_2;10}$. Show that the same result obtains from operating on ${}^A\psi^{(0)}_{n_1n_2;1,-1}$ with S_+.

(e) Use orthogonality to obtain ${}^A\psi^{(0)}_{n_1n_2;00}$. Check your answers to parts (c) through (e) with Eqs. (8.53) to (8.57).

9
The Shell Model of the Atom

"It is reasonable," he admitted. "And therefore I suspect it. Whenever you find a perfectly reasonable explanation of anything in nature or human conduct, look for something else. Things in the real universe don't all fit together like the pieces of a child's puzzle."
John Taine (Eric Temple Bell), The Time Stream, Three Science Fiction Novels *(N.Y.: Dover Publications, Inc., 1964).*

In this chapter we propose to extend the results of the model N-electron "atom" of Sec. 8.4 and to survey several important qualitative properties of real multielectron atoms. It is not possible to pursue an extensive study of atomic structure here; rather we shall try to gain an understanding of the nature and ordering of the energy levels and organization of electrons in multielectron atoms. The subject of atomic structure is pursued in more detail in the references at the end of this chapter. In Chapter 10 we shall conclude our study of atoms with an examination of the calculation of multielectron wave functions and energies.

The principal differences between actual three-dimensional N-electron atoms and the models of the last chapter are two in number. First, we must cope with three dimensions rather than one. This means, for example, that the electrons can have nonzero orbital angular momenta; this fact must somehow be incorporated into our theory. Second, we must drop the pretense that the electron-electron interaction potential energy can be arbitrarily chosen, for it is known to be the coulomb interaction

$$V'(|\mathbf{r}_i - \mathbf{r}_j|) = \frac{e^2}{r_{ij}}, \tag{9.1}$$

where r_{ij} is the distance between electrons i and j, $r_{ij} = |\mathbf{r}_i - \mathbf{r}_j|$.

9.1 THE ORBITAL APPROXIMATION FOR ATOMS

The multielectron problem is to solve the time-independent Schroedinger equation

$$\mathcal{H}\psi_\gamma = E_\gamma \psi_\gamma \tag{9.2}$$

for the eigenfunctions and eigenvalues of the Hamiltonian[1]

$$\mathcal{H} = \sum_{i=1}^{N}\left[-\frac{\hbar^2}{2m_e}\nabla_i^2 - \frac{Ze^2}{r_i}\right] + \sum_{i=1}^{N}\sum_{j>i}^{N}\frac{e^2}{r_{ij}}. \tag{9.3}$$

Each solution of Eq. (9.2) is labeled by a subscript γ, which denotes all quantum numbers necessary to specify uniquely the state of the system.

Since the electrons are indistinguishable, $\mathcal{H}$ commutes with all particle exchange operators P_{ij},

$$[\mathcal{H}, P_{ij}] = 0, \qquad i, j = 1, \ldots, N; \quad i \neq j, \tag{9.4}$$

and since electrons are fermions, the wave functions ψ_γ must be antisymmetric—that is,

$$P_{ij}\psi_\gamma = -\psi_\gamma, \qquad i, j = 1, \ldots, N; \quad i \neq j. \tag{9.5}$$

The Independent Particle Model

The reason that Eq. (9.2) is so difficult to solve is that the motions of all electrons are coupled by the terms e^2/r_{ij}. The problem would be greatly simplified if we could uncouple this motion; in such an approximation, we view each electron as moving independently of all other electrons. (Thus this picture of the atom is an "independent particle model.") Each electron is described by a single-particle function, or orbital, which spans the atom. We call this approximation *the orbital approximation.*

[1] In writing Eq. (9.2), we are, of course, neglecting many weak interactions—for example, magnetic interactions that give rise to spin-orbit coupling. Moreover, the electromagnetic field, totally absent from our Hamiltonian, cannot be ignored. In reality, most of these "stationary states" are unstable with respect to photon emission.

Let us take as our approximate Hamiltonian

$$\mathcal{H}^{OA} = \sum_{i=1}^{N}\left[-\frac{\hbar^2}{2m_e}\nabla_i^2 + V_i(\mathbf{r}_i)\right], \tag{9.6}$$

where the potential energy of the ith electron $V_i(\mathbf{r}_i)$ does not explicitly depend on the coordinates of the other $N - 1$ electrons. The eigenvalue equation for $\mathcal{H}^{OA}$ separates into a set of N equations of the form

$$\left[-\frac{\hbar^2}{2m_e}\nabla_i^2 + V_i(\mathbf{r}_i)\right]\psi_{n_im_{s_i}}(\mathbf{r}_i) = E_{n_im_{s_i}}\psi_{n_im_{s_i}}(\mathbf{r}_i), \tag{9.7}$$

where n_i denotes all spatial quantum numbers of the ith electron and m_{s_i} the spin quantum number (for all electrons $s_i = \frac{1}{2}$). Let us indicate them collectively by ϵ_i so that $\epsilon_i = (n_im_{s_i})$ and the wave function $\psi_{n_im_{s_i}}(\mathbf{r})$ is written $\psi_{\epsilon_i}(\mathbf{r})$. Each eigenfunction $\psi_{\epsilon_i}(\mathbf{r})$ is a two-component vector, which we may construct by forming a product of the spatial function obtained by solution of Eq. (9.7) and one of the spin functions α or β, depending on whether $m_{s_i} = +\frac{1}{2}$ or $-\frac{1}{2}$, respectively.

A properly antisymmetrized eigenfunction of the full Hamiltonian $\mathcal{H}^{OA}$ is given by the Slater determinant of these one-electron orbitals,

$$\psi_\gamma = |\psi_{\epsilon_1}(\mathbf{r}_1) \quad \psi_{\epsilon_2}(\mathbf{r}_2) \quad \cdots \quad \psi_{\epsilon_N}(\mathbf{r}_N)|, \tag{9.8}$$

where we recall from Eqs. (8.74) that only the diagonal terms of the determinant are explicitly shown in this notation. As a rule, there are many ways the spins can be assigned to the N electrons; the only restriction is the Pauli exclusion principle, which prohibits any two electrons from having the same set of quantum numbers $\epsilon_i = (n_i, m_{s_i})$. Consequently, there are, in general, a large number of degenerate eigenfunctions of the form (9.8). This situation is familiar from Sec. 8.4.

In order to solve the single-particle equations (9.7), we must specify the form of the ith potential energy $V_i(\mathbf{r}_i)$. One reasonable if somewhat naive form is obtained by simply ignoring all the electron-electron repulsion terms e^2/r_{ij}. In this form of the orbital approximation, each electron moves in a potential due to the coulomb interaction of the electron with the nucleus

$$V_i(\mathbf{r}_i) = V(r_i) = -\frac{Ze^2}{r_i}, \tag{9.9}$$

and is completely oblivious of the presence of any other electrons in the system. We call this the *isolated electron approximation.*

With $V_i(\mathbf{r}_i)$ given by $-Ze^2/r_i$, the single-particle Schroedinger equation for the ith electron is precisely the equation for an isolated hydrogenic atom

$$\left[-\frac{\hbar^2}{2m_e}\nabla_i^2 - \frac{Ze^2}{r_i}\right]\psi_{n_i\ell_i m_{\ell_i} m_{s_i}}(\mathbf{r}) = E_{n_i}\psi_{n_i\ell_i m_{\ell_i} m_{s_i}}(\mathbf{r}), \tag{9.10}$$

the solutions of which are well known to us. The radial dependence of $\psi_{n_i\ell_i m_{\ell_i} m_{s_i}}(\mathbf{r})$ is given by the radial function $R_{n_i\ell_i}(r)$ of Table 3.2, and the angular dependence is described by the spherical harmonic $Y_{\ell_i m_{\ell_i}}(\theta, \varphi)$ of Table 2.2. The subscripts on the single-particle orbitals in Eq. (9.10), $n_i\ell_i m_{\ell_i} m_{s_i}$, are the principal quantum number, orbital angular momentum quantum number, magnetic quantum number, and spin quantum number, respectively. Thus in the isolated electron approximation, it is an easy task to form the full eigenfunctions ψ_γ that solve $\mathcal{H}^{OA}\psi_\gamma = E_\gamma\psi_\gamma$. The energy E_γ is just[2] the sum of the hydrogenic energies of each electron; recall that these energies are given for an electron with principal quantum number n_i by [see Eq. (3.61)]

$$E_{n_i} \simeq -\frac{m_e Z^2 e^4}{2\hbar^2 n_i^2}, \qquad n_i = 1, 2, \ldots. \tag{9.11}$$

Unfortunately, this is a very poor approximation, for, as we shall see shortly, the ith electron is affected by the other $N-1$ electrons. For example, an electron that is, on the average, far from the nucleus experiences a much weaker nuclear attraction than does an electron near the nucleus because the far electron is "shielded" or "screened" by the electrons between it and the nuclear charge $+Ze$.[3]

Is there a way to take into account the effect of the electron-electron interactions in the orbital approximation without making the problem of solving the single-particle equations (9.7) unmanageable? We might be able to do so by constructing a single-particle potential energy $V_i(\mathbf{r}_i)$ for the ith electron of the form

$$V_i(r_i) = -\frac{Ze^2}{r_i} + V_i^{\text{eff}}(r_i), \tag{9.12}$$

where $V_i^{\text{eff}}(r_i)$ is an *effective potential energy* that somehow incorporates the average effect of the motion of the other $N-1$ electrons. The potential energy (9.12) has the enormous advantage of being independent of θ_i and φ_i; therefore the angular dependence of the ith orbital $\psi_{\epsilon_i}(\mathbf{r}_i)$ is given by the spherical harmonic $Y_{\ell_i m_{\ell_i}}(\theta_i, \varphi_i)$. Of course, the radial dependence of $\psi_{\epsilon_i}(\mathbf{r}_i)$ must

[2] If necessary, we could try to improve on these wave functions and the corresponding energies by including the terms $\sum_{i=1}^{N}\sum_{j>i}^{N} e^2/r_{ij}$ by perturbation theory.

[3] See Prob. 3.7.

be determined by solution of a radial equation of the form [see Eq. (2.69)]

$$\left\{\frac{1}{r_i^2}\frac{d}{dr_i}\left(r_i^2\frac{d}{dr_i}\right) - \frac{\ell(\ell+1)}{r_i^2} + \frac{2m_e}{\hbar^2}[E_{\epsilon_i} - V_i(r_i)]\right\} R_{n_i\ell_i}(r) = 0, \qquad (9.13)$$

but this task is within our capabilities (see Sec. 10.3). Because the single-particle potential energy is independent of θ_i and φ_i, this particular refinement of the orbital approximation is called the *central field approximation.* We shall have more to say about this approximation in Chapter 10; in particular, we shall look further at the calculation of $V_i^{\text{eff}}(r_i)$ and the solution of the single-particle equations.

9.2 ELECTRON SCREENING AND ATOMIC STRUCTURE

In Chapter 3 the shell structure of the hydrogenic atom was described. In the one-electron atom, a particular value of n defines a *shell* with energy E_n given by Eq. (3.63), $E_n = -Z^2(13.596)/n^2$ eV. Each shell consists of *subshells* defined by the orbital angular momentum quantum number ℓ. We thus speak of electrons in subshells $1s$, $2s$, $2p$, and so on.

This structure is easily generalized to the N-electron atom. The atom is viewed as consisting of shells and subshells into which we place the electrons in a manner consistent with the Pauli exclusion principle. (This process is similar to the filling of levels encountered in our one-dimensional model, Sec. 8.4.) Electrons with the same principal quantum number are said to be in the same shell; electrons with the same values of n and ℓ are said to be in the same subshell.

The "radius" of each shell is given roughly by the expectation value $\langle r \rangle_{n0}$ and that of each subshell by $\langle r \rangle_{n\ell}$ (see Fig. 3.4 and Example 3.1). In the hydrogenic atom (ignoring the spin-orbit interaction), quantum states with the same principal quantum number but different orbital angular momentum quantum numbers are degenerate. This is not the case in the central field approximation to the N-electron atom.

For example, for a given $n > 1$, the s orbital has a nonzero probability amplitude close to the nucleus. Consequently, it *penetrates* the $n = 1$ shell more than the corresponding p orbitals do, and the electron in the s subshell experiences more of the attractive nuclear charge than the electrons in the p subshell do. On the basis of this argument, we predict that s orbitals lie lower in energy than do p orbitals; detailed calculations reveal that such is the case. Here is one example of the phenomenon of *screening*.

In fact, screening arguments are the basis of the Aufbauprinzip, or Aufbau principle, which tells us how to put electrons into the shells of the atom. The Aufbau principle dictates that the subshells ($n\ell$) be filled in order of increasing energy, the energy being determined in part by screening arguments. This principle is most easily understood in the context of a discussion of the ground states of multielectron atoms.

9.3 GROUND STATES OF N-ELECTRON ATOMS[4]

Consider an N-electron atom. The ground state of this atom is simply the state with the lowest possible total energy. Since the single-particle Hamiltonians in the central field approximation do not single out a unique direction in space, the energy of the state is independent of the quantum numbers m_{ℓ_i}, $i = 1, 2, \ldots, N$. Similarly, it is independent of m_{s_i}, $i = 1, 2, \ldots, N$.

Hence an energy level of the system may be specified by the $2N$ quantum numbers n_i, ℓ_i for $i = 1, 2, \ldots, N$. The *electronic configuration* of a state of the atom is a statement of the quantum numbers n_i and ℓ_i for all electrons. It also indicates how the electrons fill the shells and subshells of the atom. For example, the electronic configuration of the ground state of hydrogen is $1s$, of helium $1s^2$, of lithium $1s^2 2s$. The superscript 2 on the configuration of helium indicates that there are two electrons in the $1s$ shell.

Notice that we did not write $1s^3$ for the electronic configuration of lithium. The reason is that the Pauli exclusion principle forbids more than two electrons from occupying the same s subshell. To see this we need only recall that electrons in an s subshell have quantum numbers $\ell = 0$, $m_\ell = 0$. Thus their spin quantum numbers must differ. There are only two possibilities, $m_s = \pm\frac{1}{2}$, so only two electrons can occupy the s subshell.

The shells of the atom are sometimes labeled by the letters $K, L, M, \ldots$ corresponding to principal quantum numbers $1, 2, 3, \ldots$. Each shell consists of a number of subshells.[5]

The *Aufbau principle* states that the electrons are to be placed in subshells in accordance with the Pauli exclusion principle. The most natural ordering of the subshells would seem to be

$$\boxed{1s,\ 2s,\ 2p,\ 3s,\ 3p,\ 3d,\ 4s,\ \ldots,} \tag{9.14}$$

where the energy increases from left to right. This ordering seems reasonable. Penetration arguments like those in Sec. 9.2 indicate that subshells of a

[4]See Prob. 9.1 for excited states.

[5]The subshells are denoted by spectroscopists as $K, L_1, L_2, M_1, M_2, \ldots$ corresponding to $1s, 2s, 2p, 3s, 3p, \ldots$, respectively. Thus you may see the shell structure of an atom indicated by notation like $K, L(L_1, L_2), M(M_1, M_2, M_3), \ldots$.

particular shell are ordered in energy as $s, p, d, \ldots$, and we know that the energy of a shell increases with its principal quantum number.

As stated earlier, the number of electrons that can occupy a particular subshell is determined by the Pauli exclusion principle. Thus no more than two electrons can occupy an s ($\ell = 0$) subshell (e.g., $1s$, $2s$, $3s$, . . .), and in each subshell the electrons must be oppositely paired. Six electrons can occupy a p subshell ($\ell = 1$), two with $m_{\ell_i} = -1$, two with $m_{\ell_i} = 0$, and two with $m_{\ell_i} = +1$. Similar arguments show that 10 electrons fill a d subshell, 14 fill an f subshell, and so on.

A shell that contains the full complement of electrons allowed by the exclusion principle is called a *closed shell* (or *filled shell*). Obviously, a shell that is not closed is *open* (or *partially filled*). If an atom consists of a number of closed shells and subshells plus one or more electrons in a partially filled shell, these electrons are called *valence electrons*. In lithium, $1s^2 2s$, for example, the $2s$ electron is the valence electron.

The Aufbau principle determines much of the shape and form of the periodic table of the elements. The chemical properties of various elements are largely determined by the number of valence electrons. It is chiefly these electrons that participate in bonding and chemical reactions. Because of the Aufbau principle, certain patterns in electronic configurations reappear periodically as the number of electrons increases. Thus carbon ($1s^2\, 2s^2\, 2p^2$) and silicon ($1s^2\, 2s^2\, 2p^6\, 3s^2\, 3p^2$) each have two p valence electrons; this fact implies that some of their chemical properties will be similar, and experiments indicate that such is indeed the case.[6]

Unfortunately, the simple ordering given in (9.14) breaks down in the $n = 3$ shell. For example, if the configuration of an atom consists of partially occupied $3d$ and $4s$ subshells, then, contrary to the ordering of (9.14), the $4s$ level fills before the $3d$ level. And the electronic configuration of potassium is $1s^2\, 2s^2\, 2p^6\, 3s^2\, 3p^6\, 4s$, not $1s^2\, 2s^2\, 2p^6\, 3s^2\, 3p^6\, 3d$. Qualitatively, this ordering can be understood by referring to probability densities. The small amount of probability density of the $4s$ orbital near the nucleus is enough to pull its energy level below that of the $3d$ orbital, which penetrates very little into the region of small r (see Fig. 3.4). In fact, the simple ordering of (9.14) is observed to break down in several regions of the periodic table—for example, for neutral and singly ionized transition metal ions, alkaline earths, and alkalies. The more highly ionized species tend to follow the ordering of (9.14). The details are too involved to go into here (see the Suggested Readings for this chapter).

[6]In fact, silicon-based life has been proposed as a possible alternative to the carbon-based life of which we are so fond. See Carl Sagan, *The Cosmic Connection* (New York: Anchor Press, 1973), p. 47.

The actual ordering of orbitals for neutral atoms is better described by the following convenient table.

Shell n	Increasing energy $\longrightarrow$
$n = 1, 2, 3$	$1s$ $2s$ $2p$ $3s$ $3p$ · $3d$
$n = 4$	$4s$ $4p$ · $4d$ · · $4f$ · ·
$n = 5$	$5s$ $5p$ · $5d$ ·
$n = 6$	$6s$ $6p$

(9.15)

A list of the ground-state configurations of all atoms appears in Appendix 5.

Wave Functions

Summarizing, we see that, in order to construct the ground-state configuration of a multielectron atom, we assign electrons to subshells according to (9.15) and the Pauli principle, beginning with the subshell of lowest energy. The corresponding ground-state wave function can be written down from Eq. (9.8). For example, the eigenfunction for the ground state of an atom with N even is

$$\begin{aligned}\psi_\gamma = |\psi_{1s}(\mathbf{r}_1)\alpha(1)\quad \psi_{1s}(\mathbf{r}_2)\beta(2)\quad \psi_{2s}(\mathbf{r}_3)\alpha(3)\\ \cdots\quad \psi_{n\ell m_\ell}(\mathbf{r}_{N-1})\alpha(N-1)\ \psi_{n\ell m_\ell}(\mathbf{r}_N)\beta(N)|,\end{aligned} \tag{9.16}$$

where the functions $\psi_{(n\ell m_\ell)_i}(\mathbf{r}_i)$ are single-particle orbitals obtained by solving the Schroedinger equation for the ith particle in the isolated electron approximation or the central field approximation. Using a bar to denote spin "down," we can abbreviate Eq. (9.16) as

$$\psi_\gamma = |1s(1)\quad \overline{1s(2)}\quad 2s(3)\quad \overline{2s(4)}\quad \cdots\quad n\ell m_\ell(N-1)\quad \overline{n\ell m_\ell(N)}|. \tag{9.17}$$

The energy is simply the sum of the energies of each electron.

Example 9.1
Ground State of Neon

For neon ($N = 10$), the ground-state configuration is (by the Aufbau principle) $1s^2 2s^2 2p^6$. Therefore the ground-state wave function is

$$\psi_1 = |1s(1)\,\overline{1s(2)}\,2s(3)\,\overline{2s(4)}\,2p_0(5)\,\overline{2p_0(6)}\,2p_{-1}(7)\,\overline{2p_{-1}(8)}\,2p_1(9)\,\overline{2p_1(10)}|. \tag{9.18}$$

The ground state of neon is a singlet (why?).

Neon was a particularly fortunate choice for our illustration; because its configuration is that of a closed shell, in which all the electrons are paired, we could write down a unique Slater determinant as in Eq. (9.18). If the configuration of an atom corresponds to an open shell, there will be some ambiguity about the assignment of quantum numbers to electrons. For example, the ground-state configuration of lithium is $1s^2 2s$, and the $2s$ electron can have $m_s = +\frac{1}{2}$ or $m_s = -\frac{1}{2}$. The two resulting wave functions are degenerate. Similarly, for carbon, with configuration $1s^2 2s^2 2p^2$, we can write several Slater determinants corresponding to the same energy. The $2p$ subshell contains two *equivalent electrons*, so called because their principal and orbital angular momentum quantum numbers are the same. The electronic configuration and energy depend only on the quantum numbers n_i and ℓ_i for each electron, but a wave function depends on n_i, ℓ_i, m_{ℓ_i}, and m_{s_i}, $i = 1, 2, \ldots, N$.

> **Exercise 9.1** Write the 15 different degenerate wave functions corresponding to the ground-state configuration of carbon. What is the ground-state energy of these functions in the isolated electron approximation?

Thus the presence of equivalent electrons in a partially filled shell gives rise to degenerate wave functions. As in the one-electron atom, some of this degeneracy is lifted when previously ignored interactions are considered (e.g., the spin-orbit interaction).

9.4 ANGULAR MOMENTUM COUPLING IN MULTIELECTRON ATOMS

A number of interactions have been omitted from the Hamiltonian in Eq. (9.3). The N electrons each have a spin magnetic moment and an orbital magnetic moment associated with them. All these magnetic moments can interact in various ways. The mutual coulomb repulsion of the electrons, which is related to forces that are directed not toward the nucleus but along lines between electrons, influences the individual orbital angular momenta $\mathbf{L}_i$ in such a way that they couple. Although each individual angular momentum operator does not commute with $\mathcal{H}$ (if these potential energy terms are included) and so is not a constant of the motion, the resultant total orbital angular momentum is a constant of the motion. The individual spins can also be coupled together to form a total spin angular momentum.

We see, therefore, that *electrostatic interactions* tend to couple all the orbital angular momenta and all the spin angular momenta. In addition, there are the spin-orbit interaction and other *magnetic interactions* (spin-spin, spin-other-orbit). The spin-orbit interaction gives rise to terms proportional to $\mathbf{L}_i \cdot \mathbf{S}_i$ and acts to couple the orbital angular momentum of the ith electron to its spin angular momentum so as to form a total $\mathbf{J}_i$.

Both types of interaction, electrostatic and magnetic, are present in any particular atom. However, the magnetic interactions are usually weaker. For example, in light atoms, the electrostatic repulsions are of the order of ~ 1 eV, whereas the spin-orbit potential energy is of the order of $\sim 10^{-4}$ to $\sim 10^{-3}$ eV. Hence the angular momentum coupling in a light atom is as follows:

1. The orbital angular momenta $\mathbf{L}_i$, $i = 1, 2, \ldots, N$, couple to form a total orbital angular momentum

$$\mathbf{L} \equiv \sum_{i=1}^{N} \mathbf{L}_i, \tag{9.19}$$

and the spin angular momenta $\mathbf{S}_i$ couple to form the total spin angular momentum

$$\mathbf{S} \equiv \sum_{i=1}^{N} \mathbf{S}_i. \tag{9.20}$$

This coupling is established at the level where only the electrostatic interactions are included in the Hamiltonian.

2. The weaker magnetic interactions act to couple $\mathbf{L}$ to $\mathbf{S}$ so as to form the total angular momentum of the atom

$$\mathbf{J} = \mathbf{L} + \mathbf{S}. \tag{9.21}$$

This particular coupling scheme, in which the electrostatic interactions dominate the magnetic interactions and are thereby included in the Hamiltonian first, is called *Russell-Saunders coupling*.[7]

In heavy atoms, the electrostatic repulsion terms are of the order of $\sim 10^3$ eV for the inner electrons and the magnetic or spin-orbit terms are of the order of $\sim 10^4$ eV. Thus an alternate coupling scheme is more appropriate:

1. Each electron's $\mathbf{L}_i$ and $\mathbf{S}_i$ couple via the magnetic interactions to yield

$$\mathbf{J}_i = \mathbf{L}_i + \mathbf{S}_i, \qquad i = 1, 2, \ldots, N. \tag{9.22}$$

[7]Some authors refer to this process as *LS* coupling.

2. The influence of the electrostatic interactions then couples the $\mathbf{J}_i$ together to form

$$\mathbf{J} = \sum_{i=1}^{N} \mathbf{J}_i. \tag{9.23}$$

This scheme is called *j-j coupling* and is generally less common than Russell-Saunders coupling.

Regardless of which scheme is used, the vector coupling proceeds according to the familiar rules of Chapter 7. If both electrostatic and magnetic interactions are taken into account, the individual ℓ_i, m_{ℓ_i}, m_{s_i}, j_i, m_{j_i}, L, M_L, S, and M_S are no longer good quantum numbers. Only J and M_J, corresponding to the magnitude of the total angular momentum of the atom and its projection along the $\hat{z}$ axis, are good quantum numbers of the system. The important point is that the question of which quantum numbers are good quantum numbers can be answered only in the context of a particular form of the orbital approximation. For example, in the isolated electron approximation, where each electron completely ignores its fellow electrons, L_i^2, S_i^2, $(L_z)_i$, $(S_z)_i$, J_i and $(J_z)_i$, $i = 1, 2, \ldots, N$, all commute with the Hamiltonian; therefore ℓ_i, m_{ℓ_i}, s_i, m_{s_i}, j_i, and m_{j_i} are all perfectly good quantum numbers and can be used to label states of the atom.

Exercise 9.2 Show that $(L_z)_i$ does not commute with $V'(\mathbf{r}_i - \mathbf{r}_j) = e^2/r_{ij}$ and thus does not commute with the Hamiltonian $\mathcal{H}$ of Eq. (9.3). Show also that $(L_z)_i + (L_z)_j$ does commute with $V'(\mathbf{r}_i - \mathbf{r}_j)$ so that $L_z = \sum_{i=1}^{N} (L_z)_i$ commutes with $\mathcal{H}$.

Because it is more frequently encountered, let us consider Russell-Saunders coupling in some detail.

9.5 RUSSELL-SAUNDERS COUPLING

In this case, the relevant equations are (9.19) to (9.21). The quantum numbers associated with the operators L^2, S^2, J^2, L_z, S_z, and J_z are L, S, J, M_L, M_S, and M_J, respectively.

Exercise 9.3 Draw a set of vector diagrams like the one in Fig. 7.1, illustrating Russell-Saunders coupling for a two-electron atom.

In the approximation in which spin-orbit coupling terms are neglected, eigenfunctions can be labeled by the quantum numbers $\{n, L, S, M_L, \text{and } M_S\}$ or by $\{n, L, S, J, M_J\}$. In improved approximations, in which spin-orbit interactions are included in the Hamiltonian, we must use the latter set—for M_L and M_S are no longer good quantum numbers—and in the vector model we picture the "precession of **L** and **S** about **J**" (see Prob. 9.4). Thus the appropriate set depends on two interrelated factors: (a) the strength of the spin-orbit interaction tending to couple **L** and **S** and (b) the accuracy to which we wish to treat the system.

How can we determine the new quantum numbers, given the assigned values of n_i, ℓ_i, m_{ℓ_i}, and m_{s_i} for each electron? The relationships are familiar from the treatment of angular momentum coupling in Sec. 7.3 and the one-dimensional model of Sec. 8.4. For example, the allowed values of M_L and M_S are given by

$$M_L = \sum_{i=1}^{N} m_{\ell_i} \tag{9.24}$$

$$M_S = \sum_{i=1}^{N} m_{s_i}, \tag{9.25}$$

and the eigenfunctions ψ_γ in the orbital approximation satisfy

$$L_z\psi_\gamma = M_L\hbar\psi_\gamma, \tag{9.26}$$

$$S_z\psi_\gamma = M_S\hbar\psi_\gamma. \tag{9.27}$$

Similarly, it is easy to determine the allowed values of L and S for, say, a two-electron atom. Generalizing the rules of vector coupling in Chapter 7, we have

$$L = |\ell_1 - \ell_2|, \ldots, \ell_1 + \ell_2 \tag{9.28}$$

and

$$S = 0, 1, \tag{9.29}$$

where we have assumed that $(n_1\ell_1) \neq (n_2\ell_2)$.[8] Of course, for a given L and S, M_L and M_S are given by

$$M_L = -L, \ldots, L, \tag{9.30}$$

$$M_S = -S, \ldots, S. \tag{9.31}$$

These relations are consistent with Eqs. (9.24) and (9.25). Given values for L and S, the allowed values of J and M_J are

$$J = |L - S|, \ldots, L + S, \tag{9.32}$$

$$M_J = -J, \ldots, J. \tag{9.33}$$

Notice that

$$M_J = M_L + M_S. \tag{9.34}$$

Exercise 9.4 Prove that the only allowed values of L and S for a closed-shell configuration are $L = 0$ and $S = 0$.

[8] More complicated cases are treated similarly. See Prob. 9.2.

In addition to labeling states of the system, these quantum numbers provide useful labels for the energy levels of the atom.

Atomic Terms

We expect the energy of a particular level to depend on L and S (and J if spin-orbit interactions are not ignored) but not on M_L, M_S (or M_J). Therefore we may label the levels by their *atomic terms*, which are written ${}^{2S+1}L$, where we use letters $S, P, D, \ldots$ for $L = 0, 1, 2, \ldots$, The superscript $2S + 1$ is called the *multiplicity* of the level; that is, singlet, doublet, triplet, and so on. For example, states with $S = 0$ and $L = 2$, arising from the same configuration, are collectively referred to as the ${}^1D$ term. In the approximation where spin-orbit effects are ignored, all states of a given term are degenerate. In general, a term ${}^{2S+1}L$ is $(2S + 1)(2L + 1)$-fold degenerate. For example, the ${}^3P$ term of carbon $(1s^22s^22p^2)$ is ninefold degenerate.

In the Russell-Saunders coupling scheme, the subscript J can be appended to the term designation to indicate one of the allowed values $J = |L - S|, \ldots, L + S$. For example, for $L = 1$, $S = 1$—a ${}^3P$ term—we have ${}^3P_0$, ${}^3P_1$, ${}^3P_2$. We may refer to these as *sublevels*; they are degenerate as long as the spin-orbit interaction is ignored. We shall return to this point in our discussion of Hund's rules. Each sublevel ${}^{2S+1}L_J$ is $(2J + 1)$-fold degenerate, and obviously the sum of the degeneracies of all sublevels must equal the degeneracy of the corresponding term. For example, the sublevels ${}^3P_0$, ${}^3P_1$, ${}^3P_2$ are one-, three-, and five-fold degenerate, respectively, for a total nine-fold degeneracy as required. Of course, when the spin-orbit interactions are considered, the degeneracy between different J sublevels is lifted. However, the $(2J + 1)$-fold degeneracy of *each* sublevel remains. This degeneracy can be removed by application of an external magnetic field; this is the Zeeman effect again (see Prob. 9.5).

Implied Terms[9]

It might seem a trivial matter to determine the terms for, say, the ground state of a particular atom. We merely write down the quantum numbers of the equivalent electrons in partially filled shells and couple them according to the rules of Eqs. (9.28) to (9.33). (Note that closed shells and subshells contribute nothing; see Exercise 9.4.) Doing so for carbon, which has two equivalent $2p$ electrons, we obtain the terms shown in Table 9.1. However, we have completely neglected the Pauli exclusion principle, which restricts the quantum numbers of the electrons in the atom and forces us to discard some of the terms in the table.

[9]The method outlined in this subsection follows closely the discussion in Peter O'D. Offenhartz, *Atomic and Molecular Orbital Theory* (New York: McGraw-Hill, 1970), Chap. 6.

Table 9.1

Atomic terms and sublevels obtained by applying Russell-Saunders coupling to the carbon atom. Only ${}^1S_0$, ${}^1D_2$, ${}^3P_1$, and ${}^3P_0$ remain after the Pauli exclusion principle has been taken into consideration.

L	S	J	Term	Sublevels
0	0	0	${}^1S$	${}^1S_0$
1	0	1	${}^1P$	${}^1P_1$
2	0	2	${}^1D$	${}^1D_2$
0	1	1	${}^3S$	${}^3S_1$
1	1	2, 1, 0	${}^3P$	${}^3P_2$, ${}^3P_1$, ${}^3P_0$
2	1	3, 2, 1	${}^3D$	${}^3D_3$, ${}^3D_2$, ${}^3D_1$

Table 9.2

Values of m_{ℓ_1}, m_{ℓ_2}, m_{s_1}, and m_{s_2} allowed by the Pauli exclusion principle for carbon ($1s^22s^22p^2$) and resulting values of M_L and M_S.

Entry	m_ℓ			M_L	M_S
	−1	0	+1		
1	↑ ↓			−2	0
2		↑ ↓		0	0
3			↑ ↓	2	0
4	↑	↑		−1	1
5	↑	↓		−1	0
6	↓	↑		−1	0
7	↓	↓		−1	−1
8	↑		↑	0	1
9	↑		↓	0	0
10	↓		↑	0	0
11	↓		↓	0	−1
12		↑	↑	1	1
13		↑	↓	1	0
14		↓	↑	1	0
15		↓	↓	1	−1

To illustrate, we begin by listing the allowed values of m_{ℓ_i} and m_{s_i} (valence electrons only) and the resultant values of M_L and M_S for carbon in Table 9.2. Since Exercise 9.1 showed that there are 15 degenerate Slater determinants for the ground state of carbon, we expect to find 15 entries in the table. Each entry in this table corresponds to one allowed state of the system. We now list the number of states with a particular M_L and M_S in Table 9.3a, which is

Table 9.3a
First Implied-Terms Table

M_L \ M_S	1	0	−1
2	0	1	0
1	1	2	1
0	1	3	1
−1	1	2	1
−2	0	1	0

called an *implied terms table*. We shall use it to determine the allowed terms for carbon.

First, consider the largest possible values of L and S: $L = 2$ and $S = 1$. These values give rise to the ${}^3D$ term in Table 9.1. If they were allowed, entries corresponding to $M_L = 2$ and $M_S = 1$ would have to appear in Table 9.2. A glance at the table shows that they do not. We must discard the ${}^3D$ term and the sublevels ${}^3D_3$, ${}^3D_2$, and ${}^3D_1$ as disallowed by the Pauli principle.

Now we turn to the next largest values: $L = 2$, $S = 0$. This is the ${}^1D$ term. Since there are entries in Table 9.2 with $M_L = 2$, $M_S = 0$, we know that this term is allowed. The ${}^1D$ term corresponds to five states, depending on whether $M_L = -2, -1, 0, 1, 2$. Each state is a singlet and has one of $M_L = -2, -1, 0, 1, 2$. Consequently, we have determined the nature of five of the states in Table 9.3a. Subtracting these states from the table yields Table 9.3b (we subtract 1 from each element in the second column). We look next at $L = 1$, $S = 1$—that is, the ${}^3P$ terms. Entries with $M_L = -1, 0, +1$ and $M_S = -1, 0, 1$ remain in Table 9.3b, so the ${}^3P_2$, ${}^3P_1$, and ${}^3P_0$ sublevels are allowed. They correspond to nine states altogether. Substracting them from Table 9.3b, we obtain Table 9.3c. But only one state is left in Table 9.3c. It

Table 9.3b
Second Implied-Terms Table

M_L \ M_S	1	0	−1
2	0	0	0
1	1	1	1
0	1	2	1
−1	1	1	1
−2	0	0	0

Table 9.3c
Final Implied-Terms Table

M_L \ M_S	1	0	−1
2	0	0	0
1	0	0	0
0	0	1	0
−1	0	0	0
−2	0	0	0

has $M_L = 0$, $M_S = 0$ and gives the ^1S term. Thus this term is allowed; and since no more states remain, the terms ^3S and ^1P are forbidden. We have shown that of all the atomic terms that might be generated by Russell-Saunders coupling for carbon, only the terms ^1S, ^3P, and ^1D are allowed by the Pauli exclusion principle.

Precisely the same analysis holds for any atom with two equivalent p electrons outside closed subshells. The implied-terms method is a convenient bookkeeping procedure and can be applied to any configuration (see Prob. 9.2).

To complete our picture of the quantum mechanics of multielectron atoms, we must know how the terms obtained above are ordered in energy. This is the subject of Sec. 9.6. Also unresolved at this point is the question of precisely which of the many degenerate Slater determinants constitute a particular wave function for state $\gamma = (S, L, M_L, M_S)$. To answer, we must actually construct the eigenfunction of interest. This topic is discussed in Prob. 9.3.

9.6 HUND'S RULES

It is now possible to determine the allowed atomic terms corresponding to a particular electronic configuration. For example, in carbon we found ^1S, ^3P, and ^1D; each sublevel $^1S_0$, $^1D_2$, $^3P_2$, $^3P_1$, and $^3P_0$ arising from these terms has a different energy when spin-orbit interactions are included in the Hamiltonian. We would like to know these energies. However, Chapter 7 has demonstrated that actual calculation of the shift and ordering in energy of these levels can be a time-consuming and difficult process. Fortunately, a set of rules exists that enables us to avoid such calculations if qualitative rather than quantitative results are satisfactory. Called *Hund's rules*, they provide an estimate of the ordering in energy of the levels and sublevels for the case of equivalent electrons in the ground configuration.

RULE 1: The terms(s) arising from the ground configuration with the maximum multiplicity (2S + 1) *lies lowest in energy.*

Rule 1 can be understood by generalizing the spin-pairing arguments of Sec. 8.3 to three-dimensional multielectron atoms. Recall that in the two-electron case, we found that like spins (unpaired spins) "repel" and unlike spins (paired spins) "attract." In a multielectron atom, a state with high multiplicity contains a greater number of electrons with parallel spins than does one of low multiplicity. Since these electrons all "avoid one another", the energy of the state with high multiplicity lies below that of a state with low multiplicity.

RULE 2: Of several levels with the same multiplicity, the one with the maximum value of L *lies lowest in energy.*

Rule 2 can also be crudely understood as a consequence of the electrostatic repulsion of electrons. In a state of large orbital angular momentum quantum number L, the electrons can be thought of as orbiting "in the same direction." Such electrons can remain separated from one another at all times and so have a lower energy than do electrons "orbiting in the opposite direction," which must approach one another at some time.

RULE 3: Of several sublevels with the same multiplicity and total quantum number L*:*

(*a*) *the sublevel with the minimum value of* J *lies lowest in energy if the configuration has a shell that is less than half-filled;*

(*b*) *the sublevel with the maximum value of* J *lies lowest in energy if the configuration has a shell that is more than half-filled.*

Rule 3 is primarily a consequence of the spin-orbit interaction plus the fact that the electrostatic potential energy increases as r increases.[10] If we apply Hund's rules, together with our knowledge of the effects of electrostatic and magnetic interactions for Russell-Saunders coupling, to carbon, we obtain the energy level diagram of Fig. 9.1. (The splittings in this diagram are not exactly to scale.)

As with the Aufbau principle, Hund's rules provide a simple rule of thumb for determining features of atomic structure. They, too, break down in some cases. The reason for this breakdown lies at the very core of the orbital approximation. In writing an electronic configuration, all we must specify is how electrons are placed into subshells. The actual distribution of electrons in a particular quantum mechanical state is sometimes more accurately

[10] In case (a) the sublevels are called *regular multiplets;* in case (b) they are called *inverted multiplets.*

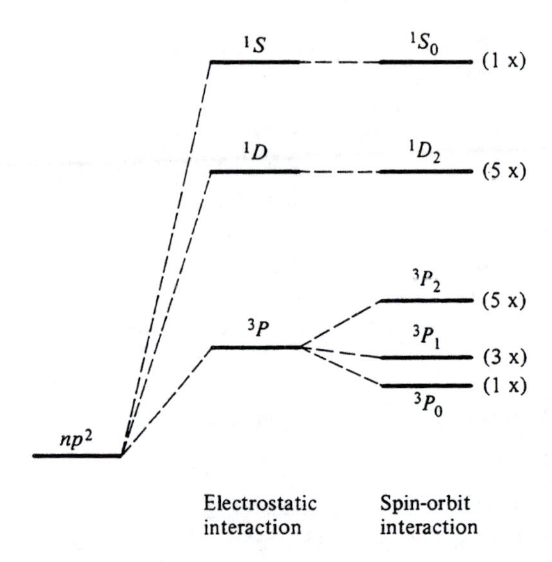

Figure 9.1 Schematic energy level diagram for the np^2 valence configuration (e.g., $n = 2$: carbon) showing the effects of electrostatic and spin-orbit interactions. Degeneracies are indicated to the side of each sublevel (Russell-Saunders coupling).

described by a mixture of more than one such configuration. A detailed discussion of this phenomenon, called "configuration interaction" (or configuration mixing), is beyond the scope of this book.

9.7 CONCLUDING REMARKS: *j-j* COUPLING

We suggested in Sec. 9.4 that Russell-Saunders coupling is not applicable to heavy atoms because the magnetic interactions, which tend to couple the spin and orbital angular momenta of the individual electrons to one another to form $\mathbf{J}_i = \mathbf{L}_i + \mathbf{S}_i$, become more important than the electrostatic interactions.

An alternative to Russell-Saunders coupling is *j-j coupling*; the relevant equations are (9.22) and (9.23). The quantum numbers of the individual electrons j_i and m_{j_i} are, for a given ℓ_i (≥ 1) and $s_i = \frac{1}{2}$,

$$j_i = \ell_i - \tfrac{1}{2}, \ell_i + \tfrac{1}{2}, \tag{9.35}$$

$$m_{j_i} = -j_i, \ldots, j_i. \tag{9.36}$$

The individual total angular momenta $\mathbf{J}_i$ then couple to give the total $\mathbf{J}$. For a two-electron atom, the quantum numbers J and M_J are simply

$$J = |j_1 - j_2|, \ldots, j_1 + j_2, \tag{9.37}$$

$$M_J = -J, \ldots, J. \tag{9.38}$$

The levels are labeled by J and (j_1, j_2) rather than by term designations as in Russell-Saunders coupling (see Prob. 9.6).

In general, the level corresponding to all electrons having the smallest values of j_i will lie lowest in energy. Thus in the case of lead (Pb), which has an np^2 valence configuration built on a closed shell (see Appendix 5), we find levels ordered as $(\frac{1}{2}, \frac{1}{2})$, $(\frac{3}{2}, \frac{1}{2})$, and $(\frac{3}{2}, \frac{3}{2})$. Usually the level with the lowest value of J for given j_1 and j_2 lies lowest in energy; however, this should not be taken as a strict rule.

Two final comments are necessary. First, for many medium-weight and heavy atoms, neither Russell-Saunders nor *j-j* coupling is precisely accurate; the electrostatic and magnetic interactions are of comparable magnitude. In such a case, a much more complicated scheme must be employed. Second, there is a relationship between the two schemes. This point is illustrated in Fig. 9.2, which shows schematically the transition from pure Russell-Saunders coupling in a light element of Group IV of the periodic table to nearly pure *j-j* coupling in the heaviest element in Group IV. Notice that all the atoms in Fig. 9.2 have the same valence configuration (np^2). The lines from left to right correlate the two extreme limits connecting levels of the same total angular momentum quantum number J.

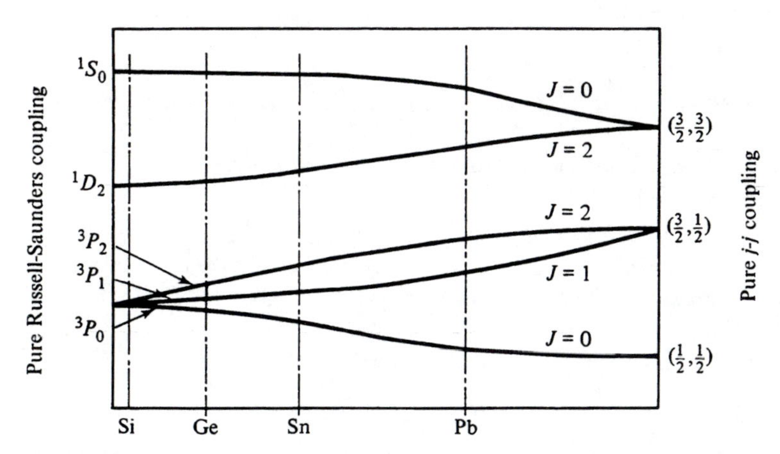

Figure 9.2 Illustration of the "correlation" of Russell-Saunders and *j-j* coupling schemes for Group IV atoms (electronic valence configuration: np^2). (From P. W. Atkins, *Molecular Quantum Mechanics,* Vol. 2. Oxford: The Clarendon Press, 1970.)

So ends our discussion of atomic structure. It would easily be possible to spend many more pages improving our understanding of atomic properties and exploring the oddities of the periodic table. We refer the interested reader to the references below instead.

SUGGESTED READINGS

A number of books contain useful treatments of atomic structure. At an introductory level, there are, for example,

EISBERG, ROBERT M., *Fundamentals of Modern Physics.* New York: Wiley, 1961. See especially Chap. 11.

PILAR, FRANK L., *Elementary Quantum Chemistry.* New York: McGraw-Hill, 1968. See especially Chap. 12.

A treatment of this subject in the context of atomic spectroscopy may be found in

KUHN, H. G., *Atomic Spectra.* New York: Academic, 1969,

or, at a slightly more advanced level, in

HERZBERG, G., *Atomic Spectra and Atomic Structure.* New York: Dover, 1944.

For further information regarding the periodic table and properties of the elements, see

RICH, R., *Periodic Correlations.* New York: W. A. Benjamin, 1965.

Finally, several of the references more appropriately placed after Chapter 10 contain well-written discussions of atomic structure.

PROBLEMS

9.1 Excited-State Configurations of the Alkali Atoms: Quantum Defect (*)

The excited-state configurations usually correspond to simply promoting one of the outer electrons to a higher single-particle state. In the case of very highly excited configurations, the energy levels are frequently found to describe a *Rydberg series* such that

$$E_{n\ell} \simeq -\frac{R}{(n - \delta_{n\ell})^2},$$

where n and ℓ are the principal and orbital angular momentum quantum numbers describing the excited orbital; $\delta_{n\ell}$ is called the *quantum defect* and gives a measure of "how hydrogenic" the excited orbital really is. The *Rydberg constant* is

$$\begin{aligned} R &= 109{,}737 \text{ cm}^{-1} \\ &= 13.605 \text{ eV}. \end{aligned}$$

The table shown contains observed energy levels (given as positive numbers in cm^{-1} units) for two alkali atoms, lithium ($Z = 3$) and sodium ($Z = 11$).

	Li				Na			
n	s	p	d	f	s	p	d	f
2	43487	28583	—	—	—	—	—	—
3	16281	12562	12204	—	41450	24484	12277	—
4	8475	7017	6864	6857	15710	11180	6901	6861
5	5187	4471	4392	4382	8249	6408	4413	4391
6	—	—	—	—	5078	4153	3062	3042

(a) Calculate the quantum defect $\delta_{n\ell}$ for each energy level of Li and Na.

(b) For a given value of n, explain the observed variation of $\delta_{n\ell}$ with ℓ.

(c) Explain the relative magnitudes of the quantum defects for Li and Na.

(d) Explain why the defects do not vanish in the limit $n \longrightarrow \infty$.

9.2 Russell-Saunders Coupling in Multielectron Atoms: Terms of p^k (**)

Consider a configuration consisting of k equivalent p electrons outside a closed shell, which we denote simply by p^k (e.g., carbon p^2, nitrogen p^3, oxygen p^4).

(a) Use the implied-terms method to determine all the terms that can arise from p^3. Which of them will have the lowest energy?

(b) Repeat part (a) for the cases p^4, p^5, and p^6 to show that, in general, the terms arising from p^k are precisely the same as those arising from p^{6-k}. [The result in part (b) is quite general and applies to d^k and d^{10-k}, f^k, and f^{14-k}, and so on. It is common practice to refer to the electrons that would be required to fill a subshell as "holes." Hence the configuration p^2 corresponds to 4 holes, p^3 to 3 holes, p^4 to 2 holes, and so on. The terms can be obtained by coupling the angular momentum of either the electrons or the holes!]

9.3 Wave functions for p^2 Configuration: Russell-Saunders Coupling (****)

We know that the single Slater determinants that correspond to particular configurations are not "good wave functions." They are, however, adequate zeroth-order functions for use in perturbation theory. Such calculations can be rather complicated for configurations like $1s^22s^22p^2$ (carbon) because the zeroth-order energy is $(36 - 6)/2 = 15$-fold degenerate (why?), and the degenerate theory of Sec. 4.6 would be required. We avoid this difficulty in the usual manner by forming new zeroth-order functions that are eigenfunctions of L^2, L_z, S^2, and S_z.

(a) Using a convenient notation, write down all 15 Slater determinants that arise in the $1s^22s^22p^2$ configuration (Exercise 9.1).

(b) For each choice of m_ℓ and $m_{\ell'}$ (quantum numbers of the $2p$ electrons), form linear combinations (summing over m_s and $m_{s'}$) as required to obtain eigenfunctions of S^2 and S_z. You should obtain 15 such functions. [Hint: You may find it convenient to consider only the diagonal element of each Slater determinant when coupling the spins. A little thought should convince you that the resulting combinations apply using the entire determinant.]

(c) In most cases, you will find that the functions obtained in part (b) are already eigenfunctions of L^2 and L_z. The exceptions are ${}^1D\,(M_L = 0)$ and ${}^1S$. Make up a table showing explicitly all the eigenfunctions for ${}^3P$ (all M_L and M_S), ${}^1D$ (all M_L), and ${}^1S$. [HINT: You may need the Clebsch-Gordan coefficients of Appendix 4.]

9.4 Spin-Orbit Coupling in Complex Atoms: Origin of Fine Structure (***)

Spin-orbit coupling is strictly an internal effect arising from the interaction between the electron spin and the effective magnetic field due to the apparent nuclear motion. By analogy with the one-electron atom, we can write for the N-electron atom

$$\mathcal{H}_{so} = \sum_{i=1}^{N} \xi_i(\mathbf{r}_i)\mathbf{L}_i \cdot \mathbf{S}_i,$$

where $\xi_i(\mathbf{r}_i)$ is defined in a manner similar to the one-electron case, assuming an effective potential field can be defined for each electron.

(a) For the case of weak spin-orbit coupling, use classical and quantum mechanical arguments to show that the first-order correction to the energy may be given as

$$E_{so}^{(1)} = \langle LSJM_J | \mathcal{H}_{so} | LSJM_J \rangle$$
$$\simeq A(L, S)\left[\frac{J(J+1) - L(L+1) - S(S+1)}{2}\right],$$

where $A(L, S)$ is a constant, independent of J. [HINT: For the case of weak spin-orbit coupling, we can picture each $\mathbf{L}_i$ as rapidly precessing about $\mathbf{L}$ and each $\mathbf{S}_i$ rapidly precessing about $\mathbf{S}$; this should give you a good start. It will help to draw vector diagrams. Next, it is helpful to recall the operator expression $J^2 = L^2 + S^2 + 2\mathbf{L}\cdot\mathbf{S}$.]

(b) The result of part (a) shows that each term $E(L, S)$ of a particular configuration is split by $\mathcal{H}_{so}$ into several different levels, one for each value of J. The separation of these so-called *fine-structure* levels is given by

$$E(L, S, J) - E(L, S, J-1) = AJ.$$

This is called the *Landé Interval Rule*. It is found that $A > 0$ ("normal multiplets") for terms arising from subshells less than half full and $A < 0$ ("inverted multiplets") for subshells over half full. For half-filled subshells, $A = 0$. Consider the ground state ${}^3P$ fine structure of carbon, oxygen, silicon, and sulfur, where the energy levels are all referred to the lowest one taken as zero; the energies in cm^{-1} units are

C	O	Si	S
0.0	0.0	0.0	0.0
16.4	158.5	77.2	396.0
27.1	226.5	223.3	573.6

(i) Write the ground-state electronic configuration for each atom and indicate whether the multiplet should be normal or inverted. Draw an energy level diagram for each atom and label the levels with the correct values of J.

(ii) Calculate empirical values of A for each interval of an atom and give the percent difference as a measure of the accuracy of the interval rule for each atom.

(iii) Briefly explain the observed variation of A with nuclear charge.

9.5 Zeeman Effect in Complex Atoms (***)

The electronic spin and orbital angular momenta in a complex atom give rise to a magnetic moment that we may write, by analogy with the one-electron atom, as

$$\mathbf{M} = \sum_{i=1}^{N} \mathbf{M}_i = -\frac{\beta}{\hbar}\sum_{i=1}^{N} (\mathbf{L}_i + 2\mathbf{S}_i).$$

In an external magnetic field $\mathbf{B} = \hat{\mathbf{z}}B_z$, the total Hamiltonian becomes

$$\mathcal{H} = \mathcal{H}^{(0)} + \mathcal{H}_{so} + \mathcal{H}_B,$$

where the term containing $\mathbf{B}$ is

$$\mathcal{H}_B = -\mathbf{M}\cdot\mathbf{B} = -M_z B_z.$$

(a) Use classical precession arguments to derive the approximate expression

$$M_z = -\left(\frac{e}{2m_e c}\right) J_z \left\{1 + \frac{J^2 + S^2 - L^2}{2J^2}\right\}$$

(assume both a weak spin-orbit interaction and a weak magnetic field interaction, but take the spin-orbit interaction to be dominant). [HINT: In the case of weak spin-orbit interaction and a weak magnetic field, we can use the precession picture of Prob. 9.4a; namely, the vectors $\mathbf{L}_i$ precess rapidly about $\mathbf{L}$ and the vectors $\mathbf{S}_i$ precess rapidly about $\mathbf{S}$. It might be helpful to draw vector diagrams to treat summations like $\sum_i (\mathbf{L}_i\cdot\mathbf{L})$ and $\sum_i (\mathbf{S}_i\cdot\mathbf{S})$.]

(b) Using the results of part (a), show that the first-order correction to the energy of a level $E(L, S, J)$ due to the perturbation $\mathcal{H}_B$ is

$$E_B^{(1)} = \left(\frac{e\hbar}{2m_e c}\right) B_z g_J M_J = g_J \beta B_z M_J,$$

where the Landé g factor is given by

$$g_J = 1 + \frac{J(J+1) + S(S+1) - L(L+1)}{2J(J+1)}.$$

Explain how we get by without "degenerate perturbation theory."

9.6 Russell-Saunders and *j-j* Coupling Schemes (*)

Consider a multielectron atom whose electron configuration is

$$1s^2 2s^2 2p^6 3s^2 3p^6 3d^{10} 4s^2 4p 4d$$

(a) To what element does this configuration correspond? Is it in the ground state or an excited state? Explain.

(b) Suppose that we apply the Russell-Saunders coupling scheme to this atom. Draw an energy level diagram roughly to scale for the atom, beginning with the single unperturbed configuration energy and taking into account the various interactions (e.g., spin-orbit coupling) one at a time in the correct order. Be sure to label each level at each stage of your diagram with the appropriate term designation, quantum numbers, and so on.

(c) Suppose that instead we apply pure *j-j* coupling to the atom. Starting again with the unperturbed $n = 4$ level, draw a second energy level diagram. [HINT: Assume that for a given level (j_1, j_2), the state with lowest J lies lowest in energy.]

10

Quantitative Approaches to Multielectron Atoms

When you cannot measure it, when you cannot express it in numbers, your knowledge of it is of a meagre and unsatisfactory kind.

Lord Kelvin 1824–1907

The previous two chapters developed a rather qualitative but reasonably accurate picture of the nature and properties of multielectron atoms. In it the atom was viewed as composed of shells (and subshells) that are filled with electrons according to the Aufbau principle, which incorporates the Pauli exclusion principle. [Excited states are generated from the ground state by promoting, or "exciting," one (or more) electrons to higher subshells.] Such a picture is consonant with the orbital approximation, in which each electron is described by a single-particle orbital. Properly antisymmetrized stationary-state wave functions of the atom are simply Slater determinants made up of these orbitals.

The atom is further described by its energy level diagram. For example, electrostatic and magnetic interactions in a light atom result in coupling of orbital and spin angular momenta and in splitting of energy levels. Hund's rules enable us to estimate the ordering of these levels but do not help in calculating the energies. Certainly our quantum mechanical theory would be less than satisfactory if it could not provide a tractable means of calculating these energies with some precision.

Indeed, much of the recent research in atomic structure (since, say, 1930) has been devoted to the development and utilization of many and diverse quantitative procedures for the calculation of energies and wave functions of multielectron atoms. In this chapter, we shall look at some calculations for heliumlike atoms ($N = 2$) and then briefly mention a widely-used approach to the problem for $N > 2$.

10.1 THE TWO-ELECTRON ATOM (PERTURBATION THEORY)

We choose to examine the case of $N = 2$ in some depth because the calculations are simple enough to be carried out without the aid of enormous computers. Moreover, we studied a one-dimensional model of precisely this atom (Sec. 8.3), and seeing how to generalize the results of that discussion will be instructive.

Let us consider, then, two-electron atoms and ions of nuclear charge Ze. Neutral He corresponds to $Z = 2$, Li^+ to $Z = 3$, Be^{++} to $Z = 4$, and so on. We shall treat this atom by perturbation theory, taking as the perturbation Hamiltonian the potential energy term due to the electron-electron coulomb repulsion; that is,

$$\mathcal{H}^{(1)} = V(r_{12}) = \frac{e^2}{r_{12}}. \tag{10.1}$$

The full Hamiltonian is, of course, just $\mathcal{H}^{(0)} + \mathcal{H}^{(1)}$, or

$$\mathcal{H} = \sum_{i=1}^{2} \left(-\frac{\hbar^2}{2m_e} \nabla_i^2 - \frac{Ze^2}{r_i} \right) + \frac{e^2}{r_{12}}. \tag{10.2}$$

Since $\mathcal{H}^{(1)}$ is independent of Z, it is smaller [compared to $\mathcal{H}^{(0)}$] for large Z than for small Z, and we would expect perturbation theory to give better results, say, for Be^{++} than for helium.[1]

From our work in Chapter 8 we know what to expect from the perturbation calculations: the energy of the ground state will be raised from the zeroth-order value, and the first excited level will split into singlet and triplet levels, the latter lying lower in energy than the former. Let us look first at the ground state.

[1]This perturbation does not appear to be "weak," however, for it becomes infinite as $r_{12} \longrightarrow 0$. Nevertheless, perturbation theory is applicable, since its validity is determined not by V alone but by the magnitude of the energy correction terms.

Ground State

The configuration of the ground state is $1s^2$; the two electrons are paired, one with $m_s = +\frac{1}{2}$, the other with $m_s = -\frac{1}{2}$, and they have identical spatial quantum numbers $n_1 = n_2 = 1$, $\ell_1 = \ell_2 = 0$, and $m_{\ell_1} = m_{\ell_2} = 0$. As we saw in Chapter 8, the ground state of a two-electron atom is a singlet with $S = 0$, $M_S = 0$. The zeroth-order eigenfunction is[2] [see Eqs. (8.58) and (8.59)]

$$\psi^{(0)}_{1s,1s;00} = \psi^{(0)}_{1s}(\mathbf{r}_1)\psi^{(0)}_{1s}(\mathbf{r}_2)\chi_{00}, \tag{10.3}$$

where χ_{00} is the antisymmetric spin function

$$\chi_{00} = \frac{1}{\sqrt{2}}[\alpha(1)\beta(2) - \beta(1)\alpha(2)] \tag{10.4}$$

and $\psi^{(0)}_{1s}(\mathbf{r}_1)$ and $\psi^{(0)}_{1s}(\mathbf{r}_2)$ are the familiar solutions of the single-particle Schroedinger equation with $n = 1$, $\ell = 0$, $m_\ell = 0$. Thus we have from Table 3.3

$$\psi^{(0)}_{1s}(\mathbf{r}_i) = \frac{1}{\sqrt{\pi}}\left(\frac{Z}{a_0}\right)^{3/2} e^{-Zr_i/a_0} \qquad (i = 1, 2), \tag{10.5}$$

where $a_0 \simeq 0.529$ Å is the first Bohr radius of the hydrogen atom.

The zeroth-order ground-state energy is just twice the energy of $\psi^{(0)}_{1s}(\mathbf{r}_i)$,

$$E^{(0)}_{1s,1s} = 2E^{(0)}_1 = -\frac{Z^2e^2}{a_0}. \tag{10.6}$$

Zeroth-order energies of several two-electron atomic systems are listed in column 4 of Table 10.1; experimentally measured ground-state energies appear in column 3. In each case, the zeroth-order energies lie "too low" (are too negative). This fact is not surprising, for the zeroth-order approximation completely ignores the electron-electron repulsion term e^2/r_{12}. The relative error due to neglect of this term is larger for small Z than for large Z.

The ground-state energy, to first order, is simply

$$E_{1s,1s} \simeq 2E^{(0)}_1 + \left\langle \psi^{(0)}_{1s,1s} \left| \frac{e^2}{r_{12}} \right| \psi^{(0)}_{1s,1s} \right\rangle, \tag{10.7}$$

or [Eq. (8.61)]

$$E_{1s,1s} \simeq 2E^{(0)}_1 + J_{1s,1s}, \tag{10.8}$$

[2] We shall omit the superscript A, it being understood that we are always dealing with antisymmetric wave functions in this chapter. When we generalize a result from an earlier chapter, you should refer back and convince yourself that the result does carry over as claimed.

Table 10.1

Comparison of experimental (E), zeroth-order (0), first-order (1), and variational (V) energies for heliumlike atomic systems. Also shown are $\Delta E(1, E)$, the difference between the first-order and experimental energies, and $\Delta E(V, E)$, the difference between the variational and experimental energies. (All numbers are in eV.)

Z	Element	(E)	(0)	(1)	$\Delta E(1, E)$	V	$\Delta E(V, E)$
2	He	−78.62	−108.8	−74.42	4.2	−77.09	1.5
3	Li^+	−197.14	−244.7	−192.80	4.3	−195.47	1.7
4	Be^{++}	−369.96	−435.0	−365.31	4.7	−367.98	2.0
5	B^{3+}	−596.4	−679.8	−591.94	4.5	−594.6	1.8
6	C^{4+}	−876.2	−987.8	−872.69	3.5	−875.4	1.8

where the coulomb (or direct) integral is, in general [see Eq. (8.37)],

$$J_{n\ell,n'\ell'} \equiv \iint |\psi^{(0)}_{n\ell m_\ell}(\mathbf{r}_1)|^2 \left(\frac{e^2}{r_{12}}\right) |\psi^{(0)}_{n'\ell' m_\ell}(\mathbf{r}_2)|^2 \, d\mathbf{r}_1 \, d\mathbf{r}_2. \tag{10.9}$$

Notice that the integration in (10.9) is taken over all coordinates of both electrons and that $J_{n\ell,n'\ell'}$ is independent of m_ℓ (see Prob. 10.2a).

Since we do not expect Eq. (10.8) to be a very good approximation to the exact ground-state energy, evaluation of the integral in Eq. (10.9) may seem pointless. However, it offers an opportunity to introduce some useful integration techniques and is examined in considerable detail in Prob. 10.1, which you should try unless these techniques are already familiar. The result obtained in that problem is

$$J_{1s,1s} = +\frac{5}{8}\frac{Ze^2}{a_0}. \tag{10.10}$$

Hence the first-order ground-state energy of a two-electron atom is

$$E_{1s,1s} \simeq -\frac{Z^2e^2}{a_0}\left(1 - \frac{5}{8}\frac{1}{Z}\right). \tag{10.11}$$

We see that the first-order effect of the electron-electron repulsion term is to increase the ground-state energy from the zeroth-order value $-Z^2e^2/a_0$.

In column 5 of Table 10.1 values of $E_{1s,1s}$ for several atoms are shown. In column 6 we have $\Delta E(1, E)$, which is the difference between the first-order approximation and the experimentally observed energy. Since the perturbation e^2/r_{12} is more important for small Z, we expect that, in general, the improvement will be most substantial for small Z. Table 10.1 reveals that this is true. In each case, He, Li^+, Be^{++}, etc., the first-order approximation is closer to the observed energy than is the zeroth-order approximation.

These encouraging results might tempt us to consider higher-order calculations; such calculations would lower the energy as required. However, we recall from Chapter 4 that high-order perturbation calculations are usually extremely complicated and unsuitable for highly accurate results. Thus a more flexible and potentially accurate approximation technique is more appropriate, and we shall return to the ground state in Sec. 10.3 and apply the variational method to it. First, however, we shall continue the generalization of the perturbation theory calculation of Sec. 8.3 by looking at excited states.

Excited States

Suppose that we excite one of the $1s$ electrons to a subshell $(n'\ell')$, $n' \neq 1$.[3] The zeroth-order wave functions for this state are [Eqs. (8.58) and (8.59)]

$$\psi^{(0)}_{1s,n'\ell'm_\ell';00} = \frac{1}{\sqrt{2}}[\psi^{(0)}_{1s}(\mathbf{r}_1)\psi^{(0)}_{n'\ell'm_\ell'}(\mathbf{r}_2) + \psi^{(0)}_{n'\ell'm_\ell'}(\mathbf{r}_1)\psi^{(0)}_{1s}(\mathbf{r}_2)]\chi_{00} \qquad (10.12a)$$

$$\psi^{(0)}_{1s,n'\ell'm_\ell';1M_S} = \frac{1}{\sqrt{2}}[\psi^{(0)}_{1s}(\mathbf{r}_1)\psi^{(0)}_{n'\ell'm_\ell'}(\mathbf{r}_2) - \psi^{(0)}_{n'\ell'm_\ell'}(\mathbf{r}_1)\psi^{(0)}_{1s}(\mathbf{r}_2)]\chi_{1M_S}, \qquad (10.12b)$$

where $\psi^{(0)}_{n'\ell'm_\ell'}(\mathbf{r})$ is the hydrogenic function for state $(n'\ell'm_\ell')$. The spin functions are

$$\chi_{00} = \frac{1}{\sqrt{2}}[\alpha(1)\beta(2) - \beta(1)\alpha(2)] \qquad (10.13a)$$

$$\chi_{11} = \alpha(1)\alpha(2), \qquad (10.13b)$$

$$\chi_{10} = \frac{1}{\sqrt{2}}[\alpha(1)\beta(2) + \beta(1)\alpha(2)], \qquad (10.13c)$$

$$\chi_{1,-1} = \beta(1)\beta(2), \qquad (10.13d)$$

corresponding to singlet states ($S = 0$, $M_S = 0$) and triplet states ($S = 1$, $M_S = +1, 0, -1$).

[3]The energies of doubly excited configurations like $(n\ell)(n'\ell')$ for n and $n' \neq 1$ are larger than the amount of energy required to completely remove one electron from the atom, the *ionization energy*. States arising from such doubly excited configurations are called *autoionizing states*, since they are unstable with respect to ionization. Thus if we were to excite an He atom to such a state (by an incident photon for example), it would subsequently ionize, ejecting one electron with a significant amount of kinetic energy. These autoionization states also show up as resonances in electron-scattering cross sections.

Typical configurations that the wave functions of Eqs. (10.12) might represent are shown below.

Configuration	Zeroth-order Energy
$1s2s$, $1s2p$	$E_1^{(0)} + E_2^{(0)}$
$1s3s$, $1s3p$, $1s3d$	$E_1^{(0)} + E_3^{(0)}$
$1s4s$, $1s4p$, ...	$E_1^{(0)} + E_4^{(0)}$
⋮	⋮

where $E_n^{(0)}$ is the single-particle hydrogenic energy. Notice that states with the same n' but different ℓ' are degenerate in the zeroth-order approximation, since we are treating each electron as if it were alone in a hydrogenic atom.

Now we take account of the electron-electron repulsion. To first order, the new energies are [Eq. (8.63)]

$$E_{1s,n'\ell'} \simeq E_1^{(0)} + E_{n'}^{(0)} + J_{1s,n'\ell'} \pm K_{1s,n'\ell'}, \tag{10.14}$$

where the $+$ sign corresponds to the singlet and the $-$ sign to the triplet states. The coulomb integral is given in Eq. (10.9). The *exchange integral* is [see Eq. (8.42)]

$$K_{n\ell,n'\ell'} \equiv \int\int \psi_{n\ell m_\ell}^{(0)}(\mathbf{r}_1)^* \psi_{n'\ell' m_\ell}^{(0)}(\mathbf{r}_1) \frac{e^2}{r_{12}} \psi_{n\ell m_\ell}^{(0)}(\mathbf{r}_2) \psi_{n'\ell' m_\ell}^{(0)}(\mathbf{r}_2)^* \, d\mathbf{r}_1 \, d\mathbf{r}_2. \tag{10.15}$$

These integrals can be evaluated;[4] the results for two typical excited configurations of helium are shown in Table 10.2, together with zeroth-order and observed results from Table 10.1. Notice that, in accordance with the arguments of Sec. 8.3, $J_{1s,n'\ell'} > 0$ and $K_{1s,n'\ell'} > 0$. As expected, the pairing of the spins in the singlet level gives it a higher energy than the triplet level.

The results of our perturbation theory calculations can be conveniently represented on an energy level diagram as in Fig. 10.1, where the effects of the coulomb and exchange integrals are explicitly shown. We have labeled

[4] This step is done by expanding $1/r_{12}$ via the expression given in Prob. 10.1 and evaluating the angular integrations by using the identity

$$\int_0^{2\pi}\int_0^{\pi} Y_{\ell m_\ell}^*(\theta,\varphi) Y_{\lambda\mu}(\theta,\varphi) Y_{\ell' m_{\ell}'}(\theta,\varphi) \sin\theta \, d\theta \, d\varphi$$
$$= \sqrt{\frac{(2\ell'+1)(2\lambda+1)}{4\pi(2\ell+1)}} \langle \ell'\lambda 00 | \ell'\lambda\ell 0\rangle \langle \ell'\lambda m_\ell' \mu | \ell'\lambda \ell m_\ell\rangle,$$

where the required Clebsch-Gordan coefficients may be found in Appendix 4.

Table 10.2

Comparison of zeroth-order (0), first-order (1), and experimental (E) energies for $1s2s$ and $1s2p$ excited states of helium. Also given are the difference between the first-order and experimental energies $\Delta E(1, E)$ and values for exchange (K) and coulomb (J) integrals. (All numbers are in eV.)

State	$1s2s$		$1s2p$	
	Singlet	Triplet	Singlet	Triplet
(0)	−68.0	−68.0	−68.0	−68.0
J	11.4	11.4	13.2	13.2
K	1.2	1.2	0.9	0.9
(1)	−55.4	−57.8	−53.9	−55.7
(E)	−58.4	−59.2	−57.8	−58.0
$\Delta E(1, E)$	3.0	1.4	3.9	2.3

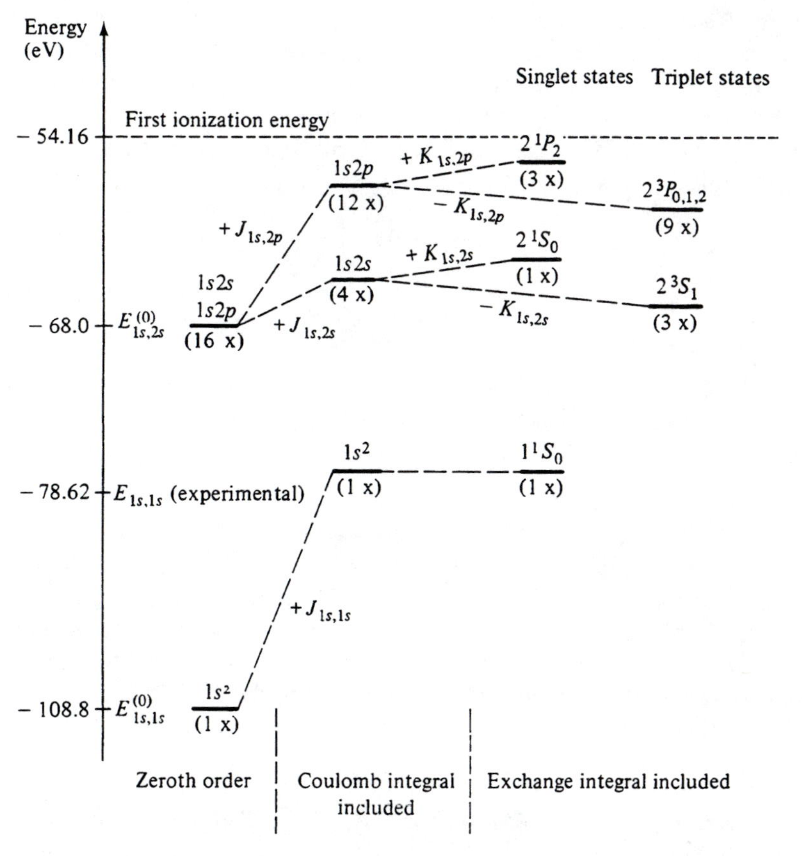

Figure 10.1 Energy level diagram for the ground and excited states of helium corresponding to $1s2s$ and $1s2p$ configurations (energies are in eV).

the ground-state term $1\ ^1S$ to distinguish it from the terms arising from the excited-state configurations ($2\ ^1S$, $2\ ^3S$, and so forth). Our quantitative calculations verify what we would have predicted using Hund's rules (Sec. 9.5). Moreover, the inclusion of the coulomb integral lifts the degeneracy of the configurations $1s2s$ and $1s2p$, reminding us yet again that this degeneracy holds only for a "purely" hydrogenic atom. Notice that the average energy of the terms arising from $1s2s$ is lower than that for $1s2p$, reflecting the influence of the $\ell = 1$ centrifugal potential discussed earlier. Although the errors in the first-order energies for the $2\ ^1S$, $2\ ^3S$, $2\ ^1P$, and $2\ ^3P$ levels appear to be rather large, the percentage error is not too bad—for example, 5% for $2\ ^1S$ and 2% for $2\ ^3S$.

We have placed the singlet and triplet states in separate columns because spectroscopic measurements indicate that transitions between singlet and triplet states occur with very small probability; in effect, they are forbidden.

In fact, not even all singlet $\longrightarrow$ singlet or triplet $\longrightarrow$ triplet transitions are allowed. Recall that in Chapter 5 the absorption and emission of electromagnetic radiation from a one-electron atom were discussed. For the excited states of helium with configurations $1sn'\ell'$, an atom in which only one electron is excited, we can assume to a good approximation that only this electron is involved in transitions. Thus we can generalize the electric dipole selection rules of Eqs. (5.43) to these atoms:

$$\begin{aligned} \Delta M_L &= 0, \pm 1 \\ \Delta L &= \pm 1 \\ \Delta S &= 0, \end{aligned} \tag{10.16}$$

where the spin selection rule follows from the fact that the electric dipole moment operator is independent of electron spin. [Very weak transitions (with low probability) that violate these rules are observed.]

Probability Densities.

Of course, perturbation theory could be used to obtain more accurate wave functions for the ground and excited states of helium. However, a good idea of the electronic structure of excited states of the atom can be obtained by examining the zeroth-order functions. In this regard, it is useful to calculate radial probability densities for, say, the states arising from a $1s2s$ configuration—namely, the singlet state, $2\ ^1S$, and the triplet state, $2\ ^3S$. Let us introduce the function $D(r_1, r_2)_\pm$, defined by

$$D(r_1, r_2)_\pm \equiv (4\pi r_1^2)(4\pi r_2^2)\tfrac{1}{2}[R_{10}(r_1)R_{20}(r_2) \pm R_{20}(r_1)R_{10}(r_2)]^2,$$

where the $+$ and $-$ signs refer to singlet and triplet $1s2s$ spatial functions, respectively. $D(r_1, r_2)_\pm\, dr_1\, dr_2$ is the probability of finding one electron in a shell of width dr_1 at a distance r_1 from the nucleus and the other electron in a shell of thickness dr_2 at a distance r_2. Figure 10.2 shows plots of curves of constant $D(r_1, r_2)_\pm$ for the singlet and triplet states. Notice that, as expected, the electrons seem to "repel one another" in the triplet state, whereas in the singlet state there is a finite probability that the electrons will be found in close proximity.

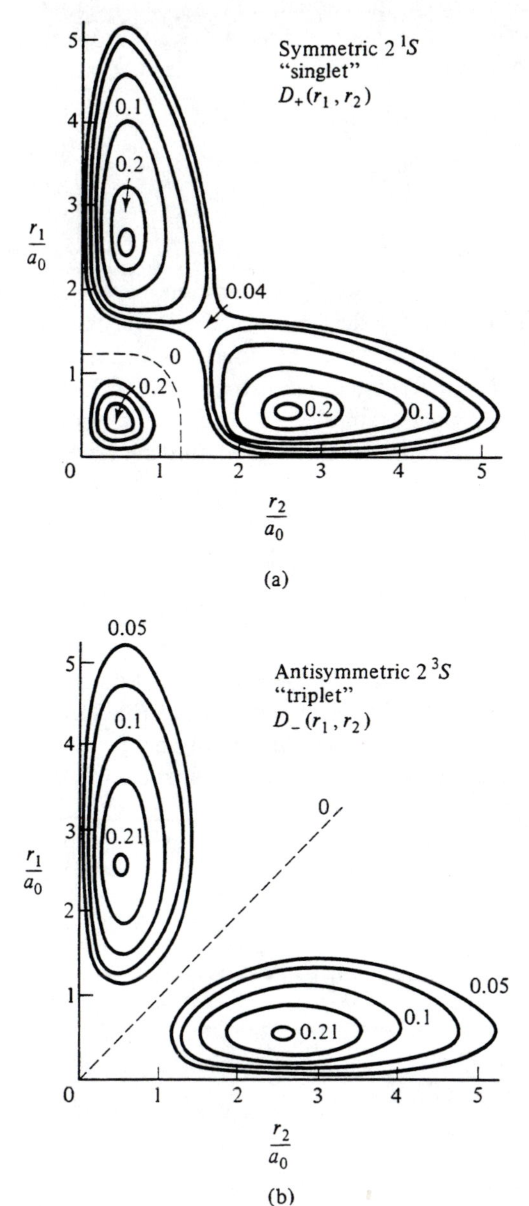

Figure 10.2 Curves of constant $D(r_1, r_2)_\pm$ for the singlet and triplet states corresponding to $1s2s$ configuration of helium. (Adapted from J. W. Linnett, *Wave Mechanics and Valency,* London: Chapman and Hall, Ltd., 1960.)

10.2 THE TWO-ELECTRON ATOM (THE VARIATIONAL METHOD)

It would be a mistake to believe that we can do no better than first-order perturbation calculations in studying heliumlike atoms. In fact, even simple variational calculations as in Sec. 4.3 give better results than those of Sec. 10.1. To illustrate, consider once more the ground state $1s^2$.

We must choose a trial wave function φ that we expect to be a reasonable approximation to the ground-state wave function for a two-electron atom. This function then leads to an approximate ground-state energy,

$$E'_{1s,1s} = \frac{\langle\varphi|\mathcal{H}|\varphi\rangle}{\langle\varphi|\varphi\rangle}, \tag{10.17}$$

which is an upper bound to the exact value.

Our work in Sec. 8.3 with the one-dimensional two-fermion system can guide us in choosing φ. We know that (a) the true eigenfunction can be written as the product of a spatial function times a spin function, and (b) the true ground state is a singlet ($S = M_S = 0$) with a symmetric spatial function. Our trial functions should have the proper symmetry. For example, we might use hydrogenic functions with a variable nuclear charge to construct a trial function of the form

$$\varphi = \varphi_{100}(\mathbf{r}_1)\varphi_{100}(\mathbf{r}_2), \tag{10.18}$$

where
$$\varphi_{100}(\mathbf{r}_i) = \frac{1}{\sqrt{\pi}}\left(\frac{Z'}{a_0}\right)^{3/2} e^{-Z'r_i/a_0}, \qquad i = 1, 2. \tag{10.19}$$

This is just the normalized product of two ground-state hydrogenic functions. We shall let Z' be the variational parameter, which is to be distinguished from Z, the fixed nuclear charge in the electron-nucleus potential energy terms $-Ze^2/r_i$. (If $Z' = Z$, then φ is just the zeroth-order wave function of Sec. 10.1.)

The evaluation of $E'_{1s,1s}$ is straightforward and very similar to the perturbation calculations of Sec. 10.1. We find that

$$E'_{1s,1s} = -\frac{e^2}{a_0}\left(2ZZ' - Z'^2 - \frac{5}{8}Z'\right). \tag{10.20}$$

The minimum principle of Chapter 4 ensures that minimization of this result with respect to the variational parameter Z' provides an upper bound to the true ground-state energy. By requiring that

$$\frac{dE'_{1s,1s}}{dZ'} = -\frac{e^2}{a_0}\left(2Z - 2Z' - \frac{5}{8}\right) = 0,$$

we obtain

$$Z' = Z - \frac{5}{16}, \tag{10.21}$$

where Z, the (actual) fixed nuclear charge is 2 for He, 3 for Li^+, and so on. Substituting this result into Eq. (10.20), we obtain

$$E'_{1s,1s} = 2E_1^{(0)}\left(1 - \frac{5}{8Z} + \frac{25}{256}\frac{1}{Z^2}\right). \tag{10.22}$$

As expected, we see that including the electron-electron repulsion term raises the ground-state energy from the zeroth-order value $2E_1^{(0)}$. Since the first two terms in the variational energy $E'_{1s,1s}$ are the same as the first-order perturbation result of Eq. (10.11), the third term may be considered as a "variational correction" to the perturbation result. (This term approximates the corrections we would obtain by higher-order perturbation calculations.) Since $E_1^{(0)}$ is negative, the effect of this variational correction is to "lower" our approximate energy, bringing it still nearer the correct value. This effect is explicitly illustrated in columns 7 and 8 of Table 10.1, where numerical values of the variational energy and the difference $\Delta E(V, E)$ between it and the experimental results are shown.

We can interpret these results by thinking again in terms of screening. One of the $1s$ electrons in the atom partially screens the other from the effect of the full nuclear charge Ze so that the screened electron behaves like an electron in the field of a nucleus of charge $Z'e$, where Z' is given by Eq. (10.21) and is clearly smaller than Z. Thus the effect of including the electron-electron repulsion term is to reduce the "effective nuclear charge" seen by each electron.

Had a more elaborate trial wave function been chosen, we would have obtained an even better (lower) energy. The variational method, especially in its linear version (Sec. 4.4), is attractive to people interested in elaborate atomic-structure calculations aimed at very accurate energies, for it is conveniently coded onto a computer.[5]

10.3 MULTIELECTRON ATOMS: THE HARTREE AND HARTREE-FOCK THEORIES

In order to treat atoms with many more than two electrons, an alternative to perturbation theory is needed, for the computations quickly become cumbersome. Moreover, first-order perturbation theory is not sufficiently

[5]For example, a linear variational calculation of the ground state of helium, using 1075 terms, gives an upper bound of −2.903724375 au or −78.95807 eV. [See C. L. Pekeris, *Phys. Rev.* **112**, 1649(1959).] This gives an idea of what is possible when sufficient computer time (i.e., money) is available to permit such large calculations.

accurate, even for helium. In Chapter 9 two ways to think about multielectron atoms were suggested: the isolated electron approximation and the central field approximation. In each case, the electrons are assigned to orbitals (single-particle functions) that satisfy single-particle equations—for example,

$$\left[-\frac{\hbar^2}{2m_e}\nabla_i^2 + V_i(\mathbf{r}_i)\right]\psi_{\epsilon_i}(\mathbf{r}_i) = E_{\epsilon_i}\psi_{\epsilon_i}(\mathbf{r}_i). \tag{10.23}$$

In the isolated electron approximation, the single-particle potential energy is simply the electron-nucleus interaction $-Ze^2/r_i$. We have seen, in the case of helium, that the effect of the electron-electron repulsion is too important to be neglected completely; for example, it has a considerable effect on the energies (see Fig. 10.1). This interaction term is included in an approximate fashion in the central field approximation, in which we take $V_i(\mathbf{r}_i)$ to be independent of θ_i and φ_i. For instance, we may use the form

$$V_i(\mathbf{r}_i) = -\frac{Ze^2}{r_i} + V_i^{\text{eff}}(r_i), \tag{10.24}$$

where $V_i^{\text{eff}}(r_i)$ is a spherically symmetric *screening potential*. This potential may be approximated by performing a spherical average of the potentials e^2/r_{ij} over the motion of all electrons except the ith (i.e., for $j = 1, 2, \ldots, N; j \neq i$).

It is probably not obvious precisely how this averaging should be performed. Intuitively, screening suggests that the average effect of electron j on electron i might be approximated by somehow averaging e^2/r_{ij} over the probability distribution of electron j:

$$\int \frac{e^2}{r_{ij}}\rho_j(\mathbf{r}_j)\, d\mathbf{r}_j.$$

This is the basic idea behind the procedure used to obtain $V_i^s(r_i)$. Notice that, in general, the screening potential will be different for each electron and that in the central field approximation the radial dependence of $\psi_{\epsilon_i}(\mathbf{r}_i)$ is not given by the simple hydrogenic radial functions of Chapter 3.

The Hartree Theory

One systematic version of the central field approximation is the *Hartree theory*. We ignore the spins of the electrons except insofar as they influence the assignment of electrons to shells and subshells via the Pauli exclusion principle. Although not strictly correct, this approach will provide a rough idea of the energies of the atom.

Thus the wave function for state γ is written

$$\psi_\gamma = \psi_{(n\ell m_\ell)_1}(\mathbf{r}_1)\psi_{(n\ell m_\ell)_2}(\mathbf{r}_2) \cdots \psi_{(n\ell m_\ell)_N}(\mathbf{r}_N), \tag{10.25}$$

where each orbital $\psi_{(n\ell m_\ell)_i}(\mathbf{r}_i)$ is an eigenfunction of the single-particle Hamiltonian

$$\mathcal{H}_i = -\frac{\hbar^2}{2m_e}\nabla_i^2 - \frac{Ze^2}{r_i} + V_i^s(r_i).$$

The wave function of Eq. (10.25) is an eigenfunction of the Hamiltonian

$$\mathcal{H} = \sum_{i=1}^{N}\left[-\frac{\hbar^2}{2m_e}\nabla_i^2 - \frac{Ze^2}{r_i} + V_i^s(r_i)\right]. \tag{10.26}$$

Suppose that we wish to solve for the orbital describing the motion of the ith electron. Obviously, we must first obtain V_i^s. This screening potential will take into account the influence on the ith electron of all the other electrons $j = 1, 2, \ldots, N$; $j \neq i$. In the Hartree method, we return to classical electromagnetic theory and postulate that the effective potential $V_i^s(r_i)$ has the same form as the potential energy of interaction of a point charge $-e$ at $\mathbf{r}_i$ with $N - 1$ other charges, spatially distributed, each with density $-e\rho_j(\mathbf{r}_j)$ and $j = 1, \ldots, N$ $(j \neq i)$:

$$V_i^s(r_i) = \sum_{\substack{j=1\\(j\neq i)}}^{N} \int \psi^*_{(n\ell m_\ell)_j}(\mathbf{r}_j)\frac{e^2}{r_{ij}}\psi_{(n\ell m_\ell)_j}(\mathbf{r}_j)\,d\mathbf{r}_j, \tag{10.27}$$

where $\mathbf{r}_j$ is a dummy variable of integration.[6] Therefore, in order to obtain the orbitals and hence the energies, we solve the *Hartree equations*:

$$\left[-\frac{\hbar^2}{2m_e}\nabla_i^2 - \frac{Ze^2}{r_i} + \sum_{j\neq i}\left\langle\psi^H_{(n\ell m_\ell)_j}\left|\frac{e^2}{r_{ij}}\right|\psi^H_{(n\ell m_\ell)_j}\right\rangle\right]\psi^H_{(n\ell m_\ell)_i}(\mathbf{r}_i) = E^H_{(n\ell m_\ell)_i}\psi^H_{(n\ell m_\ell)_i}(\mathbf{r}_i), \qquad i = 1, 2, \ldots, N. \tag{10.28}$$

The Hartree equations

Once these equations have been solved, the resulting orbitals can be substituted into the product for ψ_γ to give approximate wave functions for the atom. The expectation value of $\mathcal{H}$ with respect to this wave function is the approximate energy of the atom.[7]

[6] In cases where Eq. (10.27) is not spherically symmetric, it is usually averaged over all angles.

[7] Notice that this quantity, $\langle\psi_\gamma^H|\mathcal{H}|\psi_\gamma^H\rangle$, is not the same as the sum of the individual orbital energies $E^H_{(n\ell m_\ell)_i}$.

It is not possible to discuss the details of the solution of the Hartree equations here. The method used is called *The Method of Self-Consistent Fields* and is essentially an iterative procedure in which we initially guess a set of "reasonable" effective potentials, and then use them to obtain a set of orbitals. These orbitals are next used to determine a "better" set of effective potentials and so on; the iteration stops at step M when the $(M + 1)$st set of potentials does not differ significantly from the Mth set. (This method is discussed in more detail in the references at the end of the chapter.) Let us look now at the results of a calculation based on the Hartree method.

An Example: Cu^+

The ground-state configuration of Cu is $1s^2 2s^2 2p^6 3s^2 3p^6 3d^{10} 4s$. Thus Cu^+ corresponds to a closed shell and can be studied by using the Hartree method (which, recall, ignores spin). Hartree himself carried out these calculations originally.[8] The Hartree energies $E^H_{(n\ell m_\ell)_i}$ are presented in Table 10.3, together

Table 10.3

Energies $E^H_{(n\ell m_\ell)_i}$ obtained from Hartree calculations compared with experimental energies for Cu^+ (all energies in Rydbergs).

Orbital	$E^H_{(n\ell)_i}$	$E_{(n\ell)_i}$ (exp)
$1s$	−658.0	−661.6
$2s$	−78.45	−81.0
$2p$	−69.86	−68.9
$3s$	−8.968	−8.9
$3p$	−6.078	−5.7
$3d$	−1.195	−0.4

with experimental energies determined by X-ray measurements. We see that except for the $3d$ orbital, E^H_i and E_i (exp) agree to within 7%. Notice that $E^H_{(n\ell m_\ell)_i}$ is independent of m_{ℓ_i}.

Such calculations can be used to verify the qualitative predictions of atomic structure made in the last chapter. For example, in Fig. 10.3 we show a plot of the radial probability density of electrons as a function of distance from the nucleus for Cu^+. The figure, based on the Hartree solution of this ion, clearly reveals the K, L, and M shells introduced in Sec. 9.3.

[8]See D. R. Hartree, *Proc. Roy. Soc.* (*London*) **A141,** 282(1933).

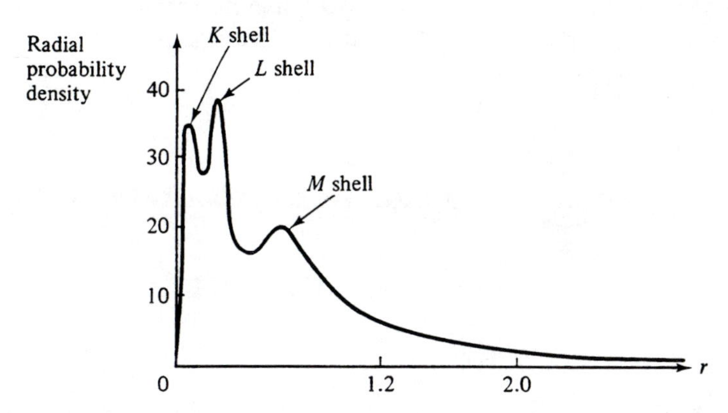

Figure 10.3 Plot of the radial distribution $4\pi r^2 \rho(r)$ for Cu^+ showing the shell structure of this atomic species.

The Hartree-Fock Theory

However useful the Hartree theory may appear, we know that it is not strictly correct and that spin can indeed affect both the orbital energies and the total energies of states of the atom. Thus we should take spin into account in any theory that is expected to give highly accurate results.

The first (and most obvious) step in the inclusion of spin into the theory is to replace the product function of Eq. (10.25) with a Slater determinant:

$$\psi_\gamma = |\psi_{\epsilon_1}(\mathbf{r}_1) \quad \psi_{\epsilon_2}(\mathbf{r}_2) \quad \psi_{\epsilon_3}(\mathbf{r}_3) \quad \cdots \quad \psi_{\epsilon_N}(\mathbf{r}_N)|, \tag{10.29}$$

where each orbital $\psi_{\epsilon_i}(\mathbf{r}_i)$ is a product of a spatial part and a spin part. We obtain these functions by solution of single-particle equations like Eq. (10.23).

Unfortunately, the situation is now so complicated that we cannot guess the form of the screening potential. We would still expect $V_i^s(\mathbf{r}_i)$ to take the form of an average of e^2/r_{ij} over some sort of charge distribution ρ_j, but the exchange effects might be expected to make this distribution rather complicated.

However, we can use our experience with the helium atom to anticipate the influence of exchange effects on the energies of the atom. In particular, we expect paired electrons to repel one another more strongly than unpaired electrons, since the spatial symmetry imposed by the Pauli principle prevents the latter from occupying the same point in space. Since this effect is left out of the Hartree theory, we should expect it to yield energies which lie too high. The Hartree-Fock theory properly accounts for spin-pairing and results in lower energies, in better agreement with measurements.

A detailed analysis of the problem, taking spin into account as suggested above, leads to the rather formidable Hartree-Fock equations

$$\begin{aligned}
&\left[-\frac{\hbar^2}{2m_e}\nabla_i^2 - \frac{Ze^2}{r_i}\right]\psi^{\text{HF}}_{(n\ell m_\ell)_i}(\mathbf{r}_i) \\
&\quad + \sum_{\substack{j=1 \\ (j\neq i)}}^{N} \left\{\psi^{\text{HF}}_{(n\ell m_\ell)_i}(\mathbf{r}_i)\left\langle \psi^{\text{HF}}_{(n\ell m_\ell)_j}(\mathbf{r}_j)\left|\frac{e^2}{r_{ij}}\right|\psi^{\text{HF}}_{(n\ell m_\ell)_j}(\mathbf{r}_j)\right\rangle\right. \\
&\quad \left. - \psi^{\text{HF}}_{(n\ell m_\ell)_j}(\mathbf{r}_i)\,\delta_{m_{s_i} m_{s_j}}\left\langle \psi^{\text{HF}}_{(n\ell m_\ell)_j}(\mathbf{r}_j)\left|\frac{e^2}{r_{ij}}\right|\psi^{\text{HF}}_{(n\ell m_\ell)_i}(\mathbf{r}_j)\right\rangle\right\} \\
&= E^{\text{HF}}_{(n\ell m_\ell)_i}\psi^{\text{HF}}_{(n\ell m_\ell)_i}(\mathbf{r}_i). \qquad (10.30)
\end{aligned}$$

These equations are presented so that you can see what they look like; the details of their derivation and solution are well beyond the scope of this book.

Hartree-Fock calculations for Cu^+ yield[9] the orbital energies shown in Table 10.4 and the radial functions of Fig. 10.4. Comparison with Fig. 3.2 reveals that these radial functions are similar to the corresponding hydrogenic functions (for the appropriate value of Z) in their qualitative behavior but that they differ in their detailed structure.

Table 10.4
One-electron energies $E^{\text{HF}}_{(n\ell m_\ell)_i}$ for Cu^+ obtained by Hartree-Fock calculations. (Energies are given in Rydbergs.)

Orbital	$E^{\text{HF}}_{(n\ell)_i}$
$1s$	−658.4
$2s$	−82.30
$2p$	−71.83
$3s$	−10.651
$3p$	− 7.279
$3d$	− 1.613

We have done little more than suggest that rather accurate and detailed atomic-structure calculations can be performed on multielectron atoms. However, armed with the picture of the atom developed in Chapter 9, you

[9]D. R. Hartree and W. Hartree, *Proc. Roy. Soc.* (*London*) **A157**, 490(1936).

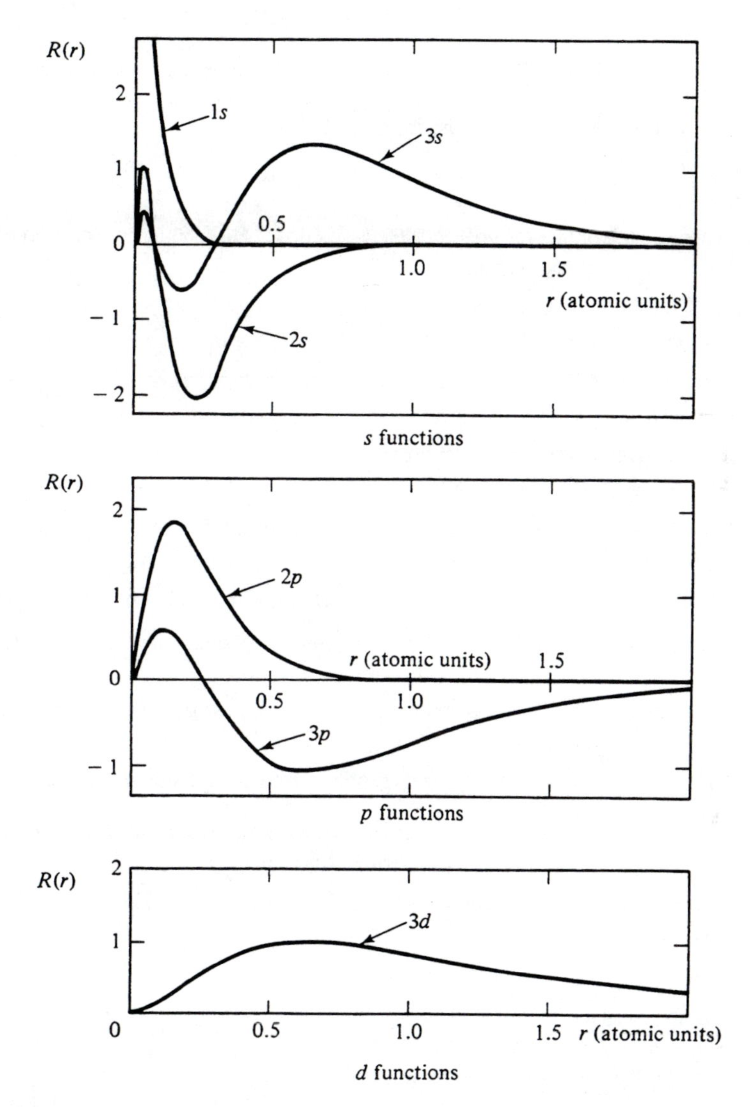

Figure 10.4 Radial wave functions $P_{n\ell}(r)$ for Cu^+ obtained by Hartree-Fock calculations performed by Hartree and Hartree. (From John C. Slater, *Quantum Theory of Atomic Structure,* Vol. 1. New York: McGraw-Hill, 1960.)

should be able to pursue without too much difficulty any one of a number books that discuss the physics of multielectron atoms at a more advanced level.

SUGGESTED READINGS

Treatments of atomic-structure calculations can be found in

KAUZMAN, WALTER, *Quantum Chemistry*. New York: Academic, 1957. See Chap. 10.

SLATER, JOHN C., *Quantum Theory of Matter*. 2nd ed. New York: McGraw-Hill, 1968. See Chaps. 12, 16, 17, and 18.

An excellent discussion of the mathematical aspects of orbital theory, which should be accessible to the reader of our text, is presented in

OFFENHARTZ, PETER O'D., *Atomic and Molecular Orbital Theory*. New York: McGraw-Hill, 1970.

Chapter 6 of this reference deals with atomic orbitals. The electronic structure of complex atoms is treated in Chap. 8 of

MIZUSHIMA, MASATAKA, *Quantum Mechanics of Atomic Spectra and Atomic Structure*. New York: W. A. Benjamin, 1970.

Two books of interest, which are more on the level of the researcher in this field, are

HARTREE, DOUGLAS R., *The Calculation of Atomic Structure*, New York: Wiley, 1957.

HERMAN, F., and S. SKILLMAN, *Atomic Structure Calculation*, Englewood Cliffs, N.J.: Prentice-Hall, 1963.

The latter volume contains tables of atomic potentials as well as detailed discussions of their calculation.

Finally, we should mention that a very thorough bibliography of atomic structure studies through 1958 is contained in Appendix 16 of

SLATER, JOHN C., *Quantum Theory of Atomic Structure*, Vol. I, New York: McGraw-Hill, 1960.

PROBLEMS

10.1 Evaluation of a Direct Integral (**)

In this problem we shall verify Eq. (10.10),

$$J_{1s,1s} = +\frac{5}{8}\frac{Ze^2}{a_0}.$$

(a) Using the law of cosines,

$$r_{12}^2 = r_1^2 + r_2^2 - 2r_1r_2\cos\Theta,$$

where Θ is the angle between $\mathbf{r}_1$ and $\mathbf{r}_2$, and the properties of Legendre polynomials, show that $1/r_{12}$ can be expanded as

$$\frac{1}{r_{12}} = \sum_{\lambda=0}^{\infty} \left(\frac{r_<^\lambda}{r_>^{\lambda+1}}\right) P_\lambda(\cos\Theta),$$

where $r_<$ is the lesser of r_1 and r_2 and $r_>$ is the greater of r_1 and r_2. Thus a different expansion is presented for each region (see Figure 10.5).

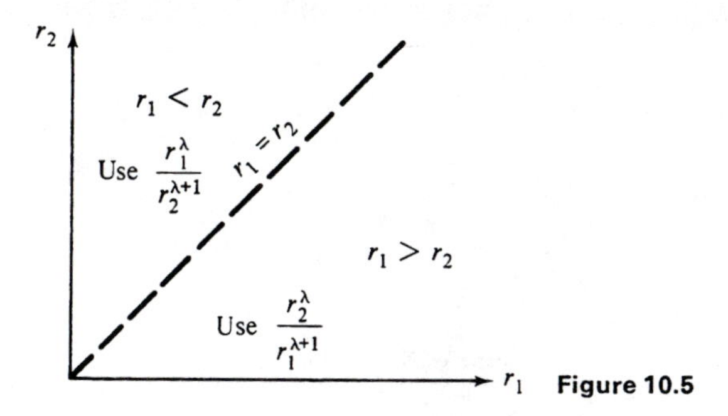

Figure 10.5

(b) Use the Addition Theorem of spherical harmonics,[10]

$$P_\lambda(\cos\Theta) = \frac{4\pi}{2\lambda+1} \sum_{\mu=-\lambda}^{+\lambda} Y_{\lambda\mu}^*(\theta_1, \varphi_1) Y_{\lambda\mu}(\theta_2, \varphi_2)$$

in this expansion to write $J_{1s,1s}$ as the product of a radial integral times an angular integral. Using properties of the spherical harmonics, show that your result reduces to

$$J_{1s,1s} = e^2 \int_0^\infty \int_0^\infty [R_{10}(r_1)]^2 \frac{1}{r_>} [R_{10}(r_2)]^2 r_1^2 r_2^2 \, dr_1 \, dr_2.$$

(c) Some skill is required to evaluate this radial integral, since $r_>$ depends on whether $r_1 > r_2$ or $r_1 < r_2$. To see how to handle the problem, let us rewrite $J_{1s,1s}$ as

$$J_{1s,1s} = \int_0^\infty \left[\int_0^{r_1} f(r_1 \geq r_2)\, dr_2 + \int_{r_1}^\infty f(r_1 \leq r_2)\, dr_2\right] dr_1,$$

where $f(r_1 \geq r_2)$ is written for the integrand as defined in the region $(r_1 \geq r_2)$. In Figure 10.6 we illustrate this choice of limits. The shaded region is an r_2-integration area for a fixed value of r_1. The region below the line $r_2 = r_1$ corresponds to the first of the inner integrals, and the region above this line corresponds to the second. The final integration, over r_1 from 0 to ∞, is pictorially represented by sweeping the vertical ribbon toward the right from 0 to ∞.

[10]See M. E. Rose, *Elementary Theory of Angular Momentum* (New York: Wiley, 1957), pp. 59–60 for a derivation of this result.

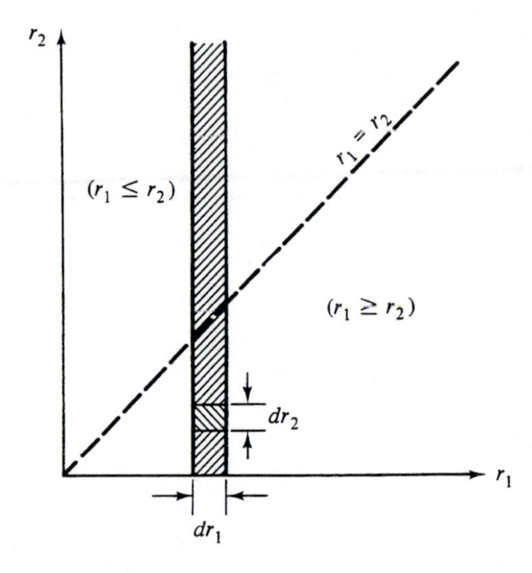

Figure 10.6

Thus we have

$$J_{1s,1s} = e^2 \int_0^\infty \left\{ \frac{1}{r_1} \int_0^{r_1} [R_{10}(r_2)]^2 r_2^2 \, dr_2 + \int_{r_1}^\infty [R_{10}(r_2)]^2 \frac{1}{r_2} r_2^2 \, dr_2 \right\} [R_{10}(r_1)]^2 r_1^2 \, dr_1.$$

Evaluate this expression, obtaining $J_{1s,1s} = (5/8)(Ze^2/a_0)$.

10.2 First Excited Configurations of Helium: Coulomb versus Exchange Energy (**)

Consider the terms ^1S, ^3S, arising from the $1s2s$ configuration and ^1P, ^3P arising from the $1s2p$, configuration of helium.

(a) Derive expressions for the coulomb integrals $J_{1s,2s}$ and $J_{1s,2p}$ and exchange integrals $K_{1s,2s}$ and $K_{1s,2p}$ all in terms of one-electron radial integrals. Carry out all angular integrations. You need not evaluate the radial integrals. Show that $J_{1s,2p}$ and $K_{1s,2p}$ are independent of m_ℓ.

(b) The numerical values for these integrals are found in Table 10.2. Think carefully about the nature of the integrals and give the best physical arguments (charge distributions, nodes, etc.) for each inequality below:

(i) $J_{1s,2s} \gg K_{1s,2s}$ and $J_{1s,2p} \gg K_{1s,2p}$.

(ii) $J_{1s,2s} < J_{1s,2p}$ (but same order of magnitude).

(iii) $K_{1s,2s} > K_{1s,2p}$ (but same order of magnitude).

[HINT: You may find it more convenient to discuss some of these integrals in their original form, before reduction to one-electron radial integrals.]

10.3 Second Excited Configurations of Helium: Short-Range Nature of Exchange (**)

Consider the terms ^1S, ^3S, ^1P, ^3P, ^1D, and ^3D arising from the three configurations $1s3s$, $1s3p$, and $1s3d$.

(a) Write the energies of these terms, correct to first order, in terms of the appropriate coulomb and exchange integrals $J_{1s,3s}$, $K_{1s,3s}$, $J_{1s,3p}$, and so on.

(b) Use the table of observed energy levels (given below) to obtain empirical, numerical values for all the coulomb integrals (relative to $J_{1s,3s}$) and exchange integrals. Give the relative ordering and provide a physical explanation (as in Prob. 10.2) of this ordering for

(i) $J_{1s,3s}$, $J_{1s,3p}$, $J_{1s,3d}$,
(ii) $K_{1s,3s}$, $K_{1s,3p}$, $K_{1s,3d}$,
(iii) $J_{1s,3s}$ versus $K_{1s,3s}$, $J_{1s,3p}$ versus $K_{1s,3p}$,
$J_{1s,3d}$ versus $K_{1s,3d}$ (order of magnitude only).

Table of Energies[a]

	S	P	D
$S = 1$	0	2328	2865
$S = 0$	1628	2973	2868

[a]All energies in cm^{-1} relative to the lowest level ^3S, taken as zero.

(c) Exchange is often referred to as a "short-range" effect. Comment, using the results of this problem (and Prob. 10.2 if you wish).

(d) Using hydrogenic $1s$ and $3d$ wave functions ($Z = 2$), calculate a numerical value for $K_{1s,3d}$. Discuss your result and compare with the empirical result. Explain possible causes of the discrepancy, if any.

10.4 Review Question on Helium (*)

(a) Consider a helium atom with principal quantum numbers $n = 1$ and $n' = 3$ for the two electrons. What are the possible electronic states, characterized by quantum numbers L, S, and J, for this atom? For each of these states, write the atomic term and the electronic configuration.

(b) Discuss semiquantitatively, in order of decreasing importance, the interactions and perturbations that cause the energies of the 10 electronic states found in part (a) to be different from each other. For a given state, how might the M_J degeneracy be removed?

(c) Sketch an energy level diagram (see Fig. 10.1), indicating for each electronic state of part (a) how the energy degeneracies are removed by the various interactions and perturbations of part (b). Label all states with the appropriate quantum numbers. Indicate what happens to each state if a magnetic field is applied.

APPENDICES

APPENDIX 1 FUNDAMENTAL CONSTANTS AND CONVERSION FACTORS[1]

Fundamental Constants

Planck's constant	$h = 6.6262 \times 10^{-27}$ erg-sec
	$= 4.1357 \times 10^{-15}$ eV-sec
Rationalized Planck's constant	$\hbar = 1.0546 \times 10^{-27}$ erg-sec
Boltzmann constant	$k_B = 1.3807 \times 10^{-16}$ erg/°K
Rest mass of the electron	$m_e = 9.1095 \times 10^{-31}$ kg
Bohr radius	$a_0 = 0.5292 \times 10^{-8}$ cm
Speed of light (in vacuum)	$c = 2.9979 \times 10^{10}$ cm/sec
Elementary charge	$e = 4.8032 \times 10^{-10}$ esu
Rest mass of the proton	$m_p = 1.6726 \times 10^{-27}$ kg

Conversion Factors

	bohr	angstrom
1 bohr	1	0.529177
1 angstrom	1.88973	1

	eV	Hartree	cm^{-1}
1 eV	1	3.6748×10^{-2}	8065.48
1 Hartree	27.212	1	219477.84
1 cm^{-1}	1.23985×10^{-4}	4.55627×10^{-6}	1

1 angstrom $= 10^{-8}$ cm
1 Rydberg $= \frac{1}{2}$ Hartree $= 13.606$ eV
1 eV $= 1.6022 \times 10^{-12}$ erg

[1]"Review of Particle Properties," *Rev. Mod. Phys.* **45** (Supplement), 529(1973). Also see B.N., Taylor, W. H. Parker, and D. N. Langenberg, *Rev. Mod. Phys.* **41,** 375(1969), and "Symbols, Units, and Nomenclature in Physics, *Physica* **93A**, 1 (1978).

APPENDIX 2 ATOMIC UNITS

In many advanced quantum theory textbooks and in some published literature, a special system of units, called *atomic units*, is used. In this system we set

$$\hbar = m_e = e = a_0 = 1. \tag{1}$$

In practical terms, this means that we define the unit of length to be the Bohr radius, the unit of charge to be the charge of an electron, and the unit of mass to be the mass of an electron. In atomic units, Schroedinger's equation for the hydrogen atom becomes

$$\left(-\frac{1}{2}\nabla^2 - \frac{1}{r}\right)\psi(\mathbf{r}) = E\psi(\mathbf{r}). \tag{2}$$

With these definitions, the unit of energy turns out to be twice the ground-state energy of the hydrogen atom. This unit of energy is called a *Hartree*:

$$1 \text{ Hartree} = 27.212 \text{ eV}. \tag{3}$$

Table 1 presents several important physical quantities in atomic units.

Also frequently encountered is a special energy unit, the Rydberg. One Rydberg is equal to $\frac{1}{2}$ Hartree, so that

$$1 \text{ Rydberg} = 13.606 \text{ eV}. \tag{4}$$

In Rydbergs, we write the Schroedinger equation for a central-force problem

$$[\nabla^2 + V(r)]\psi(\mathbf{r}) = E\psi(\mathbf{r}),$$

where the energy and the potential energy are in Rydberg units.

Table 1[a]

Constant	Symbol	Value in Atomic Units[b]
Electronic charge	e	1
Electronic mass	m_e	1
Bohr radius	a_0	1
Planck's constant	$\hbar$	1
Proton mass	m_p	1836.11
Boltzmann constant	k_B	3.1668×10^{-8}
Speed of light	c	137.039

[a] *Source: Physics Today* **17** (2), 48(1964).

[b] The unit of time in atomic units is $t_0 = 2.4189 \times 10^{-17}$ sec., the time required for an electron in the first Bohr orbit of a hydrogen atom to describe one radian. The unit of velocity is $v_0 = 2.1877 \times 10^8$ cm/sec.

APPENDIX 3 EXPECTATION VALUES OF SEVERAL IMPORTANT OPERATORS

The following apply to observable properties of stationary states of the one-electron atom discussed in Chapter 3. We adopt the notation

$$\langle G \rangle_{n\ell m} = \langle \psi_{n\ell m} | G | \psi_{n\ell m} \rangle$$

for operator G.

1. Powers of r

$$\langle r \rangle_{n\ell m} = \frac{a_0}{2Z}[3n^2 - \ell(\ell + 1)]$$

$$\langle r^2 \rangle_{n\ell m} = \frac{n^2 a_0^2}{2Z^2}[5n^2 + 1 - 3\ell(\ell + 1)]$$

$$\langle r^{-1} \rangle_{n\ell m} = \frac{Z}{a_0 n^2}$$

$$\langle r^{-2} \rangle_{n\ell m} = \frac{Z^2}{a_0^2 n^3 (\ell + \frac{1}{2})}$$

$$\langle r^{-3} \rangle_{n\ell m} = \frac{Z^3}{a_0^3 n^3 \ell(\ell + \frac{1}{2})(\ell + 1)}$$

2. Potential Energy

$$\langle V \rangle_{n\ell m} = -(Ze^2)\langle r^{-1} \rangle_{n\ell m} = -\frac{\mu Z^2 e^4}{\hbar^2 n^2}$$

3. Square of Linear Momentum

$$\langle p^2 \rangle_{n\ell m} = \left(\frac{\mu Z e^2}{n\hbar}\right)^2$$

This suggests a definition for the RMS (root mean square) speed of the electron:

$$\sqrt{\langle v^2 \rangle_{n\ell m}} = \frac{Ze^2}{n\hbar}$$

4. Kinetic Energy

$$\langle T \rangle_{n\ell m} = \frac{1}{2\mu}\langle p^2 \rangle_{n\ell m} = \frac{\mu Z^2 e^4}{2\hbar^2 n^2}$$

APPENDIX 4 CLEBSCH-GORDAN COEFFICIENTS[a]

$(j_1\,\tfrac{1}{2}\,m_1 m_2 \,|\, j_1\,\tfrac{1}{2}\,j\,m)$

$j=$	$m_2=\frac{1}{2}$	$m_2=-\frac{1}{2}$
$j_1+\frac{1}{2}$	$\sqrt{\dfrac{j_1+m+\frac{1}{2}}{2j_1+1}}$	$\sqrt{\dfrac{j_1-m+\frac{1}{2}}{2j_1+1}}$
$j_1-\frac{1}{2}$	$-\sqrt{\dfrac{j_1-m+\frac{1}{2}}{2j_1+1}}$	$\sqrt{\dfrac{j_1+m+\frac{1}{2}}{2j_1+1}}$

$(j_1\,1\,m_1\,m_2 \,|\, j_1\,1\,j\,m)$

$j=$	$m_2=1$	$m_2=0$	$m_2=-1$
j_1+1	$\sqrt{\dfrac{(j_1+m)(j_1+m+1)}{(2j_1+1)(2j_1+2)}}$	$\sqrt{\dfrac{(j_1-m+1)(j_1+m+1)}{(2j_1+1)(j_1+1)}}$	$\sqrt{\dfrac{(j_1-m)(j_1-m+1)}{(2j_1+1)(2j_1+2)}}$
j_1	$-\sqrt{\dfrac{(j_1+m)(j_1-m+1)}{2j_1(j_1+1)}}$	$\dfrac{m}{\sqrt{j_1(j_1+1)}}$	$\sqrt{\dfrac{(j_1-m)(j_1+m+1)}{2j_1(j_1+1)}}$
j_1-1	$\sqrt{\dfrac{(j_1-m)(j_1-m+1)}{2j_1(2j_1+1)}}$	$-\sqrt{\dfrac{(j_1-m)(j_1+m)}{j_1(2j_1+1)}}$	$\sqrt{\dfrac{(j_1+m+1)(j_1+m)}{2j_1(2j_1+1)}}$

$(j_1\,\tfrac{3}{2}\,m_1\,m_2 \,|\, j_1\,\tfrac{3}{2}\,j\,m)$

$j=$	$m_2=\frac{3}{2}$	$m_2=\frac{1}{2}$
$j_1+\frac{3}{2}$	$\sqrt{\dfrac{(j_1+m-\frac{1}{2})(j_1+m+\frac{1}{2})(j_1+m+\frac{3}{2})}{(2j_1+1)(2j_1+2)(2j_1+3)}}$	$\sqrt{\dfrac{3(j_1+m+\frac{1}{2})(j_1+m+\frac{3}{2})(j_1-m+\frac{3}{2})}{(2j_1+1)(2j_1+2)(2j_1+3)}}$
$j_1+\frac{1}{2}$	$-\sqrt{\dfrac{3(j_1+m-\frac{1}{2})(j_1+m+\frac{1}{2})(j_1-m+\frac{3}{2})}{2j_1(2j_1+1)(2j_1+3)}}$	$-(j_1-3m+\frac{3}{2})\sqrt{\dfrac{j_1+m+\frac{1}{2}}{2j_1(2j_1+1)(2j_1+3)}}$
$j_1-\frac{1}{2}$	$\sqrt{\dfrac{3(j_1+m-\frac{1}{2})(j_1-m+\frac{1}{2})(j_1-m+\frac{3}{2})}{(2j_1-1)(2j_1+1)(2j_1+2)}}$	$-(j_1+3m-\frac{1}{2})\sqrt{\dfrac{j_1-m+\frac{1}{2}}{(2j_1-1)(2j_1+1)(2j_1+2)}}$
$j_1-\frac{3}{2}$	$-\sqrt{\dfrac{(j_1-m-\frac{1}{2})(j_1-m+\frac{1}{2})(j_1-m+\frac{3}{2})}{2j_1(2j_1-1)(2j_1+1)}}$	$\sqrt{\dfrac{3(j_1+m-\frac{1}{2})(j_1-m-\frac{1}{2})(j_1-m+\frac{1}{2})}{2j_1(2j_1-1)(2j_1+1)}}$

$j=$	$m_2=-\frac{1}{2}$	$m_2=-\frac{3}{2}$
$j_1+\frac{3}{2}$	$\sqrt{\dfrac{3(j_1+m+\frac{3}{2})(j_1-m+\frac{1}{2})(j_1-m+\frac{3}{2})}{(2j_1+1)(2j_1+2)(2j_1+3)}}$	$\sqrt{\dfrac{(j_1-m-\frac{1}{2})(j_1-m+\frac{1}{2})(j_1-m+\frac{3}{2})}{(2j_1+1)(2j_1+2)(2j_1+3)}}$
$j_1+\frac{1}{2}$	$(j_1+3m+\frac{3}{2})\sqrt{\dfrac{j_1-m+\frac{1}{2}}{2j_1(2j_1+1)(2j_1+3)}}$	$\sqrt{\dfrac{3(j_1+m+\frac{3}{2})(j_1-m-\frac{1}{2})(j_1-m+\frac{1}{2})}{2j_1(2j_1+1)(2j_1+3)}}$
$j_1-\frac{1}{2}$	$-(j_1-3m-\frac{1}{2})\sqrt{\dfrac{j_1+m+\frac{1}{2}}{(2j_1-1)(2j_1+1)(2j_1+2)}}$	$\sqrt{\dfrac{3(j_1+m+\frac{1}{2})(j_1+m+\frac{3}{2})(j_1-m-\frac{1}{2})}{(2j_1-1)(2j_1+1)(2j_1+2)}}$
$j_1-\frac{3}{2}$	$-\sqrt{\dfrac{3(j_1+m-\frac{1}{2})(j_1+m+\frac{1}{2})(j_1-m-\frac{1}{2})}{2j_1(2j_1-1)(2j_1+1)}}$	$\sqrt{\dfrac{(j_1+m-\frac{1}{2})(j_1+m+\frac{1}{2})(j_1+m+\frac{3}{2})}{2j_1(2j_1-1)(2j_1+1)}}$

Clebsch-Gordan Coefficients[a] (Continued)

$(j_1\,2\,m_1\,m_2\,|\,j_1\,2\,j\,m)$

$j=$	$m_2=2$	$m_2=1$
j_1+2	$\sqrt{\dfrac{(j_1+m-1)(j_1+m)(j_1+m+1)(j_1+m+2)}{(2j_1+1)(2j_1+2)(2j_1+3)(2j_1+4)}}$	$\sqrt{\dfrac{(j_1-m+2)(j_1+m+2)(j_1+m+1)(j_1+m)}{(2j_1+1)(j_1+1)(2j_1+3)(j_1+2)}}$
j_1+1	$-\sqrt{\dfrac{(j_1+m-1)(j_1+m)(j_1+m+1)(j_1-m+2)}{2j_1(j_1+1)(j_1+2)(2j_1+1)}}$	$-(j_1-2m+2)\sqrt{\dfrac{(j_1+m+1)(j_1+m)}{2j_1(2j_1+1)(j_1+1)(j_1+2)}}$
j_1	$\sqrt{\dfrac{3(j_1+m-1)(j_1+m)(j_1-m+1)(j_1-m+2)}{(2j_1-1)2j_1(j_1+1)(2j+3)}}$	$(1-2m)\sqrt{\dfrac{3(j_1-m+1)(j_1+m)}{(2j_1-1)j_1(2j_1+2)(2j_1+3)}}$
j_1-1	$-\sqrt{\dfrac{(j_1+m-1)(j_1-m)(j_1-m+1)(j_1-m+2)}{2(j_1-1)j(j_1+1)(2j_1+1)}}$	$(j_1+2m-1)\sqrt{\dfrac{(j_1-m+1)(j_1-m)}{(j_1-1)j_1(2j_1+1)(2j_1+2)}}$
j_1-2	$\sqrt{\dfrac{(j_1-m-1)(j_1-m)(j_1-m+1)(j_1-m+2)}{(2j_1-2)(2j_1-1)2j_1(2j_1+1)}}$	$-\sqrt{\dfrac{(j_1-m+1)(j_1-m)(j_1-m-1)(j_1+m-1)}{(j_1-1)(2j_1-1)j_1(2j_1+1)}}$

$j=$	$m_2=0$	$m_2=-1$
j_1+2	$\sqrt{\dfrac{3(j_1-m+2)(j_1-m+1)(j_1+m+2)(j_1+m+1)}{(2j_1+1)(2j_1+2)(2j_1+3)(j_1+2)}}$	$\sqrt{\dfrac{(j_1-m+2)(j_1-m+1)(j_1-m)(j_1+m+2)}{(2j_1+1)(j_1+1)(2j_1+3)(j_1+2)}}$
j_1+1	$m\sqrt{\dfrac{3(j_1-m+1)(j_1+m+1)}{j_1(2j_1+1)(j_1+1)(j_1+2)}}$	$(j_1+2m+2)\sqrt{\dfrac{(j_1-m+1)(j_1-m)}{j_1(2j_1+1)(2j_1+2)(j_1+2)}}$
j_1	$\dfrac{3m^2-j_1(j_1+1)}{\sqrt{(2j_1-1)j_1(j_1+1)(2j_1+3)}}$	$(2m+1)\sqrt{\dfrac{3(j_1-m)(j_1+m+1)}{(2j_1-1)j_1(2j_1+2)(2j_1+3)}}$
j_1-1	$-m\sqrt{\dfrac{3(j_1-m)(j_1+m)}{(j_1-1)j_1(2j_1+1)(j_1+1)}}$	$-(j_1-2m-1)\sqrt{\dfrac{(j_1+m+1)(j_1+m)}{(j_1-1)j_1(2j_1+1)(2j_1+2)}}$
j_1-2	$\sqrt{\dfrac{3(j_1-m)(j_1-m-1)(j_1+m)(j_1+m-1)}{(2j_1-2)(2j_1-1)j_1(2j_1+1)}}$	$-\sqrt{\dfrac{(j_1-m-1)(j_1+m+1)(j_1+m)(j_1+m-1)}{(j_1-1)(2j_1-1)j_1(2j_1+1)}}$

$j=$	$m_2=-2$
j_1+2	$\sqrt{\dfrac{(j_1-m-1)(j_1-m)(j_1-m+1)(j_1-m+2)}{(2j_1+1)(2j_1+2)(2j_1+3)(2j_1+4)}}$
j_1+1	$\sqrt{\dfrac{(j_1-m-1)(j_1-m)(j_1-m+1)(j_1+m+2)}{j_1(2j_1+1)(j_1+1)(2j_1+4)}}$
j_1	$\sqrt{\dfrac{3(j_1-m-1)(j_1-m)(j_1+m+1)(j_1+m+2)}{(2j_1-1)j_1(2j_1+2)(2j_1+3)}}$
j_1-1	$\sqrt{\dfrac{(j_1-m-1)(j_1+m)(j_1+m+1)(j_1+m+2)}{(j_1-1)j_1(2j_1+1)(2j_1+2)}}$
j_1-2	$\sqrt{\dfrac{(j_1+m-1)(j_1+m)(j_1+m+1)(j_1+m+2)}{(2j_1-2)(2j_1-1)2j_1(2j_1+1)}}$

[a] *Source:* E. U. Condon, and G. H. Shortley, *The Theory of Atomic Spectra* (London: Cambridge University Press, 1951), pp. 76–77.

APPENDIX 5 TABLE OF ATOMIC PROPERTIES[a]

Z	Atom	Configuration	Ground state	First ionization energy[b]	Orbital radius[c]
1	H	$1s$	$^2S_{1/2}$	13.505	0.529
2	He	$1s^2$	$^1S_0$	24.580	0.291
3	Li	[He]$2s$	$^2S_{1/2}$	5.390	1.586
4	Be	[He]$2s^2$	$^1S_0$	9.320	1.040
5	B	[He]$2s^22p$	$^2P_{1/2}$	8.296	0.776
6	C	[He]$2s^22p^2$	$^3P_0$	11.264	0.620
7	N	[He]$2s^22p^3$	$^4S_{3/2}$	14.54	0.521
8	O	[He]$2s^22p^4$	$^3P_2$	13.614	0.450
9	F	[He]$2s^22p^5$	$^2P_{3/2}$	17.42	0.396
10	Ne	[He]$2s^22p^6$	$^1S_0$	21.559	0.354
11	Na	[Ne]$3s$	$^2S_{1/2}$	5.138	1.713
12	Mg	[Ne]$3s^2$	$^1S_0$	7.644	1.279
13	Al	[Ne]$3s^23p$	$^2P_{1/2}$	5.984	1.312
14	Si	[Ne]$3s^23p^2$	$^3P_0$	8.149	1.068
15	P	[Ne]$3s^23p^3$	$^4S_{3/2}$	11.00	0.919
16	S	[Ne]$3s^23p^4$	$^3P_2$	10.357	0.810
17	Cl	[Ne]$3s^23p^5$	$^2P_{3/2}$	13.01	0.725
18	Ar	[Ne]$3s^23p^6$	$^1S_0$	15.755	0.659
19	K	[Ar]$4s$	$^2S_{1/2}$	4.339	2.162
20	Ca	[Ar]$4s^2$	$^1S_0$	6.111	1.690
21	Sc	[Ar]$4s^23d$	$^2D_{3/2}$	6.56	1.570
22	Ti	[Ar]$4s^23d^2$	$^3F_2$	6.83	1.477
23	V	[Ar]$4s^23d^3$	$^4F_{3/2}$	6.74	1.401
24	Cr	[Ar]$4s3d^5$	$^7S_3$	6.76	1.453
25	Mn	[Ar]$4s^23d^5$	$^6S_{5/2}$	7.432	1.278
26	Fe	[Ar]$4s^23d^6$	$^5D_4$	7.896	1.227
27	Co	[Ar]$4s^23d^7$	$^4F_{9/2}$	7.86	1.181
28	Ni	[Ar]$4s^23d^8$	$^3F_4$	7.633	1.139
29	Cu	[Ar]$4s3d^{10}$	$^2S_{1/2}$	7.723	1.191
30	Zn	[Ar]$4s^23d^{10}$	$^1S_0$	9.391	1.065
31	Ga	[Ar]$4s^23d^{10}4p$	$^2P_{1/2}$	6.00	1.254
32	Ge	[Ar]$4s^23d^{10}4p^2$	$^3P_0$	8.13	1.090
33	As	[Ar]$4s^23d^{10}4p^3$	$^4S_{3/2}$	10.00	1.001
34	Se	[Ar]$4s^23d^{10}4p^4$	$^3P_2$	9.750	0.918
35	Br	[Ar]$4s^23d^{10}4p^5$	$^2P_{3/2}$	11.84	0.851
36	Kr	[Ar]$4s^23d^{10}4p^6$	$^1S_0$	13.996	0.795
37	Rb	[Kr]$5s$	$^2S_{1/2}$	4.176	2.287
38	Sr	[Kr]$5s^2$	$^1S_0$	5.692	1.836
39	Y	[Kr]$4s^24d$	$^2D_{3/2}$	6.6	1.693
40	Zr	[Kr]$5s^24d^2$	$^3F_2$	6.95	1.593

Table of Atomic Properties[a] (Continued)

Z	Atom	Configuration	Ground state	First ionization energy[b]	Orbital radius[c]
41	Nb	$[Kr]5s4d^4$	$^6D_{1/2}$	6.77	1.589
42	Mo	$[Kr]5s4d^5$	$^7S_3$	7.18	1.520
43	Tc	$[Kr]5s^24d^5$	$^6S_{5/2}$	–	1.391
44	Ru	$[Kr]5s4d^7$	$^5F_5$	7.5	1.410
45	Rh	$[Kr]5s4d^8$	$^4F_{9/2}$	7.7	1.364
46	Pd	$[Kr]4d^{10}$	$^1S_0$	8.33	0.567
47	Ag	$[Kr]5s4d^{10}$	$^2S_{1/2}$	7.574	1.286
48	Cd	$[Kr]5s^24d^{10}$	$^1S_0$	8.991	1.184
49	In	$[Kr]5s^24d^{10}5p$	$^2P_{1/2}$	5.785	1.382
50	Sn	$[Kr]5s^24d^{10}5p^2$	$^3P_0$	7.332	1.240
51	Sb	$[Kr]5s^24d^{10}5p^3$	$^4S_{3/2}$	8.64	1.193
52	Te	$[Kr]5s^24d^{10}5p^4$	$^3P_2$	9.01	1.111
53	I	$[Kr]5s^24d^{10}5p^5$	$^2P_{3/2}$	10.44	1.044
54	Xe	$[Kr]5s^24d^{10}5p^6$	$^1S_0$	12.127	0.986
55	Cs	$[Xe]6s$	$^2S_{1/2}$	3.893	2.518
56	Ba	$[Xe]6s^2$	$^1S_0$	5.210	2.060
57	La	$[Xe]6s^25d$	$^2D_{3/2}$	5.61	1.915
58	Ce	$[Xe](6s^24f5d)$	$(^3H_5)$	6.91	1.978
59	Pr	$[Xe](6s^24f^3)$	$(^4I_{9/2})$	5.76	1.942
60	Nd	$[Xe]6s^24f^4$	$^5I_4$	6.31	1.912
61	Pm	$[Xe](6s^24f^5)$	$(^6H_{5/2})$	–	1.882
62	Sm	$[Xe]6s^24f^6$	$^7F_0$	5.6	1.854
63	Eu	$[Xe]6s^24f^7$	$^8S_{7/2}$	5.67	1.826
64	Gd	$[Xe]6s^24f^75d$	$^9D_2$	6.16	1.713
65	Tb	$[Xe](6s^24f^9)$	$(^6H_{15/2})$	6.74	1.775
66	Dy	$[Xe](6s^24f^{10})$	$(^5I_8)$	6.82	1.750
67	Ho	$[He](6s^24f^{11})$	$(^4I_{15/2})$	–	1.727
68	Er	$[Xe](6s^24f^{12})$	$(^3H_6)$	–	1.703
69	Tm	$[Xe]6s^24f^{13}$	$^2F_{7/2}$	–	1.681
70	Yb	$[Xe]6s^24f^{14}$	$^1S_0$	6.2	1.658
71	Lu	$[Xe]6s^24f^{14}5d$	$^2D_{3/2}$	5.0	1.553
72	Hf	$[Xe]6s^24f^{14}5d^2$	$^3F_2$	5.5	1.476
73	Ta	$[Xe]6s^24f^{14}5d^3$	$^4F_{3/2}$	7.88	1.413
74	W	$[Xe]6s^24f^{11}5d^4$	$^5D_0$	7.98	1.360
75	Re	$[Xe]6s^24f^{14}5d^5$	$^6S_{5/2}$	7.87	1.310
76	Os	$[Xe]6s^24f^{14}5d^6$	$^5D_4$	8.7	1.266
77	Ir	$[Xe]6s^24f^{14}5d^7$	$^4F_{9/2}$	9.2	1.227
78	Pt	$[Xe]6s4f^{14}5d^9$	$^3D_3$	8.96	1.221
79	Au	$[Xe]6s4f^{14}5d^{10}$	$^2S_{1/2}$	9.223	1.187
80	Hg	$[Xe]6s^24f^{14}5d^{10}$	$^1S_0$	10.434	1.126

Table of Atomic Properties[a] (Continued)

Z	Atom	Configuration	Ground state	First ionization energy[b]	Orbital radius[c]
81	Tl	$[Xe]6s^2 4f^{14} 5d^{10} 6p$	$^2P_{1/2}$	6.106	1.319
82	Pb	$[Xe]6s^2 4f^{14} 5d^{10} 6p^2$	$^3P_0$	7.415	1.215
83	Bi	$[Xe]6s^2 4f^{14} 5d^{10} 6p^3$	$^4S_{3/2}$	7.287	1.295
84	Po	$[Xe]6s^2 4f^{14} 5d^{10} 6p^4$	$^3P_2$	8.43	1.212
85	At	$[Xe](6s^2 4f^{14} 4d^{10} 6p^5)$	$(^2P_{3/2})$	–	1.146
86	Rn	$[Xe]6s^2 4f^{14} 5d^{10} 6p^6$	$^1S_0$	10.746	1.090
87	Fr	$[Rn](7s)$	$(^2S_{1/2})$	–	2.447
88	Ra	$[Rn]7s^2$	$^1S_0$	5.277	2.042
89	Ac	$[Rn]7s^2 6d$	$^2D_{3/2}$	6.9	1.895
90	Th	$[Rn]7s^2 6d^2$	$^3F_2$	–	1.788
91	Pa	$[Rn](7s^2 5f^2 6d)$	$(^4K_{11/12})$	–	1.804
92	U	$[Rn]7s^2 5f^3 6d$	$^5L_6$	4	1.775
93	Np	$[Rn](7s^2 5f^4 6d)$	$(^6L_{11/12})$	–	1.741
94	Pu	$[Rn](7s^2 5f^6)$	$(^7F_0)$	–	1.784
95	Am	$[Rn](7s^2 5f^7)$	$(^8S_{7/2})$	–	1.757
96	Cm	$[Rn](7s^2 5f^7 6d)$	$(^9D_2)$	–	1.657
97	Bk	$[Rn](7s^2 5f^9)$	$(^6H_{15/2})$	–	1.626
98	Cf	$[Rn](7s^2 5f^{10})$	$(^5I_8)$	–	1.598
99	Es	$[Rn](7s^2 5f^{11})$	$(^4I_{15/2})$	–	1.576
100	Fm	$[Rn](7s^2 5f^{12})$	$(^3H_6)$	–	1.557
101	Md	$[Rn](7s^2 5f^{13})$	$(^2F_{7/2})$	–	1.527
102	No	$[Rn](7s^2 5f^{14})$	$(^1S_0)$	–	1.581
103	Lw	$[Rn](7s^2 5f^{14} 6d)$	$(^2D_{3/2})$	–	–

[a]*Source:* Data from C. E. Moore, *Atomic Energy Levels*, National Bureau of Standards (U.S.), Circular No. 467 (U.S. Government Printing Office, Washington, D.C.), 1949, and Waber and Cromer, *J. Chem. Phys.* **42,** 4116(1965).

[b]Energy in electron volts.

[c]Radius of the maximum in radial probability density of the outermost orbital (in angstroms).

Bibliography

This bibliography lists useful references that are not mentioned in the Selected Readings. It is not meant to encompass all valuable texts on quantum mechanics, atomic physics, molecular physics, and solid-state physics but rather to acknowledge other volumes drawn on at one time or another in preparation of this book. Each reference is keyed: E = elementary, I = intermediate, A = advanced.

ARYA, ATAM P., *Fundamentals of Atomic Physics*. Boston: Allyn and Bacon, 1971. [E]

BAYM, GORDON, *Lectures on Quantum Mechanics*. New York: W. A. Benjamin, 1969. [A]

BEARD, DAVID B., and G. B. BEARD, *Quantum Mechanics with Applications*. Boston: Allyn and Bacon, 1970. [E]

BOHM, DAVID, *Quantum Theory*. Englewood Cliffs, N.J.: Prentice-Hall, 1951. [I]

CONDON, E. U., and G. H. SHORTLY, *The Theory of Atomic Spectra*. New York: Cambridge University Press, 1935. [A]

DAVYDOV, A. S., *Quantum Mechanics*. New York: Pergamon Press, 1965. [A]

DEKKER, A. J., *Solid State Physics*. Englewood Cliffs, N.J.: Prentice-Hall, 1957. [E]

DICKE, R. H., and J. P. WITTKE, *Introduction to Quantum Mechanics*. Reading, Mass.: Addison-Wesley 1960. [E]

DIRAC, P. A. M., *The Principles of Quantum Mechanics*, 4th ed., New York: The Clarendon Press, 1958. [A]

FANO, U., and L. FANO, *Physics of Atoms and Molecules*. Chicago: The University of Chicago Press, 1972. [A]

FERMI, ENRICO, *Notes on Quantum Mechanics*. Chicago: The University of Chicago Press, 1961. [E]

HOLDEN, A., *The Nature of Solids*. New York: Columbia University Press, 1965. [I]

KOMPANEYETZ, A. S., *Basic Concepts in Quantum Mechanics*. New York: Reinhold Publishing Corp., 1966. [E]

LEIGHTON, R. B., *Principles of Modern Physics*. New York: McGraw-Hill, 1959. [E]

LUDWIG, G., *Wave Mechanics*. Oxford: Pergamon Press, 1968. [I]

MESSIAH, A., *Quantum Mechanics*. New York: Wiley, 1966. [A]

POWELL, J. L., and B. CRASEMANN, *Quantum Mechanics*. Reading, Mass.: Addison-Wesley, 1961. [E]

RAPP, D., *Quantum Mechanics*. New York: Holt, Rinehart and Winston, 1971. [I]

RICHTMYER, F. K., E. H. KENNARD, and J. N. COOPER, *Introduction to Modern Physics*, 6th ed. New York: McGraw-Hill, 1969. [E]

SCHIFF, L. I., *Quantum Mechanics*, 3rd ed. New York: McGraw-Hill, 1968. [A]

SPOSITO, G., *An Introduction to Quantum Physics*, New York: Wiley, 1970. [E]

WANNIER, G. H., *Elements of Solid State Theory*. New York: Cambridge University Press, 1959. [A]

WEINREICH, G., *Solids: Elementary Theory for Advanced Students*. New York: Wiley, 1965. [A]

WICHMANN, E. H., *Quantum Physics* (*Berkeley Physics Course, Vol. 4*). New York: McGraw-Hill, 1971. [E]

WOODGATE, G. K., *Elementary Atomic Structure*. London: McGraw-Hill, 1970. [E]

ZIOCK, K., *Basic Quantum Mechanics*, New York: Wiley, 1969. [E]

Index